精明营建：可持续的体育建筑

Smart Planning and Building: Sustainable Sports Facility

孙一民 著

SUN Yimin

中国建筑工业出版社

孙一民

长江学者特聘教授
首届广东省工程勘察设计大师
中国建筑学会常务理事
中国体育科学学会体育建筑专业委员会副主任委员
中国建筑学会城市设计分会副理事长
建设部城市设计专家委员会成员
现任华南理工大学建筑学院院长

SUN Yimin

Changjiang Scholar Distinguished Professor
Executive Committee, Architectural Society of China
Deputy Director, Sports Architecture Committee, Architectural Society of China, China Sports Science Society
Deputy Director, Urban Design Committee, Architectural Society of China
Member, Urban Design Expert Committee, Ministry of Housing and Urban-Rural Development of China
Dean, School of Architecture, South China University of Technology

序　一

华南理工大学建筑学院的孙一民教授在多年教学、研究和设计实践的基础上，厚积薄发，出版了他关于体育建筑研究的新作《精明营建：可持续的体育建筑》，我有幸先睹了书稿，从中得到许多启发。

体育一直是我国十分热门的行业，尽管我们在竞技水准、运动普及程度、运动员数目等方面和世界一流水准还有差距，正在从体育大国向体育强国努力，但其社会影响力和人们关注度却是不可小视的。如巴西奥运会的女排决赛就牵动着多少亿国人的目光。这里也体现了一个浅显的道理：国家的强盛和发展，才能带来竞技体育和全民健身事业的发展。即如重返亚运会后，1982年新德里亚运会上一举打破日本运动员独霸亚运会的局面；重返奥运会后，在1984年洛杉矶奥运会上，中国运动员获32枚奖牌，金牌总数列第四位，实现了“零的突破”；2008年的北京奥运会上，又以51枚金牌居金牌总数首位，成为金牌总数首位的第一个亚洲国家。与此同时健身运动的社会化、体育人口的大众化、社区体育的多样化也日益成为构建和谐社会的重要目标。体育事业的发展也必然推动我国各种类型的建设，所以体育建筑设计自然而然地成为设计行业的显学，呈现出百花争艳的局面。

我是在一民读研时候知道他的，审读过他有关高校多功能厅堂设计研究的博士论文。在学期间他在梅季魁先生的指导下专攻体育建筑设计，更为幸运的是他也赶上了我国体育建筑蓬勃发展的大好时机，得以在这个难得的舞台上一展自己的才能与智慧，并取得了可喜的成就。记得早在1985年前后，他就跟随梅先生参与了亚运会朝阳体育馆和石景山体育馆的设计。1991年取得博士学位并到华南理工大学任教以后，视野更为开阔，设计的建筑类型更多，但大型体育设施的设计和研究仍是他的重要主攻方向，在我国一些大型体育赛事如奥运会、全运会、大运会等体育设施的竞投和建设，国内体育设施的论证和研究，与国外著名设计事务所的合作等都可以见到他的身影，取得了很好的社会效益和经济效益，除了获得国内的各类奖项外，还获得了许多国际性的奖项，如：北京奥运会摔跤馆获2011年“IPC/IAKS国际体育建筑杰出功勋奖”，亚运武术馆获2011年“IOC/IAKS国际体育建筑铜奖”，等等。这些不但扩大了我国体育建筑设计在国际体坛的影响，同时也向世界展现了中国建筑师的雄厚实力。一民是在这方面表现突出的建筑师。

由于一民的学术经历和他的关注重点，除了在设计项目上展现了才华之外，他还勤于思考，勇于从实践过程中发现问题，大胆剖析。众所周知，由于我国体育运行机制上的缺陷，体育事业社会化、产业化的不成熟，在轰轰烈烈的设施和场馆建设中，许多矛盾和弊病已经和正在显现出来，为人们所诟病。一民在他的新作中，就这种跃进式的建设，普遍的急功近利、缺乏理性的研究，提出了当前场馆建设中所存在的问题：总体发展失衡，重复建设；科学决策缺失，建与养矛盾突出；选地不当，与城市缺乏良性互动；功能定位单一，灵活适应性不足；高投入低效益；运营成本高，能耗大，等等。于是从自己所经手的十项建筑作品所做的探索，加上在长期研究过程中对国外实例的经验和做法的了解，提出通过城市理性、功能理性和技术理性的方法，达到可持续的体育建筑。在体育建筑的研究上提出了一个比较系统的理论框架。

一民的分析主要是针对体育建筑在现代化环境下如何在观念上更加理性化或者运用更加理性的判断来分析的社会学命题。德国的社会学家和历史学家韦伯在20世纪初提出一种新的研究方法，以研究人的社会活动的意义和目的为研究对象，他所提出的理性主义包括人们通过内心思索引发的思想层面意义关联的系统化，通过计算和分析来支配事物的科学技术的理想行为，通过意义关联及利害关系制度化而形成的系统。当时他认为这种理性体系是历来社会发展中最理想的体系。在这种理性体系的指引下，可能创造大量的财富，对大自然无尽的征服和探索，获得更多的自由发展空间。但韦伯在反省欧美地区的现代化时又发现理性的发展又使得追求自由和解放的人们在这一过程中变成了理性的奴隶。他又提出了“理性之铁笼”的著名隐喻，在科技的理性，计算的理性和官僚体制的理性之下，将跌入物质和权利的控驭，导致社会的等级化、官僚化、法律化。人们认为，在工具理性和价值理性两种理性的观点中，工具理性是十分重要的，现代化的大部分内容都是工具理性的。但如果在工具理性的指引下，片面强调功利的取向，同样也会陷入困境，这时需要价值理性的内容来加以平衡，需要人们有一种价值提升的力量来使现代化的过程更加健康。即如我们城市建设中出现的“追求视觉冲击的奇奇怪怪的建筑”“盲目崇洋”“追逐第一，豪华奢侈，盲目攀比”等乱象。就需要通过“适用、经济、绿色、美观”的方针，考虑我国人口、资源、环境的国情，考虑可持续发展的主流价值判断来加以认识。我们在现代化过程中不仅要考虑“如何去实现”，还要考虑“为什么要如此”。

一民的新作提出了很重要的问题，对于工具理性和价值理性的问题我知之甚少，只是在学习新作的过程中提出一点粗浅的体会，以此来求教于一民和其他方家。

2017 年 4 月 12 日

马国馨（1942 年—），原籍上海，出生于山东济南，中国工程院院士、全国工程勘察设计大师、“梁思成建筑奖”获得者。担任北京市建筑设计研究院有限公司顾问总建筑师，主持完成多个国家重点项目，在建筑理论、建筑设计、建筑历史、建筑评论、环境设计等领域均有开创性贡献。其中对体育设施建设更有着深刻的研究与实践积累，是国家行业标准《体育建筑设计规范》（2003）的主要负责人。20 世纪 80-90 年代，主持完成的北京亚运中心体育场馆获得国际体育建筑最高奖项——IAKS 奖，成为首位获此殊荣的中国建筑师。21 世纪初，带领团队完成北京申办奥运会的体育场馆设计方案，为我国重大体育工程的顺利开展和建成做出了重要贡献。

Forword I

Professor SUN Yimin at the School of Architecture, South China University of Technology, has published his new work on sports buildings 'Smart Planning and Building: Sustainable Sports Facility', based on years of his accumulation and profound knowledge of teaching, research, and design practice. I have had the privilege of reading the manuscript and gain a lot of inspiration from it.

Sports have always been very popular in China, although there is still a gap between the standards of sports in our country and the world-class standards in terms of the level of competition, the popularity of sports, the number of athletes, etc. Although we are on our way from being a 'big country' to becoming a 'strong country' in sports, the social influence of sports and people's attention it attracts are not to be underestimated. For example, women's volleyball finals of the Brazilian Olympic Games attracted the attention of hundreds of millions of Chinese people. These facts also embody a simple truth: only the prosperity and development of the country can bring about the growth of competitive sports and national fitness. On returning to the Asian Games, China ended Japanese athletes' dominance in the 1982 New Delhi Games. Further, on returning to the Olympic Games, Chinese athletes won 32 medals in the 1984 Los Angeles Olympic Games, and the total number of their gold medals ranked fourth in medal tally, achieving a 'zero breakthrough'. In the 2008 Beijing Olympic Games, China, with 51 gold medals, ranked first in medal tally in terms of the number of gold medals, becoming the first Asian country to achieve this feat in the history of Olympics. At the same time, the socialization of fitness, the popularization of sports participants, and the diversification of community sports have increasingly become important goals in building a harmonious society. The development of sports industry will inevitably promote various types of construction in China, so the design of sports facilities naturally becomes a famous study of the design industry, showing a blooming situation.

I learned about Yimin during his postgraduation, and reviewed his doctoral thesis on the design of multi-functional halls in colleges and universities. He specialized in sports building design under the guidance of Mr. Jikui MEI. It was even more fortunate that he caught up with great opportunities in design that resulted from vigorous development of sports buildings in China, and was able to display his talent and wisdom on this rare stage, accomplishing gratifying achievements. I remember that as early as 1985, he followed Mr. MEI to participate in the designing of the Chaoyang Gymnasium and Shijingshan Stadium for the Asian Games. After obtaining his Ph.D. in 1991 and teaching at South China University of Technology, he has a broadened vision since he has worked on several types of building design. However, the design of large-scale sports facilities and related research are still his dominant focuses. He is actively involved in some major sports events in China, such as the bidding and construction of sports facilities for the Olympic Games, the National Games, the Universiade, the demonstration of domestic sports facilities and related research, and cooperation with famous foreign design firms; achieves good social and economic benefits. In addition to various domestic awards, he has also won many international awards, such as the 'IPC/IAKS Distinctions' for the 2008 Olympic wrestling hall (China Agricultural University Gymnasium) in 2011 and the 'IOC/IAKS Award Bronze' in 2011 for the 2010 Asian Games Martial Arts Hall (Nansha Gymnasium), which is the highest international award for sports architecture, jointly sponsored by the International Olympic Committee (IOC), the International Paralympic Committee (IPC) and the International Association for Sports and Leisure Facilities (IAKS). These achievements have not only expanded the influence of China's sports architectural design in the international sports arena, but also demonstrated to the world the strength of Chinese architects; and Yimin is an outstanding architect in this respect.

Due to Yimin's academic experience and focus, in addition to his talents in design projects, he is also a diligent thinker, brave to find problems from the practice, and bold enough to analyze. As we all know, due to the defects in China's sports operation mechanism, and immature socialization and industrialization of sports industry, many contradictions and shortcomings have been showing up in the vigorous construction of facilities and venues, which is being criticized by people. In his new work, focusing on situations of leap-forward construction universal eagerness for instant success and quick profits, and lack of rational research, Yimin has raised the following contradictions in the current stadium construction: between partial repeated construction and overall unbalanced development; between construction and operation; between self-emphasis and urban-integration; between single function and diverse usage; between high capital investment and low economic and energy efficiency, etc. Further, based on the exploration of the ten design works he has handled, and the understanding of the experience and practices of foreign examples in the long-term research process, he proposes that sustainability can be achieved in sports facilities through urban, functional, and technical rationality. This is a relatively systematic theoretical framework proposed in sports architecture studies.

Yimin's analysis is mainly aimed at the sociological propositions of how sports buildings can be more rational in conception or how more rational judgments can be used in the modern environment. German sociologist and historian Weber proposed a new research method at the beginning of the last century to study the meaning and purpose of human social activities. His rationalism includes the systematic connection of meaning and inner thought at the ideological level, the ideal behavior of science and technology that controls things through calculation and analysis, and a system formed through the association of meaning and institutionalization of interests. He believed that this rational system was the most ideal system for social development. He believed that under the guidance of this rational system, it would be possible to create immense wealth, accomplish endless conquest and exploration of nature, and gain more space for further development. But when Weber reflected on the modernization of Europe and the United States, he realized that rational development made people pursuing freedom and liberation rational slaves in the process. Then, he put forward the famous metaphor of 'rationality cage'. Under the rationality of science and technology, the rationality of calculation, and the rationality of bureaucratic system, the system will fall into the control of material and rights, leading to the hierarchical organization, bureaucratization, and legalization of society. It is believed that in the two viewpoints of rationality, namely the instrumental rationality and the value rationality, the former is very important, and most of the content of modernization is based on instrumental rationality. However, under the guidance of instrumental rationality, one-sided emphasis on utilitarian will also lead to trouble. Therefore, there is need for the value rationality balance, which requires people to be guided by a value-enhancing force to make the modernization process healthier. That is to say, in the construction of our cities, as a response to the chaos of 'freakish architecture that pursues visual impact,' 'blind worship', 'the first, luxury, blindly comparing', etc., it is necessary to adopt a policy that emphasizes 'applicability, economy, greenness, and beauty,' and considers China's national conditions of population, resources and environment, as well as the mainstream value judgment of sustainable development. In the process of modernization, we must not only consider 'how to achieve' something but also 'why should it be achieved'.

The new work by Yimin raises very important questions. I don't know much about the issue of instrumental rationality and value rationality. I have just presented my elementary experience from the process of learning through this new work, so as to seek advice from Yimin and other masters.

MA Guoxin
April 2017

MA Guoxin (1942–), Shanghai by origin, born in Jinan Shandong, Academician of CAE (Chinese Academy of Engineering), National Master of Engineering Survey and Design, Winner of the 'Liang Sicheng Architecture Prize' . As the chief architect consultant of BIAD (Beijing Institute of Architectural Design), he has hosted numerous national key projects. Among the comprehensive pioneering contributions in the fields of architectural theory, design, history and review, as well as environmental design, he has profound research and practice accumulation in the construction of sports facilities, and was the main person in charge of the national industry standard 'Design Code for Sports Building' (2003). In the 1980s and 1990s, his masterpiece, the Beijing Asian Games Center venues, won the IAKS (International Association for Sports and Leisure Facilities) Award, the highest international award for sports facilities, which made him became the first Chinese architect to receive this honor.At the beginning of the 21st century, he led a team and completed the venues design for Beijing's bid to host the Olympic Games, which made as ignificant contribution to the successful development and completion of these major sports projects in China.

序　二

与孙一民教授及其设计作品的相遇很不一般。在这里，我想两方面都说一说。

十多年来，我们事务所和我个人作为大学的客座教授，每年都要去几次中国，因此碰到了很多搞建筑的人。我们非常惊讶于中国目前人口已经十分稠密的省份每年新建建筑量的巨大，远远超出了欧洲人的经验，有时甚至让我们感到十分不可思议（比如说，它们一年的建筑量就相当于意大利自古典时期以来的所有建成量）。如果只向后看十年，只看看新建成的公共建筑的数量，如火车站、机场、博物馆、工业区和体育设施——国际大型盛会肯定对此也有推波助澜的作用，如 1999 年在北京举行的 UIA 世界建筑师大会，2008 年的奥林匹克运动会和 2010 年的上海世博会——对建筑量如此之大的惊奇就不由得让人产生很多疑问，如规划条件是怎么样的，如何征到用地，材料如何获取与加工，以及巨型施工工地是如何相互协调并保证后勤供应的，等等。

另一方面，这也体现了投资商对新建建筑的普遍定位，即最新奇、最张扬的建筑外观，或是独一无二的建筑形象或建造技术，它们不仅必须在最短的时间内设计出来，还必须建成。由此产生了许多奇形怪状的建筑，也造成了许多短期的建造结果，等到下一个同类的、也设计得奇形怪状的竞争对手出现时，原先的建筑就失去了吸引力。这无疑具有极大的娱乐效应和新闻效应。

但这也造成一种当下的、持续的建筑文化不可能有机会出现。这一建筑文化在中国已延续千年，至今仍有很多杰出的建筑物得以保留，它们在建造技术和空间表达的成熟度上达到了最高的美学水准，是深沉的人类智慧长期专注和深思熟虑的结晶，它们将建筑物的存在视为空间与物体、自然环境与人类创造的文化环境之间和谐的条件 。

这应该就是那些“可持续的”“智慧型的”“绿色的”或“生态气候型的”建筑。它们作为伟大的技术产物是长期建造的结果，因此远期看来十分经济划算。今天的建筑应在此基础上，不仅在建造阶段使用有限的资源细心地、带有环境意识地进行规划，在能源使用上也应该尽可能降低能源需求，并采用零排放、无危险的环境能源来替代化石能源。

正是在此目标设定下，我看到了孙一民教授的独特贡献。他属于这类建筑师和规划师，他们对于这一问题了然于心，并多年以来在自己的工作中一步步地努力实现这一目标。但他并不将问题简化，将某一规划项目仅看作是与环境毫不相干的独立物体。从他开始接受土木工程和建筑教育以来，以及随后在他的博士论文中（这是中国首篇关于体育建筑的博士论文），他一直有意识地在他的整个工作中探讨巨型建筑项目（常常是体育设施）的社会意义。

他研究体育建筑的社会功能，对城市结构产生的影响，体育设施场地中建筑与建筑之间，以及建筑与周围公共空间之间的交通联系等。

他一直致力于体育建筑的规划设计，并据其特点发展出体育建筑的原型。功能和技术上的处理在规划阶段会根据其适宜性有所不同。这不是依自己情感的喜好所产生的形式上的突变，而是随着科学知识的不断进步，自我见解和能力的不断提升，在建筑设计表现上的不断进步前行；是根据地段的具体情况对建筑设计质量的不断改善——即明智城市设计在地形和气候设计上的要求。

对他来说，每一个建筑项目都不是为了树立自我风格。如果说他在很多项目中重复得采用了一些的母题，如超级大厅，大跨度的结构支架与顶入日光的利用相结合等，这也不是为了让人把它们当作是他的“设计标签”。为了避免对运动员和观众造成炫光——这已成为体育建筑设计中功能理性的一条普遍标准——孙一民教授在他设计建造的项目中多次采用大型的、带有形式感的建筑支撑结构，但这也仅是他找到的一种建筑解决方案而已。这与密斯 • 凡德罗的建造原则和逻辑不谋而合，密斯认为“每个星期一早上发明一种新的建筑”毫无意义也毫无必要。

孙一民教授的工作与一般的大型建筑项目形成了鲜明的对比，因此具有跨区域的意义。不是个人的形式喜好，而是作为建筑师所应有的、源于对社会的责任感而产生的思考理性决定着他的建筑设计。那些下一代的年轻学人能有他这样具有清晰理念、立场坚定的建筑师为师，实乃三生有幸。他的成功，他的众多具有类似设计要求的建造项目，充分说明了在日益复杂的建筑世界里，对贴近眼、手的细节的关注，对于人们对真实的建筑尺度、比例、形式、色彩和质地的理解，以及在大尺度上，成千上万的观众对“体育宫殿”的巨型尺度的集体性、同时性体验，提供了广泛的可能；对建筑最终质量的有意识关注将会极大地促进建筑业的发展。他既拥有广博的理论知识，又拥有丰富的实践经验（几十年来，他设计过约六十座体育建筑，其中三分之一得以建成），这使得本书远远超出了一本仅介绍建筑项目或描述其特点的作品集。他的理论建构和自身立场的构建基础，充分展现了他在没有体育赛事场地设计先驱的情况下，积极投身于自己国家的发展现状的热情。中国近二十年建筑业的蓬勃发展以提升现状质量为主，其中不乏广泛的争议。他对发展的语境式注脚，一方面是作为教师传道授业的义务；另一方面，他也通过本书中图像资料的展示逻辑，进一步强调了自己对于发展的观点。特别是当决定性的设计和规划条件出于作者的无知而被强烈忽视，而建筑体块看上去又很形式主义，往往会造成其试图表现的概念缺失应有的艺术潜质。

孙一民教授特别强调这一领域的知识要具有专业水准，他不是只提出几句口号，而是在他的表述中以一种清晰的、充满逻辑的方式建构了这一领域中关于城市理性、功能理性以及技术理性的几个研究重点。他探讨那些引发许多问题或疑问的建筑，如它们在技术和功能上的面向是什么？它们结构特征是什么？有时是特别重要的关于可持续性的问题，如在节点设计中的建筑物理和建造结构问题等。因此他对问题的解决方案也多种多样，既来自于不同领域对建筑设计的不可计数的要求，也建立在上文提到的他的研究重点上——无论是源于所希冀的建造目标，还是将建造看作是处于内部联系中的复杂过程。

不管怎样，他在做决策时不会因为有所偏好而回避矛盾。 更多的是——这也表现了他的特别用心——决策会被有意识地看作是矛盾中的最优选择，这表现在本书理论部分中经系统考虑和表达的标志上。因此，可取得的优点会被表达出来，但可能的缺点也会被毫不掩饰地指出。仅就他处理这些议题的方式，已使孙一民教授作为建筑师的工作在许多点上可被很好地理解，对于其他从事规划和建筑事业的研究机构和个人来说，也能很好地理解他的工作。他的建造成果体现了一种高度的理性，它们表明，明确的目标定位（众所周知，处于城市边缘的商业区的一个主要特征是，它让人们觉得它们好像没有任何美学追求）能够形成令人信服的建筑设计表达。最后，这也就是说，对建筑和建筑之间空间的设计要求必须成为文化能力的普遍特征，对于这些要求存在多种选择方式，在相应的工作条件具备的前提下，它们应该具有模范带头的作用。

正是在此语境下，孙一民教授在体育建筑方面的工作在我看来十分具有特色，尤其是其建成结果的多样性具有特别的意义。这也体现在那些为实现建筑所采用的建筑产品上，基于今天长途运输的可能，它们来自世界各地，根据最佳功能特征而被选取。这也充分说明，这些项目具有高度的个性和可区别性，体现了地区自身的特点。这可能是通过项目处理中不同阶段和不同尺度上的专门工作形式，以及城市空间尺度上对建筑体及建筑细节在 1∶1 的人体尺度上的规模和位置进行考虑而达到的。

托马斯·赫尔佐格

2016 年 3 月

托马斯 • 赫尔佐格（Thomas Herzog，1941 年—），出生于德国慕尼黑，德国注册建筑师，国际建筑科学院、德国艺术科学院、巴伐利亚艺术科学院、法国建筑科学院、俄罗斯艺术与科学院院士。以关注技术、注重生态的建筑设计享誉世界，获得过包括德国建筑学会金奖、密斯凡德罗奖、德国钢结构建筑奖、全球可持续建筑奖、国际建协应用技术奖、法国建筑科学院金质奖章在内的众多奖项。曾任慕尼黑工业大学建筑学院教授、院长，清华大学、瑞士洛桑联邦理工及丹麦皇家艺术学院等校客座教授，在大学中组织并领导了多学科交叉、精密的建筑技术研究，与建筑物理及太阳能技术等领域合作实现建筑设计的科学性。所从事的建筑科学研究和设计实践，带动了德国乃至整个欧洲可持续建筑的发展，使其处于世界领先地位。

Forword II

Die Begegnung mit Yimin Sun als Person und mit seiner Arbeit als Architekt ist im besten Sinne des Wortes ungewöhnlich. Von Beidem soll hier die Rede sein.

Seit gut einem Jahrzehnt sind wir von Seiten unseres Büros und bin ich selbst in China als Gastprofessor vor Ort mehrmals im Jahr konfrontiert mit vielem, was sich baulich ereignet; wir sind im hohen Maße beeindruckt von dem ganz unglaublichen Tempo mit dem in derzeit bereits dichtest besiedelten Provinzen Quantitäten an jährlichen Neubauten entstehen, die für Europäer weit außerhalb der eigenen Erfahrung, ja gelegentlich des Vorstellbaren liegen (wo z.B. der gesamte existente „building stock" Italiens seit der Antike in nur einem Jahr an Hochbauten entsteht).

Blickt man nun nur ein Jahrzehnt zurück und sieht die schiere Menge an neuen öffentlichen Bauten wie Bahnhöfen, Airports, Museen, Industrieanlagen und Sportstätten – sicherlich auch befeuert durch internationale Großveranstaltungen, wie den Weltkongress der Architekten UIA in Peking 1999, die Olympischen Spiele 2008 und die Weltausstellung in Shanghai 2010, so schließen sich an das verwunderte Erstaunen ob dieser unglaublichen quantitativen Leistungen Fragen an, welche zum einen den Entstehungsprozess angehen, die Planungsbedingungen, die Zugriffsmöglichkeiten auf das erforderliche Land, die Gewinnung von Material sowie seine Verarbeitung, die Koordination und Logistik der gigantischen Baustellen und vieles mehr.

Was sich aber andererseits auch zeigt, ist die vor allem unter Investoren weit verbreitete Zielsetzung, nach der in der optischen Wirkung Außergewöhnliches, Extravaganz, formale Erstmaligkeiten und bauliche Einmaligkeiten, die in allerkürzester Zeit nicht nur geplant, sondern auch realisiert sein müssen. Dies geschieht mit der als hohes Ziel gewollten Auffälligkeit bis hin zu modischen Eskapaden, bei denen auch vielfach bauliche Resultate entstanden die auf Kurzzeitwirkung setzen und die nur so lange beeindrucken, bis der nächste Konkurrent in gleicher Sache auftaucht, der wieder andere gestalterische Merkmale des Baukörpers zu bizarrer Wirkung bringt. Unterhaltsam und medienwirksam ist dies – gewiss.

Doch wird eines dabei wenig Chancen haben: Das Entstehen einer heutigen, konsistenten baulichen Kultur, die in Jahrtausende alten, noch erhaltenen großartigen Beispielen von höchstem Rang nach wie vor in China existiert und dies in einer konstruktiven und räumlichen Reife auf höchstem ästhetischem Niveau, wie sie als Ergebnis von Konzentration und Besonnenheit von tiefem Wissen um die Bedingungen für das Entstehen von Architektur als Harmonie von Raum und Objekt, von natürlicher und der von Menschen geschaffenen kulturellen Umwelt Bedingung ist.

Es sind Bauten, die als „sustainable", „smart", „green" oder auch „bioclimatic" gelten. Sie sind als technische Großgegenstände eindeutig Langzeitprodukte und damit im höchsten Maße auf Dauer effizient. Bauten von heute müssen darüberhinaus sowohl in der Phase ihres Entstehens mit beschränkten Ressourcen sorgfältig und bewusst konzipiert sein und in ihrer Energieversorgung bei möglichst geringem Bedarf emissionsfrei und risikolos in maximal möglichem Umfang durch den Einsatz von Umweltenergie an Stelle von fossilen Energieträgern betrieben werden können.

An dieser Stelle und in dieser Zielsetzung sehe ich die so besondere Leistung von Prof. Yimin Sun. Er gehört zu den Architekten und Planern, die diese Zusammenhänge kennen und seit Jahren in ihrer eigenen Arbeit schrittweise zu realisieren bemüht sind. Er vermeidet aber Vereinfachung, indem er nur ein zu planendes Bauwerk als isoliertes Thema sehen würde. Er analysiert und diskutiert im Bewusstsein der Bedeutung der riesigen Bauaufgaben wie Sportstätten es häufig sind, seit seiner Ausbildung im Bereich des Civil Engineering und der Architektur sowie in der Folge seiner Dissertation über diese Bauaufgaben (der ersten Promotion zu diesem Thema in China) sein bevorzugtes Thema in seiner ganzen Breite gesamtheitlich.

Dabei behandelt er die gesellschaftliche Funktion für die städtische Struktur, die Erschließung von Arealen in der Relation von Gebäude und umgebendem öffentlichen Raum.

Er bleibt am Thema und entwickelt die Typologie von Sportbauten entlang ihrer architektonischen Hauptmerkmale. Funktionale und technische Lösungen werden im Planungsprozess auf ihre Tauglichkeit hin variiert. Es geht dabei nicht um formale Sprünge nach eigenem emotionalen Gusto sondern um Entwicklungsschritte bei laufender wissenschaftlicher Kontrolle als dem Mittel zur Vertiefung der eigenen Einsichten und wachsende Kompetenz; um Verbesserung unter den Gegebenheiten des jeweiligen Ortes – topografisch und klimatisch entsprechend den Kriterien eines klugen Städtebaus.

Doch es geht ihm bei den Projekten selbst ganz offensichtlich nicht um die Etablierung eines eigenen Stils, der womöglich zum Erkennungszeichen oder „Label“ seiner Entwürfe werden könnte. Wenn sich Motive wie beispielsweise die große Hallen überspannenden hohen Träger in Verbindung mit der Nutzung von Tageslicht vom Zenit in unterschiedlichen Projekten wiederholen. Wenn zur Vermeidung von Blendung für Akteure und Zuschauer als eines der universell und generell maßgeblichen Kriterien funktionaler Ratio mehrfach bei den Projekten und Bauten, die Yimin Sun konzipiert und realisiert hat in markanter, Form bestimmender Weise die großen Tragstrukturen sichtbar werden, so ist dies schlichtweg eine gefundene Lösung ganz nach der baulichen Disziplin und Logik Mies van der Rohes, der es als weder sinnvoll noch notwendig erachtete „jeden Montag morgen eine neue Architektur zu erfinden“.

Professor Yimin Suns Arbeit kontrastiert mit dem allgemeinen Geschehen bei Großbauten von überregionaler Bedeutung. Nicht persönliche Formpräferenzen dominieren den Entwurf sondern Rationalität in der übergeordneten, in seiner Leistung von Architekten weltweit einzufordernden Verantwortung gegenüber der Gesellschaft. Es ist ein Glücksfall, dass er als Hochschullehrer eine klare, starke und prägende Haltung gegenüber jungen Leuten als der nächsten beruflichen Generation einnimmt. Seine Erfolge, seine vielen Bauten und Projekte mit ähnlicher Aufgabenstellung verweisen als überzeugende Beispiele darauf, dass in der hochkomplexen Welt der Architektur die sorgfältige Befassung mit den Einzelheiten, die nahe am Auge und nahe an der Hand sind, für die individuelle Wahrnehmung von realer Größe, Proportion, Form, Farbe und Textur bis zur Groß-Dimension des „Sportpalastes“ für das kollektive, synchrone Erlebnis in der Masse von Zehn- ja Hunderttausenden ein enormes Spektrum bietet und dass die bewusste Konzentration auf letztlich beispielgebende Qualität von Bauten zur Entwicklung der Architektur letztlich die maßgeblichen Beiträge leistet. Seine ausgeprägte Position in Verbindung mit dem umfangreichen theoretischen Wissen und praktischer Erfahrung aus jahrzehntelanger Tätigkeit (rund 60 Projekte zu Sportbauten, von denen etwa ein Drittel realisiert ist) veranlassen ihn in dieser Buchpublikation sich nicht auf einen Werkbericht über eigene Bauten und Darstellung von deren besonderen Merkmalen zu beschränken. Sein theoretischer „Unterbau“ und die Basis für die eigene Positionierung legen es nahe, dass er sich – nach einem Abriss geschichtlicher Vorläufer der Arenen für sportliche Veranstaltungen – zunächst ausführlich mit dem Status quo im eigenen Land nach knapp zwei Jahrzehnten des enormen Baubooms in Form einer Art Bestandserhebung befasst, die auch nicht mit deutlicher Kritik spart. Seine textlichen Kommentare zur Entwicklung bewirken einerseits eine Klärung im Sinne der für Lehrende verpflichtenden Weitergabe eindeutiger Aussagen. Andererseits belegt er naheseine Kommentare durch Bildmaterial, das in der Logik des Gezeigten deutlich seine eigene Bewertung unterstreicht. Dies gilt insbesondere, wenn die maßgeblichen Entwurfs- und Planungsgrundlagen aus Unkenntnis der Autoren zum Teil erheblich vernachlässigt werden und häufig zugunsten von Formalismen Baukörper realisiert wurden, die offenkundig geradezu artistisches Potenzial bei der Verwirklichung der Konzepte aufzeigen.

Sun fordert mit großem Nachdruck den Stand des Wissens auf diesem Sektor als professionelle Bedingung einzubringen und beschränkt sich dabei nicht auf einige verbale Monita sondern bildet in seiner Darstellung in klarer, logischer Folge Schwerpunkte zu den Bereichen einer urbanen Rationalität, einer funktionalen Rationalität und schließlich zur technischen Rationalität. Er bespricht Bauten, die eine Vielzahl von Fragen – zumindest Zweifel – was die technisch-funktionale Seite, ihre konstruktive Identität und die gelegentlich durchaus brisante

Frage der Dauerhaftigkeit, beispielsweise bei den Detailausbildungen in bauphysikalischer und baukonstruktiver Hinsicht nahelegt. Sehr differenziert stellt er die Vielfalt von Entscheidungen dar, die auf Grundlage zahlreicher Anforderungen aus den unterschiedlichsten Bereichen, aber auch innerhalb der erwähnten Schwerpunkte anstehen – sei es aus Gründen der gewollten Zielsetzung oder aus dem Bauen als komplexem Vorgang innewohnenden Zusammenhängen.
Keineswegs werden dabei Konflikte aufgrund einer Präferenz bei den Entscheidungen zunächst vermieden. Vielmehr – und darin liegt ein besonderer Wert seiner Ausarbeitungen – werden durch systematisch erfasste und dargestellte Kennzeichnungen im umfangreichen Theorieteil des Buches die Entscheidungen als Optimierungsvorgänge zwischen Widersprüchlichkeiten bewusst gemacht, erreichbare Vorteile dargestellt, aber auch mögliche oder zu erwartende Nachteile unverblümt benannt. Allein schon durch diese Art der Behandlung des Themas wird die Arbeit von Yimin Sun als Architekt zu einer in vielen Punkten nachvollziehbaren und damit auch für andere mit dem Planungs- und Bauvorgängen aufgrund ihrer Rolle verwobenen Institutionen und Individuen nachvollziehbar gemacht. Ein hohes Maß an Rationalität wird aber auch vorgeführt durch eigene bauliche Ergebnisse, bei denen intensiv vermittelt wird, dass deutliche Zweckorientierung (bekanntlich ein Hauptmerkmal von Gewerbegebieten, die oft am Rand der Städte liegen und den Eindruck vermitteln, als handele es sich um ästhetisch völlig belanglose Konglomerate) sehr wohl zu einem überzeugende architektonische Gestaltung führen können. Letztlich heißt dies, dass für die Ansprüche, welche bei der Gestaltung von Bauten und den zwischen ihnen liegenden Freiräumen durchgängiges Merkmal von kultureller Kompetenz sein müssen in vielfältiger Weise Präferenzen bestehen, die – entsprechende Arbeitsbedingungen vorausgesetzt – Vorbildcharakter haben.

In diesem Zusammenhang erscheint mir der Beitrag von Yimin Sun, den am Beispiel der Konzentration auf einen funktionalen Bereich – eben den des Sports – charakterisiert, gerade die Unterschiedlichkeit der Resultate von ganz besonderer Bedeutung. So sehr auch für die Realisierungen Produkte verwendet werden, die aufgrund heutiger Transport- und Distributionsmöglichkeiten jeweils nach besten funktionalen Merkmalen ausgewählt weltweit zum Einsatz kommen mögen, so ist doch deutlich, dass die Bauten selbst ein hohes Maß an Individualität, Unterscheidbarkeit und Prägung des Ortes erreichen. Möglich ist dies durch die spezielle Form der verschiedenen Stufen und Maßstäbe der Bearbeitung der Projekte und der übergreifenden stadträumlichen Überlegungen über die sorgfältige Dimensionierung und Positionierungen der Baulichkeiten bis hin zu Einzelheiten der Gebäude in den von den Menschen wahrgenommenen und erlebten Maßstab 1:1, also ihrer natürlichen Größe.

März 2016

Thomas Herzog (1941–), born in Munich Germany, architect BDA, member of International Academy of Architecture (Sofia, Bulgaria), the Academy of Arts (Berlin); Académie d' Architecture (Paris, France); Bavarian Academy of Fine Arts (Munich) and Russian Academy of Arts and Sciences (St Petersburg, Russia). He is an internationally successful architect with focus on technology and sustainability. He has received numerous honors and awards for his buildings, including Gold Medal from the Association of German Architects (BDA), Mies van der Rohe Award, Global Award for Sustainable Architecture, Auguste Perret Prize from the International Union of Architects, Fritz-Schumacher-Prize for Architecture, Grande médaille d' or from the Académie d' Architecture. He was professor and dean at the Technische Universität Munich and guest professor at Tsinghua University, École Polytechnique Fédérale de Lausanne and Royal Academy of Copenhagen. Thomas Herzog demonstrated how to make useful, sustainable structures contextual and adaptable through interdisciplinary research. This pioneering research and the quality of the buildings brought him international fame and promote the development of sustainable architecture worldwide.

前　言

《精明营建：可持续的体育建筑》撰写初衷在于为我国体育建筑的建设决策提供科学支持，并进一步形成系统的可持续设计方法与理论。笔者在研习前辈经验，反观学界现状的过程中意识到，我国体育建筑学在20世纪中、下叶曾有过产研并重的理性开端，但进入新世纪后逐渐呈现理性研究日渐匮乏、主观决策愈加普遍的不良趋势。我国目前体育建筑人均占有率极低，与之形成强烈反差的是追求形式、成本畸高的现象司空见惯，大型体育场馆赛后高成本、运行困难的问题屡见报端。体育建筑建设迫切需要回归科学定位、理性决策的轨道中来。

笔者有幸在20世纪80-90年代师从梅季魁教授，毕业于哈尔滨建筑工程学院，成为我国培养的第一位体育建筑学博士。毕业至今工作于广州的华南理工大学建筑学院，专注体育建筑为代表的大型公共建筑、城市设计科学研究，又适逢国家经济社会快速发展、体育事业逐渐繁荣，主持完成了包括奥运、亚运、大运及全运会在内的多项体育场馆设计实践。本书是笔者对多年体育建筑科学研究与设计实践成果的总结，凝练为“精明营建”的理念。这一理念的提出，试图突破专业设计的习惯界限，在更加宽广的领域探讨解决问题的思路。所谓“精明”，指向了决策的全面性、科学性，而“营建”则突破“建设”的狭隘范畴，突出了基于建造的“筹划”。

本书包含“论述”和“作品”两大部分，前者是围绕“精明营建”的理论论述，在总结我国体育建筑发展历程、建设趋势、使用现状，并进行问题剖析的基础上，提出可持续体育建筑的决策思想和设计方法。

1. 注重“全过程”的决策思想。强调全过程，重视早期科学决策，突破传统研究局限，从初始成本、维护成本和更新成本三个方面控制和降低全寿命周期成本；以集约建设为原则，提高体育设施规划布局的灵活机动性，研究符合国情、低成本、低损耗的体育建筑设计策略。

2. 基于整体环境，强调“城市理性”的可持续设计方法。从整体性出发，在学界率先提出“基于城市的体育建筑设计”理念，建立与整体环境影响相关联的科学分析，结合城市设计方法，从功能整合和空间整合两方面完善体育建筑效能，实现体育建筑赛后融入整体环境的可持续运营。

3. 基于灵活、适应性，坚持“功能理性”的可持续设计方法。根据体育建筑的功能构成关系，从功能的可持续性出发，结合计算机辅助技术，提出基于参数选择的建筑功能灵活适应性设计方法，通过空间设计和设施利用的优化与控制，为赛时机能转换，赛后功能应变获得极大的灵活效能。

4. 基于集约、适宜性，满足“技术理性”的可持续设计方法。基于体育建筑的技术需求特点，对大跨结构选型、设备系统设计、容积体积控制、自然通风与采光等进行多目标综合优化研究，强调对成熟“适宜技术”创新性应用与先进建筑体系实施的互相补充，实现低成本的可持续运营。

本书的图文资料涉及到设计团队完成的22项体育建筑工程，“作品”部分选取了其中的10项代表性案例，包含北京奥运会场馆2项、广州亚运会场馆3项、深圳大运会场馆1项、省市级体育场馆4项。这些作品突破了既有建筑范式，形成新的城市空间与建筑复合功能，以低成本方案成功解决技术难点和关键问题，赛时满足复杂功能转换，赛后多年运行良好。其中奥运摔跤馆是北京奥运最低造价、最节约材料的场馆，亚运武术馆技术先进、构思独特，两者均获得国际体育建筑最高奖项——IAKS奖，分别成为北京奥运和广州亚运获奖工程中唯一的中国建筑师自主创新成果。国际合作设计工程深圳宝安体育场和佛山世纪莲体育中心也分别获得IAKS奖。

然而笔者深知，要形成完善的体育建筑可持续理论体系尚需大量基础研究和应用实践，笔者本人及所带领的学生和设计师团队一起，亦不断结合新的工程实践，开展国际合作、计算机技术开发与应用、使用后评价等工作，持续地完善这一知识系统。本书希望能在该领域抛砖引玉，给出初步轮廓，引起学界讨论。书中错误在所难免，恳请广大读者能够给予批评指正。

2018年12月

广州，中国

Preface

The original intention of writing SMART PLANNING AND BUILDING - Sustainable Sports Facility is to provide scientific supports for decision-makings in constructing sports facilities in China and to generate a system of sustainable design approaches and theory. In the process of studying previous experience and reflecting on the status quo in the academia, the author found that sports architecture in China had had a rational start which attached importance to industry and research in the middle and latter half of the 20th century. While in the 21st century, it gradually stepped into a less attractive pace with decreasing scientific researches and increasing subjective decision makings. Currently, the per capita occupancy of sports facility in China is extremely low, nevertheless, in stark contrast to this, the phenomenon of pursuing styles and excessively high cost is pervasive. Problems which involve high cost and tough operation after sports events for large scale venues are frequently reported by the media. It is urgent to steer the building of sports facilities back into a track that puts emphasis on scientific positioning and rational decision making.

The author had the privilege to study under the guidance of Professor MEI Jikui during the 1980s to 1990s. He graduated from Harbin University of Civil Engineering and Architecture and is the first person who was trained and awarded a Ph.D. degree of Sports Architecture in China. From the day of his graduation, he has been working in the School of Architecture in South China University of Technology. He has been engaged in scientific researches for large-scale public buildings featuring sports facilities and urban design. With China's rapid growth of economy and society, the sports undertaking increasingly thrives. In this period of great development, the author presided over the design of multiple sports venues which serves for sports events ranging from the Olympic Games, the Asian Games, the Universiade to the National Games. This book is a summary of the author's scientific researches and design practices in sports facilities through the years, which can be conceptualized as SMART PLANNING AND BUILDING. The concept is put forward by the author in an attempt to push the common limits of professional design and probe into approaches for problems in a wider realm. SMART demands the comprehensiveness and scientific nature of a decision, while PLANNING AND BUILDING isn't confined by the limited scope of BUILDING, instead, it highlights the PLANNING based on building.

The book is divided into two parts: Discourse and Works. The first part focuses on expressing the concept of SMART PLANNING AND BUILDING and, by summing up the history, trend and status quo of China's sports facilities as well as analyzing its problems, it proposes decision-making philosophy and design approaches for sustainable sports facilities.

1. The decision-making philosophy attaches great importance to whole-course involvement. By whole-course involvement, it means to put emphasis on rational decision-making in the initial period and to break through the limits of traditional research, checking and cutting the life cycle cost in three aspects: initial cost, operating expense and renewal cost. The practice principle is intensive building so as to promote the flexibility of sports facility planning and work out design strategies for sports architecture that are desirable for the country, cost-effective and with low loss.

2. Based on the integrated environment, it adopts sustainable design approaches that highlight 'Urban Rationality'. Keeping the integrality in mind, the author takes the lead in proposing the concept of 'Sports Architecture Design from the Perspective of the City', which carries out scientific analysis that are related to the influence of integrated environment. With urban design approaches, it perfects the sports building efficiency through functional integration and spatial integration, hence realizes sustainable operation of sports facilities after sports events.

3. Based on flexibility and adaptability, it sticks to sustainable design approaches that highlight 'Functional Rationality'. Correlations are formed in line with the functions of sports facilities. Taking functional sustainability into consideration and adopting computer-aided technology, we offer design approaches for flexible adaptability of architectural functions that are based on parameters. By optimizing and controlling space design and facility utilization, it creates great flexibility in terms of functional transformation in sports event season and functional operations after the games.

4. Based on intensiveness and suitability, it advocates sustainable design approaches that conform to 'Technical Rationality'. In line with technical features of sports building, researches are conducted to integrate and optimize multiple aspects ranging from the model selection of large-span spatial structure, the design of equipment system, the control of volume and capacity, natural ventilation and lighting. It stresses the complementary implementation of novel application of appropriate existing technology and advanced building system, in this way, sustainable operation at low cost will be realized.

The textual and graphic files in this book involves 22 sports building projects carried out by the designer team, of which 10 representative cases are selected in the Works part, including two projects for the Beijing Olympics Games, three for the Guangzhou Asian Games, one for the Shenzhen Universiade and four for provincial and municipal level sports events. These works got rid of the existing building paradigm and gave rise to new functions that integrate urban space and buildings. They are successful in overcoming technical obstacles and solving critical problems at low cost. The sports facilities in these projects cater for complicated functional transformation in sports event season and maintain sound operation for years after the games. The Olympic Wrestling Hall is the lowest costly and most material-effective one in the Beijing Olympic Games while the Asian Games Wushu Hall boasts advanced technology and unique design conception. These two projects won the IAKS Architecture and Design Awards, the most recognized international award for sports facility. They respectively represent the only innovative work independently designed by Chinese architects among all the awarded projects in the Beijing Olympic Games and Guangzhou Asian Games. In addition, the Bao'an Stadium in Shenzhen and Century Lotus Stadium in Foshan, which are cooperatively designed by teams both at home and from abroad, obtained the IAKS Architecture and Design Awards as well.

Nevertheless, the author knows well that massive fundamental researches and application practices remain to be done in order to perfect the sustainable theory for sports building. The author, together with his students and designer team, will continuously enhance international cooperation, promote development and application of computer technology and make post occupancy evaluations so as to better the theory. The book is expected to be a catalyst that provides a preliminary outline and initiates discussions in the academia. It is inevitable to avert every error in this book, if any, please offer us your criticism and feedback.

SUN Yimin
December 2018
Guangzhou, China

目　录

Content

论述
Discourse

一、现代体育与体育建筑

I. MODERN SPORTS AND SPORTS ARCHITECTURE

体育运动的现代发展得益于现代奥林匹克运动的缔造者顾拜旦（1863 ~ 1937），他让体育与现代社会融汇在一起。

The development of sports in modern times owes credits to the founder of the modern Olympics, Pierre de Coubertin (1863~1937), who had woven sports into modern society.

……

啊，体育，你就是进步！
为人类的日新月异，
身体和精神的改变要同时抓起，
你规定良好的生活习惯，
要求人们对过度行为引起警惕。
你告诫人们遵守规则，
发挥人类最大能力，
而又无损健康的肌体。
……

——《体育颂》皮埃尔 · 顾拜旦

......
O Sport, you are progress!
To serve you,
a man must improve himself both physically and spiritually.
You force him to abide by a greater discipline;
you demand that he avoid all excess. You teach him wise rules
which allow him to exert himself with the maximum of intensity
without compromising his good health.
......
-'Ode to Sport', by Pierre de Coubertin

图 1-1　波赛东海神庙
Fig.1-1　Temple of Poseidon

图 1-2　雅典竞技场
Fig.1-2　Athens Stadium

图 1-3　1936 年柏林奥运会的德意志帝国体育场
Fig.1-3　Deutsches Stadium in 1936 Berlin Olympic Games

现代奥林匹克运动的发展同样对体育设施的发展起到了决定性的推动作用。100 多年前，现代奥林匹克运动在希腊重生。尽管古希腊建筑留下了经典的辉煌，雅典建筑更加是极尽雄伟（图 1-1），但第一届奥林匹克运动会设施却极为简单（图 1-2）。作为运动者的舞台，突出的是运动员，这在某种程度上反映了顾拜旦的体育理想——“警惕过度行为”“发挥最大能力”“无损健康肌体”。

The progression of the modern Olympics has been playing a decisive role in promoting the development of sports facilities. When the modern Olympics was reborn in Greece over 100 years ago, though despite of the architectural brilliance and magnificence of Classical Greece (Fig. 1-1), the facilities of the first Olympic Games in Athens were extremely simple (Fig. 1-2). Sports is a stage that gives prominence to athletes, which, to a certain extent, reflects the ideals of Coubertin's 'avoid all excess', 'maximum of intensity', and 'without compromising his good health'.

然而，现代奥林匹克质朴、健康的历史并没有很久，这项人类竞争与欢庆的盛会规模越来越大，逐渐受到政府的重视。很快，将奥运会作为政治炫耀、民族自豪的想法开始出现在德国。1936 年的柏林奥运会，纳粹德国凭借奥运会的举办而宣扬德国自第一次世界大战战败后的复兴，体育设施建设与民族、政治的炫耀首次相连。体育中心的建设也达到空前的规模（图 1-3）。

However, the pristine and healthy history of the modern Olympics did not last for long, The grand gathering for contest and festivity of mankind grew in scale and on the government. It did not take long before Germany considered the Olympic Games as a carrier of political showoff and national pride. The Nazi Germany propagandized the revival of Germany from the WWI after its defeat in the 11th Olympic Games in Berlin in 1936. It was the first time the construction of sports facilities was associated to national sports and political parade. The construction of sports centers became an unprecedented campaign (Fig. 1-3).

意大利于 1935 年筹备“罗马环球博览会”（Universal Exhibition of Roma，EUR），在古罗马的西南开发新的城区，借此向世界展示“新

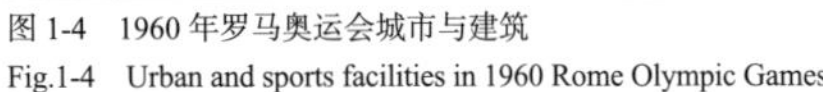

图 1-4　1960 年罗马奥运会城市与建筑

Fig.1-4　Urban and sports facilities in 1960 Rome Olympic Games

罗马帝国”的辉煌。然而，初具规模的现代花园城市规划开放被战争打断。二次大战后的，罗马的城市发展已大大落后于欧洲其他大都市。同时，战后的东西方阵营都十分重视奥林匹克运动会的社会政治意义。战败国意大利试图通过奥运宣示自己痛改前非，从过去的法西斯国家脱胎换骨，重返国际大家庭。1960 年罗马奥运会的举办，让墨索里尼时代的罗马新建筑与新城区融合复兴，意大利靓丽回归西方阵营。1960 年的罗马奥运会，不仅仅为世界展示了经典的体育建筑，而且分开设置的体育中心，更好地根据所在城市区位进行规划设计，很好地均衡了标志性与社区感，大型体育设施与城市良好地融为一体（图 1-4）。

1936 年，刚刚从 1924 年关东大地震中恢复过来的日本，国力日渐雄厚，受德国和意大利的启发，也开始关注申办奥运会。当时军国主义体制下，日本在东亚的扩张屡屡得手，越来越心高气傲，开始一切向纳粹看齐，瞄准 1940 奥运会，以此展示帝国荣耀，而其中最

In 1935, Italy located the Universal Exhibition of Roma, EUR at a new urban area in the southwest of Roman to show the world the brilliant brandish to the world the ‘New Roman Empire’. However, the rudiments of the planning of the modern garden city were interrupted by the war. After WWII, the urban development of Rome had significantly lagged behind other European cities. Meanwhile, the post-war East and West camps have attached great importance to the social and political significance of the Olympic Games. When the Olympic Games interrupted by WWII resumed, the vanquished Italy declared its repentant attempt to mend its ways at the Olympic Games, signifying the amendment for the once-fascist state and its return to the international family. The 1960 Rome Olympic Games marked the fusion and renaissance of the Roman architecture dated back to the Mussolini era, which gave Italy a stunning resurface when it was taken back by the Western camp. The Games not only demonstrated the classic sports architecture to the world, but also experimented with design and planning on independent and centralized sports facilities in accordance with urban spatial analysis. The solution found a healthy balance between iconicity and familiarity, between large-scale sports facility and urban condition (Fig. 1-4).

In 1936, encouraged by the increasingly strong national strength after its recovery from thc 1924 Great Kanto Earthquake and inspired by Germany and Italy’s revival, Japan begun to rise with interest in the Olympic bidding campaign. Under the militarism system, Japan’s success in its expansion in East Asia boosted the nation’s confidence and arrogance. Under the Nazi’s influence, Japan aimed at the 1940 Olympic Games to demonstrate the glory of the empire, of which the most important reason was that the 2600th anniversary of the enthronement of the first Mikado fell right on 1940, making the 12th International Olympic Games of that year the most desirable event in commemoration.

图 1-5　神宫外苑奥运体育场构想
Fig.1-5　Design concept of the Olympic stadium at the Meiji Shrine Outer Garden

图 1-6　驹泽体育公园
Fig.1-6　Komazawa elympic Park

为重要的原因是为纪念以第一代天皇即位为标志的所谓日本开国 2600 年，即 1940 年，恰好这一年举办的第 12 届国际奥运会成了最为理想的目标。

1936 年 3 月日本申奥计划大纲提出在东京靠近市中心的神宫外苑兴建能容纳 12 万人的新体育场（图 1-5）。

1936 年 7 月日本申奥成功。1938 年 4 月，东京主办者放弃了在神宫外苑地区建设大型体育场的计划，确定在距离市区稍远的驹泽兴建体育中心。并很快发表了驹泽体育设施的设计构想（图 1-6），随后，侵华战争全面爆发，中国人民坚强抵抗让日本快速结束战争的希望破灭，很快日本放弃了 1940 年奥运会的举办。

图 1-7　1964 年东京代代木国立综合体育馆
Fig.1-7　Tokyo Yoyogi National Gymnasium in 1964

战后的日本也怀着与意大利同样的思维，努力恢复正常国家的身份，积极筹备承办国际体育赛事。1955 年，为筹备亚运会和准备奥运会，建设了规模为 5.2 万人的国立竞技场，举办了 1958 年亚运会、1964 年东京奥运会。而新建的代代木、驹泽体育公园不仅在奥运期间承担了比赛，也在奥运建筑史上首次写下东亚人辉煌的一笔（图 1-7）。

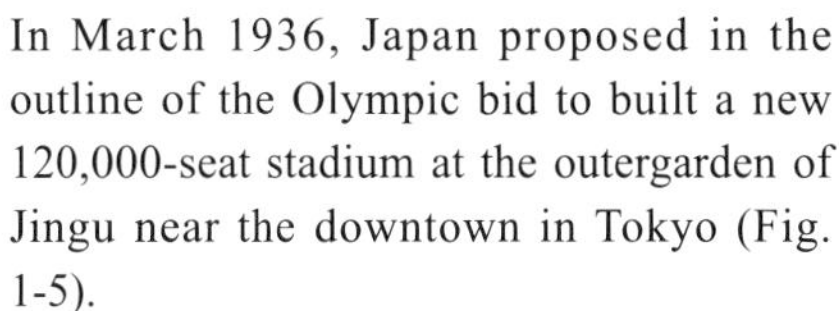

In March 1936, Japan proposed in the outline of the Olympic bid to built a new 120,000-seat stadium at the outergarden of Jingu near the downtown in Tokyo (Fig. 1-5).

After Japan won the Olympic bid in July 1936, the Tokyo organizers abandoned the outer garden for Komazawa in April 1938, which was a far away from the downtown. Soon, the design concept of Komazawa sports center was laid on the table (Fig. 1-6). But the strong impact of Chinese resistance force against the Japanese invasion smashed the Japanese dream of ending the war quickly, seeing which, Japan gave up the 1940 Olympic Games.

The postwar Japan, nourishing the same thoughts as Italy, strove to restore the identity of a normal country, and actively prepared to host international sporting events. In 1955, in preparation for the Asian Games and Olympic Games, Japan built a 52,000-seat 'National Stadium' where the 1958 Asian Games and the 1964 Tokyo Olympic Games were staged. The new Yoyogi and Komazawa Sports Parks recorded East Asian people on a glorious page of the history of Olympic architecture (Fig. 1-7).

Luxury culminated at the 1976 Montreal Olympic Games to an extent that some sports facilities were not completed when the Games started and the highly indebted cities were reduced to the verge of bankruptcy thereafter. The aftermath raised public introspection on the true meaning of Olympic Games. But the organizers did not come around until the 1984 Los Angeles Olympic Games, which set a completely new example for the design and construction of sports facilities.

Meanwhile, large-scale construction activities in developed Western countries have diminished since the late 20th century, along with a shift of focus in architectural researches. While the construction of large

1976年蒙特利尔奥运会，则让奥林匹克的奢华夸张达到了顶点，以至于赛时体育设施无法完工，赛后城市负债几近破产边缘，人们对奥林匹克的真正含义开始反思。直到1984年的美国洛杉矶奥运会，举办者思想理念彻底转变，为体育设施的建设做出了全新的范例。

与此同时，20世纪后期开始，西方发达国家大型建设活动已趋减少，建筑相关科研重心日渐转移。大型体育设施的建设逐渐沉寂，除了各类体育盛会对体育建筑的发展起到推动作用——如世界杯成为仅次于奥运会的大型赛事——其他大型体育场馆建设项目多结合职业运动的产业要求进行。比如美国的冰球、棒球、橄榄球、职业篮球则形成了完善的职业体育产业与相应的体育建筑体系。职业运动的商业化，将体育建筑的公共性意义进行了本质上的转化，尽管职业化的体育产业带来了技术、设备与观念的更新，但就建筑而言，在分析借鉴的同时需要加以区分，避免差异化的理解。建筑师尤其应该加以重视。

sports facilities gradually ebbed, most large projects started to take into account the requirements of professional training, notwithstanding the building industry was still propelled by international events such as the World Cup. For example, in the USA, a complete industry for professional training and a correspondent system for sports architecture have formed for ice hockey, baseball, football, and professional basketball in the USA. The commercialization of professional sports has changed the nature of the public significance of sports architecture. Despite the update introduced by professionalized sports industry to technology, equipment and ideas, architecturally speaking, it is necessary to separate sports architecture from the rest with analysis and reference to avoid differentiated understanding. This is important to architects in particular.

图1-8 1953年建成的广州越秀山体育场（上）和1951年建成的重庆大田湾体育场（下）

Fig.1-8 Guangzhou Yuexiu Mountain Stadium in 1953 (up) and Chongqing Datianwan Stadium in 1951 (down)

体育建筑在我国的发展

1. 理性开端，产研并重

1950年代开始，体育建筑受到政府重视。新中国成立初年就在北京、重庆、广州、长春等地建设大型体育馆。同时国际大型体育赛事的申办也提上日程。而后，北京工人体育场、体育馆的设计和兴建，在技术与理念上都在国际上有一定的地位。即使是在文革前后，北京和上海依然兴建了两座万人体育馆（图1-8）。

“文革”期间，体育建筑的研究仍有积累，1979年召开的体育建筑会议和出版的论文集代表了当时的学术水平。首次赢得的综合性国际体育盛会给中国的体育建筑带来了第一个春天。1984年中国体育科学学会、中国建筑学会会同建筑师、教授、体育管理人士联合成立了“体育建筑专业委员会”。其后，体育建

Development of Sports Facilities in China

1. The rational beginning with both study and practice

Sports architecture began to capture governmental attention since the 1950s. Large venues were erected in Beijing, Chongqing, Guangzhou, Changchun and other cities in the founding year of the foundation of the PRC. Bids for major international sporting events were also on the agenda. Design concepts and building technologies of large-scale venues of the time were globally renowned. Even before and after the Cultural Revolution, two gymnasiums, both capable of seating tens of thousands of people were erected in Beijing and Shanghai (Fig. 1-8).

With countinued researches of sports architecture throughout the Cultural Revolution, an assembly of sports architecture examplified of the general

图 1-9　1990 年北京亚运会场馆
Fig.1-9　Venues of the 1990 Beijing Asian Games

筑专业委员会形成针对工程建设的研究准备，1985、1986 年数次研讨交流、出国考察都有针对性。很快，体育建筑专业委员会委员们就为 1990 年北京亚运会做出了令人刮目相看的贡献。亚运会的所有场馆均由国内建筑师完成，不仅成功地举办了当时国内最重要的体育赛事，而且在体育建筑多功能研究、结构技术创新等多方面展现出中国建筑师的实力。马国馨院士主持的北京亚运中心体育场馆获得国际体育建筑奖（IAKS），成为首位获得国际体育建筑最高奖项的中国建筑师（图 1-9）。

academic level was held in 1979, accompanied by symposium publications. The first successful bid for a international multi-sport event brought to China's sports architecture the first flourishing season. In 1984, China Sports Science Society and Architectural Society of China, joining up with architects, professors, sports officials, established the Sports Building Committee, which organized targeted discussions and oversea investigation in 1985 and 1986 in preparation for the research of engineering construction. Soon the members on the committee made impressive contributions to the 1990 Beijing Asian Games, of which all venues were completed exclusively by domestic architects. With the most awared multi-sport event of Chinam held inside, the venues dsemonstrated Chinese architects' skill in multi-funtional and technical innovations. The academician Ma Guoxin chaired the Center Stadium of the Beijing Asian Games that won the International Sports Architecture Award (IAKS), and became the first Chinese architect to win the highest award for international sports architecture (Fig. 1-9).

1987 年全国运动会在广州召开，天河体育中心成为当时国内首次集中兴建的大型体育中心，开创了“一场两馆”的综合体育中心模式，也为体育建筑带动城市发展做了全新的尝试。

To host the 1987 National Games, Guangzhou built the Tianhe Sports Center, the first sports center of a masterplan scale. The project established the archetype of 'one-stadium-two-gymnasium' plan for future sports center designs. It was proved to be a groundbreaking attempt to encourage urban development by constructing sports centers.

2. 理性研究日渐匮乏，主观决策愈加普遍

1990 年代以后，特别是进入新世纪，从奥运会、亚运会到全运会、省运会，在各种运动会申办的推动下，国内城市体育设施建设的积极性不断高涨。特别是北京奥运体育建筑的国际招标，极大地推动了我国体育建筑的发展。同时也应该看到，由于各级政府行政意志不断催化建设的标志性，伴随着跃进式的建设

2. Rational research increasingly losing ground to subjective decision-making

Ever since the 1990s, especially in the 20th century, domestic cities have become obsessed with Olympic Games, Asian Games, National Games, Provincial Games and a variety of sports events; enthusiasm for the construction of sports facilities has been on the rise. In particular, the international tender for the sports venues

速度和普遍的浮躁与急功近利，违背体育建筑基本规律的错误设计经常重复出现，体育建筑的相关研究明显滞后。体育建筑追求怪异的形态成为定式，助长了体育建筑设计的非理性趋势。体育建筑设计出现了广泛的异化现象：追求怪异的形态、浮华的表皮，导致造作的结构和重复、夸张的表皮构造，体育建筑设计标准不断升高，建设成本不断攀升。当体育建筑的评价标准已建立在美丽、虚无的故事和无病呻吟的惆怅之上时，其创作必然热衷于浮华、夸张的表皮。花样翻新的表皮暴露出的却是对体育建筑基本设计原则的无视，体育建筑内在的灵活性和适应性，以及节能降耗、提高公共开放使用率的核心价值已被遗忘。

以建筑功能类型为基础的设计研究迅速成为过去，作为大型工程设计项目，体育建筑技术复杂、项目设计费含金量高，越来越受到设计单位的重视，但作为最复杂的建筑类型之一，其相关研究却没有赢得应有的地位。体育建筑曾经作为现代社会技术进步的代表，伴随现代建筑的发展，尽显华芳。时至今日，当体育产业日益贴近市民生活，已被冷落多时的体育建筑研究突然变得迫切了。

体育事业由国家兴办、体制单一，制约了城市体育设施的发展。随着体育体制向社会化方向的逐步转变，体育产业化的趋势日趋明显。然而，限于学科分类，体育院校无法对体育建筑进行深入研究；建筑和城市规划学科也没有从根本上予以重视；设计单位发表的论文又多限于以工程实践总结为主的评论与分析；设计师思想被动，研究范围狭小，缺乏深度与广度——上述现状导致体育设施建设中存在的普遍问题是：不了解体育工艺要求、决策主观，导致建成后使用不便，形成二次浪费；城市规划布局不尽合理，片面强调体育建筑的体量宏伟与标志性，与周围环境严重冲突，致使体育馆的建设与养护矛盾突出。另一方面，

of the Beijing Olympic Games was a great boost of the development of sports facilities in China. It should also be noted that, as administrative wills imposed the importance of iconicity on urban constructions, in the context of social impetuosity and anxiety, flawed designs against the basic principle of sports architecture occur frequently, while referential studies drop significantly. Consequently, a growing preference for sports facilities for bizarre forms fueled the irrational trend of sports architectural design. As a result, bizarre elements have taken over sports architectural designs: weird shape and flashy surface lead to pretentious infrastructure and repetitive, exaggerated surface structure. Hence the standards and costs for sports architecture soar unstoppably. When the criteria for sports architecture are based on nihilized narratives and pretended melancholy, they expedite the production of flamboyant cocoons of emptiness. What revealed by flashy surfaces is the ignorance of the basic principles of sports architectural designs, the inherent flexibility and adaptability of sports facilities, as well as the core values of energy saving, consumption reduction and better public usage have been forgotten.

Today, when the design research based on the types of architectural function quickly became out of time, sports facility has won over the attention of architectural firms due to its complex engineering and abundant payoff. However, its research did not receive its due. Sports facility, once the representative of technological advances of the modern society, has had its days along with the development of modern architecture. While the sports industry shifts increasingly close towards the life of citizens, the research of sports facility, which has been neglected for a long time, suddenly becomes in urgent demand.

Organization of sports events by the government in a unitary system has slowed down the development of urban sports facilities. As the sport system gradually

体育投入渠道正在多样化，运动、休闲、娱乐等多功能的体育设施不断涌现。全民健身运动的开展和体育产业化对体育设施提出了许多新的要求，客观上需要加强体育建筑研究，为丰富多样的体育活动提供高效灵活的设施。

北京奥运会申办成功以及全运会的主办权改由各省市竞争申请，激发了各城市兴建体育设施的热情，全国各地体育设施建设蓬勃发展，兴建规模大、标准高、配套完善的竞技体育设施已成为各地的目标。然而由于缺乏科学研究，导致主观决策，忽略了体育建筑基本的功能要求，建筑设计盲目求新求变，不仅许多新的体育功能无法满足，一些过去的研究结论也得不到应有的重视。虽然建设成本日益升高，设施灵活性、适应性却没有明显改观。2004年，在北京奥运主要设施已确定实施方案后，开展了长达半年的设施“瘦身”讨论，这样前后颠倒的过程，凸显体育建筑研究的缺位，证实了早期建设决策缺乏科学研究的支撑。北京奥运建设迎来了许多国际建筑师，然而国际著名体育建筑专业设计公司的方案却大多落选。一些境外公司，体育建筑的研究与设计经验积累少得可怜，却因体育建筑之外的设计构思中标项目。在如此情形下，这些建筑的建设“成就”就在于对其建造复杂性的高投入、高成本的成功解决，但对体育建筑科学研究的促进却十分有限（图 1-10，图 1-11）。

3. 呼唤理性，需要精明

作为大型公共建筑和城市重大公共投资项目，体育建筑不仅功能复杂，而且承担着许多社会责任，其建设迫切需要回归科学定位、理性决策的轨道中来。进入后奥运时代，体育建筑需要面对和解决的科学问题依然很多。首先是对体育场馆的科学定位。国际上，公共体育场馆的建设主体是以国家和公营事业投资为主，民营投资参加的多是职业化商业运营较为成功的领域。而我国，体育建筑建设与养护矛

turns gradually of socialization, the industrialization trend of sports has become apparent. However, the categorization of subject has hindered sports institutions from conducting in-depth study of sports facilities; nor did architecture and urban planning disciplines take sports facility seriously. The papers published by design units are mostly limited to review and analysis of summary of engineering practices, let alone passive thoughts of designers, narrow scope of research, and the lack of depth and breadth. The phenomena listed above has brought widespread problems to the construction of sports facilities: a) poor understanding of sport technological requirements, subjective decision-making, resulting in inconvenient use and secondary waste after completion; b) unreasonable urban planning, undue emphasis on mass and posture of sports facility in a serious conflict with the surrounding environment, resulting in the prominent contradiction between the construction and maintenance of venues. On the other hand, channels for sports activities diversify. Sports, leisure, entertainment and other multi-purpose sports facilities constantly rise in number. The nationwide development of civil fitness and the industrialization of sports entail many new requirements on sports facilities. The objective reality requires us to continue with the research of sports facility so as to provide efficient and flexible facilities for various sports activities.

The successful bid for the Beijing Olympic Games and the release of national bidding campaign for the National Games aroused the cities' enthusiasm to construct sports facilities. The vigorous nationwide constructions have set the decision-makers' mind to building large scale sports facilities of high standards with complete accessories. However, the lack of scientific studies have resulted in subjective decisions and ignorance of the basic functional requirements of sports facility, so that architectural designs become a blind pursue

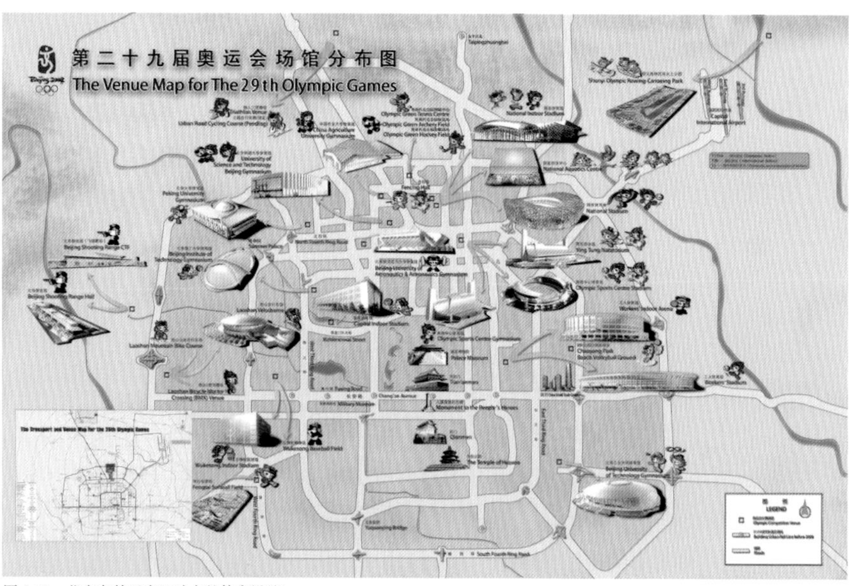

图 1-10　北京奥林匹克运动会的体育设施

Fig.1-10　The sports facilities of the Beijing Olympic Games

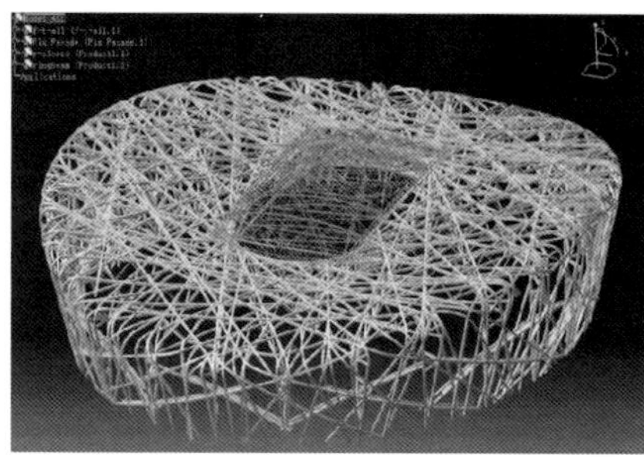

图 1-11　北京奥运主体育场瘦身

Fig.1-11　The downsizing of the main facilities of the Beijing Olympic Games

盾突出的主要原因在于定位混淆，存在前期公共投入控制不严、后续支持力度不够等问题。其次是我国体育场馆建设的盲目性。由于许多城市将大型体育场馆作为标志性工程建设，项目决策由主观肇始，建设初始缺乏科学论证，建筑标准定位不当，导致建设主体内容不准确、规模确定随意、项目策划不科学，重复建设、恶性竞争严重。因此，科学决策的前提是全过程地关注体育场馆的建设，由此开展的科学定位至关重要。同时需要特别注意的还包括提高场馆布局的科学性、加强策划与可行性研究的论证与分析。

of innovation and novelty unqualified for function demands and ignorant to previous studies. Despite the escalating construction costs, the flexibility and adaptability of facilities has not been significantly improved. A six-month discussion of the downsizing of the main facilities of the Beijing Olympic Games, whose final plans have already been already determined was held in 2004. This reversed cause-effect is a telling evidence that the study of sports facilities is neglected and the early decisions of designs are not based on scientific study. The Beijing Olympic bidding had attracted many international architects; however, most of the designs from the world's leading sports facilities firms were voted out. Some foreign companies with poor research and design experience in sports facilities won the bid merely for concepts irrelevant to the sports facility. Under such circumstances, the only achievement of the final construction is the high-input and cost solution to the complexity of sports facilities without substantial contribution to the study of sports facilities (Fig. 1-10, Fig. 1-11).

3. Call for rationality and smartness

As a large-scale public building and major urban public investment projects, sports facilities bear not only complex functions, but also many social responsibilities, therefore their construction urgently needs to return to the scientific positioning and rational decision-making. In the post-Olympic era, sports architecture is still facing a lot of scientific issues suspended for solutions. The first is the scientific targeting of sports venues. Outside China, the construction entity of large public sports venues are mainly invested by the government and public enterprises. Private investment goes to the fields of successful commercial operation of professionalized sports. In China, the obvious contradiction between the construction and the maintenance of sports facilities can be primarily imputed to confused targeting, negligent preliminary control of public input, inadequate follow-up support, etc.

The secondary cause is the aimlessness of design and construction. In constructing large sports venues as landmark projects, subjective decision-making, insufficient justification and improper criterion-targeting result in ambiguous subject-matter, inconsiderate scale and irrational project planning. Repetitive construction and vicious competition prevail. Therefore, the precondition to scientific decision-making is the attention paid to stadiums construction throughout the whole process, and to prioritize scientific targeting. Also, special attention should be paid to scientific designs of venues, careful justification and analysis of planning and feasibility study.

我国体育建筑建设的新趋势

1. 投资运营主体趋向多元

传统的国内体育建筑项目多由政府投资，并组织附属单位负责运营管理，这类项目具有公益性和非营利性特点、新时期随着社会经济发展、政府财政收入提高，体育建筑项目仍维持政府投资为主的模式，与此同时尝试投资社会化的创新实践，发展项目融资模式，这一过程中利用项目自身未来的现金流量为基础进行融资。北京奥运会场馆建设即采用了多种投资方式，如射击馆由政府投资，赛后转变为国家射击队训练的场所；国家体育馆由北京国有资产经营有限公司和中信联合体组成项目公司，负责建设、维护和运营；国家游泳中心由华人华侨捐建，赛后移交国家运营。

目前国内体育建筑项目存在的投资运营模式主要有政府投资、事业或企业运营的模式，和以 BOT（建造—运营—移交）为代表的项目融资模式。深圳湾体育中心即采用 BOT 模式，将项目整体交给华润开发集团进行投资、建设和运营管理，在合同期满后再移交政府。BOT 在实际应用中可进一步产生 BT（建设—移交）、BTO（建设—移交—运营）、TOT（移交—运营—移交）、BOL 模式（建

The New Trend of the Construction of Sport Facilities in China

1. Diversification of the investment and operation mode

Traditionally, the construction of domestic sports facilities is mostly invested by the government, who organizes affiliated institutions that are responsible for the operation and management. Such projects are of public welfare and non-profit. In the new era, with the socio-economic development and the increase in government revenues, the government investment-based mode still maintains dominant in the construction of sports facilities. At the same time, the innovation and practices of investment socialization are tried out to develop the project financing mode, which uses the future cash flow of the project itself as the base for financing. The construction of the venues for the Beijing Olympic Games had involved a variety of investment modes. For example, the shooting hall was invested by the government and transformed into a training venue for the national shooting team after the Games; The National Stadium was handled by the joint project company composed of the Beijing State-owned Assets Management Co., Ltd. and the CITIC Construction Co., Ltd., who were responsible for the construction, maintenance and operation; The National

设一运营一租赁)和BLT(建设一租赁一移交)等具体方式。这些新型方式有利于降低政府负担,协调政府和社会机构之间的责权关系,提高项目运作效率。

在BOT模式的基础上,江门滨江新城体育中心发展出DBOT(设计一建设一经营一转让)的投资运营模式。这一模式允许社会机构在前期介入项目的策划开发,直至项目运营收回成本乃至获得收益。DBOT模式中,政府可以通过合同规定以保证项目一定的公益性,而负责建设运营的社会机构从中获得更多的决策权,有助于在市场需求预测的前提下,控制项目投资,节约开发成本以及降低未来的管理运营成本,从而实现政府、社会和企业三者共赢,对体育设施项目建设运营而言是一种创新尝试(表1-1)。

投资和运营要素与一方面与体育建筑的定位和规模要素存在从属关系,另一方面与其功能和空间要素存在互动关系。投资和运营模式的多元化趋向,为体育建筑建设和运营的可持续性提供多样化的解决方案。但这也意味着决策主体、服务对象、利益诉求的复杂化,这给场馆规模定位、功能空间、技术手段的设计和研究带来新的挑战。

体育建筑投资运营模式比较　　表1-1

投资运营模式	特点	优势	劣势	案例
传统模式	多由政府投资,并组织附属单位负责运营管理	政府负责投资和运营,项目公益性具有保障	行政意识和手段的不理性,容易导致场馆运营困境	北京奥运会射击馆、国家游泳中心
BOT模式	"建造—运营—移交",企业进行投资、建设和运营管理	有利于降低政府负担,协调政府和社会机构之间的责权关系,提高项目运作效率	经济效益的不合理追逐,容易导致社会公益性的不足	深圳湾体育中心
DBOT模式	"设计—建设—经营—转让",在BOT模式基础上,允许社会机构在前期介入项目的策划开发	社会机构获得更多的决策权,有助于控制项目投资,节约开发成本及降低管理运营成本	对社会机构提出更高的要求	江门滨江新城体育中心

Swimming Center was donated by the overseas Chinese and was handed over to government for operation after the Games.

At present, the investment and operation for domestic sports construction projects mainly include two modes. The first one is investment by government, and operation by business or enterprise, and the second one is the project financing mode represented by BOT (build-operate-transfer). For example, the Shenzhen Bay Sports Center adopted the BOT model that firstly it transferred the entire project to China Resources Development Group for investment, construction and operation management, and then handed over to the government after the contract expired. In practical applications, the BOT can further be derived to models such as BT (build-transfer), BTO (build-transfer-operate), TOT (transfer-operate-transfer), BOL (build-operate-lease) and BLT (build-lease-transfer), and so on. These new modes are conducive to reducing the government burden, coordinating the relationship between the government and social organizations, and improving the efficiency of project operations.

Based on the BOT mode, the Jiangmen Binjiang New City Sports Center has developed an investment operating mode, which is called DBOT (Design-Construction-Operation-Transfer). This mode allows social organizations to intervene in the planning and development of the project in the early stage, until the project being operated, costs being recovered and even profit being gained. In the DBOT model, the government can stipulate contracts to ensure certain public welfare of the project, and the social organizations that are responsible for construction and operation can obtain more decision-making authority, which is beneficial for controlling the project investment, saving the development costs on the premise of market demand forecast, as well as reducing the future management and operating costs, therefore achieving a

2. 场馆建设重点发生转移

奥运会、亚运会、大运会以及全运会等大型体育赛事的成功举办，促使一线城市大中型体育场馆建设的基本完善，对既有场馆赛后利用成为关注重点，场馆服务对象由竞技比赛转移到面向群众和学校的日常体育，并兼顾文艺演出和展览等活动类型。

建设分布重点由一线城市向二三线城市转移。随着整体经济发展，这些城市逐渐具备举办全国、全省性的大中型体育赛事，为此建立市级体育设施，并借此机遇带动新区开发建设。与此同时，县区级中小型体育设施也得到补充和完善。

Comparison among the investment and operation modes of sports facilities — Tab. 1-1

Investment and operation modes	Feature	Advantage	Disadvantage	Case
Conventional mode	The government investes and organizes affiliated institutions for the operation and management	Invested and operated by government, so the public welfare of the project is guaranteed	Irrational administrative consciousness and means can easily lead to the dilemma of venue operations	Beijing Olympic Shooting Hall, National Swimming Center
BOT	'Build-Operate-Transfer', enterprise investes, builds, operates and manages	Helps to reduce government burden, coordinate the relationship between the government and social organizations, and improve project operational efficiency	Unreasonable pursuit of economic benefits easily leads to the lack of social welfare	Shenzhen Bay Sports Center
DBOT	'Design-Construction-Operation-Transfer', based on BOT, allows social organizations to intervene the project in early stage	Social institutions gain more decision-making rights, help to control project investment, reduce development, management and operating costs	Have higher demands on the social institutions	Jiangmen Binjiang Sports Center

win-win situation among the government, society, and the enterprises. This is an innovative attempt for the construction and operation of sports facilities (Tab. 1-1).

The investment and operation are related to the positioning and scale of the sports facilities on the one hand, and interact with the function and space on the other hand. The diversification of investment and operating modes provides diversified solutions for the sustainability of sports construction and operation. However, this also means that the decision makers, the service objects, and the interest demands are becoming complicated, which brings new challenges to the design and research of the scale and positioning, function and space, and technical means.

2. Shift of the venues construction focus

The successful hosting of the major sports events, such as the Olympic Games, Asian Games, Universiade and the national Games, has prompted the improvement of large- and medium-sized sports venues in first-tier cities. The post-game comprehensive utilization of the existing venues has become the focus of attention. The venue's service targets have shifted from competitive competition to the daily sports of the masses and schools, and taken into account those activities such as theatrical performances and exhibitions.

The construction distribution focus has shifted from the first-tier cities to the second- and third-tier cities. With the overall economic development, these cities have gradually become competent to host the national and provincial, large and medium-sized sports events. To this end, they have established city-level sports facilities and taken the opportunity to promote the development and construction of new urban districts. At the same time, the small and medium-sized sports facilities on the district levels have also been supplemented and improved.

3. 场馆经营环境逐步改善

目前和未来可预见的一段时间内，我国体育场馆建设仍以公益性为主，承担非营利性的公益活动和训练比赛。随着体育产业的发展和居民体育消费能力的提高，体育场馆所具备的独特功能，为其开展营利性产业经营提供条件。

一方面，我国目前体育事业迎来高速发展，体育运动的职业化和全民化的推进使得“体育产业”成为资本追逐的一项热点。如中国足球超级联赛（CSL）、中国男子篮球职业联赛（CBA）等为代表的体育职业化，带动职业体育市场的逐渐成长；健康意识和消费能力的提高，成为“全民健身”群众体育市场发展的基础；北京张家口冬奥会、各单项世锦赛等体育事件的陆续发生，有望促进滑冰等单项运动成为新的消费热点。

另一方面，体育场馆提供的大空间，为大型文艺演出、展览活动提供条件。文艺演出和展览市场的逐渐繁荣，对大空间提出质和量的要求，这为体育场馆提高使用率、实现可持续运营提供需求动力。

图 1-12 “鸟巢”里的文艺演出

Fig.1-12 Theatrical performance in the 'Bird's Nest'

3. Gradual improvement of the operating conditions

For the foreseeable period of the present and the future, the construction of the sports venues in our country will still be based on public welfare, and assume non-profit charitable activities, training and competitions. However, with the development of the sports industry and the improvement of the residents' sports consumption capacity, the unique functions of sports venues provide conditions for the development of profitable industry operations.

On the one hand, the current sports industry of China has ushered in rapid development. The professionalization of sports and the popularization advancement have made the ‘sports industry’ a hot spot for capital chasing. For example, the sports professionalization represented by the Chinese Football Association Super League (CSL) and the China Basketball Association (CBA) is gradually driving the growth of the professional sports market; the improvement of health awareness and consumption capacity is promoting the development of the ‘nationwide fitness’ in mass sports market; the Beijing Zhangjiakou Winter Olympics, various individual world championships and other sports events have occurred in succession, which is expected to promote the individual sports such as skating to become a new consumer hot spot.

On the other hand, the large space provided by the sports venues provides conditions for large-scale theatrical performances and exhibition activities. The gradual prosperity of theatrical performances and exhibition market puts forward quality and quantity requirements for the large space, which provides a driving force for the venues to increase the utilization rate and achieve sustainable operations.

The improvement of the external operating environment will help the sports venues to carry out business activities under the

外部经营环境的改善有助于体育场馆在满足公益性的基础上开展经营活动，以求运营的收支平衡。对于场馆的经营管理，需要协调体育馆场公益性社会目标和营利性经济目标之间的矛盾；对于场馆建设，需要在满足赛事要求的前提下，减少不必要的投入，预备相应的配套设施，进行灵活适应性设计，保证场馆由赛时比赛功能向赛后运营功能的转变（图 1-12）。

premise of satisfying the public welfare, so as to balance their operations. For the venue management, it is necessary to coordinate the conflict between the social goals and the profit-making economic goals of the sports facilities. For the venue construction, it is required to reduce unnecessary input, prepare corresponding supporting facilities, and perform flexible and adaptive design to meet the requirements of the competition and ensure the smooth function transition of the venues from the during-game to the post -game operation (Fig. 1-12).

我国体育建筑的使用现状

目前，我国体育场馆主要服务于举办国内外大型综合运动会、单项竞技赛事和职业联赛等商业比赛，以及服务于群众锻炼的大众体育、服务于师生日常教学和锻炼的学校体育，此外还包括文艺演出及展览等非体育活动。几大主要使用需求对于场馆建设及其可持续运营具有决定性影响，其中竞技体育、群众体育和学校体育构成现代体育的三大组成部分，此外职业体育和文艺演出及展览的使用需求，在我国产业现状背景下值得特别关注（表 1-2）。

1. 竞技体育

1990 年北京亚运会是中国第一次举办的综合性国际体育大赛，此后我国连续在北京、广州和深圳成功举办奥运会、亚运会和大运会等重大体育赛事，为满足赛事需求，兴建体育中心、主体育场、主体育馆、游泳跳水馆、媒体中心和运动员村等竞技体育设施。如 2010 年广州亚运会促使奥林匹克体育中心、广州体育馆、跳水游泳馆、武术武道馆、运动员村和媒体村等设施的新建和完善。2011 年深圳大运会新建 22 个场馆、维修改造 28 个场馆、临时搭建 10 个场馆。

Current Usage of Sports Facilities in China

At present, the sports venues in China mainly serve the holding of domestic and international large-scale comprehensive sports events, single event competitions and commercial competitions such as professional leagues, as well as the popular sports for the masses, school sports for teachers and students in daily teaching and exercise, and in addition the non-sports activities such as theatrical performances and exhibitions. These major usage requirements have a decisive impact on the construction of venues and their sustainable operations. Among them, competitive sports, mass sports and school sports constitute the three major components of modern sports. In addition, the use of professional sports, theatrical performances and exhibitions deserve special attention in the context of the current sports industry in China (Tab. 1-2).

1. The competitive sports

The 1990 Beijing Asian Games was the first international comprehensive sports competition held in China. Since then, China has successfully hosted major Olympic sports events such as the Beijing Olympic Games, Guangzhou Asian Games and the Shenzhen Universiade. To meet the needs of the event, sports centers including the main stadiums, main gymnasiums and natatoriums, media centers, athletes'

体育设施的活动需求类型比较　　表 1-2

体育类型	使用需求	案例	活动规模	活动频率
竞技体育	国内外大型综合运动会	奥运会、亚运会、大运会、全运会、省运会	大	低
	单项竞技赛事	单项锦标赛	大	低
群众体育	群众体育锻炼	健身、日常锻炼	小	高
	其他社会服务	企业活动、专业培训	中 / 小	中
学校体育	师生教学和锻炼	师生日常锻炼、教学课程	小	高
职业体育	体育职业联赛	中超、欧洲足球、美国 NFL、MLB、NBA、NHL 等联赛	大 / 中	中
	其他体育比赛	商业比赛	大 / 中 / 小	中 / 低
其它功能	文娱活动	节日庆典、文艺演出、演唱会	大	高 / 中
	会议和集会	讲座、报告、开学典礼	大 / 中	低
	展览	贸易展览、消费展览	大 / 中	中
	其它商业服务	配套酒店、餐饮、商店	大 / 中 / 小	高

以全运会、省运会为代表的国内重大体育赛事是地区体育设施建设的重要推力。2001 年，国务院发函取消全运会由北京、上海和广东轮流举办的限制，全运会得以陆续在江苏、山东、辽宁和天津等省市开展。可以认为，未来全运会将会在更多的省会城市陆续举办，这将为整个城市的建设和发展提供全方位的推动。由于全运会是国内水平最高、规模最大的综合性运动会，比赛项目基本与奥运会相同，因此其每一次开展都会强有力地促进主办城市的体育设施建设。

作为各省份最大的综合性运动会，省运会的举办同样带来体育设施的大批投入，如广东省运动会，素有“小全运会”之称，自 1956 年以来已经成功举办 14 届，且从改革开放后，举办地点从广州拓展到佛山、江门、深圳、湛江等其他城市。惠州为举办 2010 年第十三届广东省运会，投入约 10 亿新建 10 个场馆，修、改和扩建 7 个场馆；湛江为举办 2015 年第十四届广东省运会，投资约 22.5 亿建设主场馆及配套工程。

villages and other competitive sports facilities have been built. For example, the 2010 Guangzhou Asian Games promoted the construction and improvement of the facilities such as the Olympic Sports Center, City Gymnasium, Natatoriums, Martial Arts Hall, Athlete Village and Media Village of Guangzhou. The 2011 Shenzhen Universiade built 22 new venues and 10 temporary venus, maintained and renovated 28 venues.

The major domestic sports events represented by the National Games and the Provincial Games are an important thrust for the construction of sports facilities in certain region. In 2001, the State Council decided to cancel the restriction that the National Games could only be held in Beijing, Shanghai, and Guangdong. Afterwards the National Games has been held in Jiangsu, Shandong, Liaoning, and Tianjin in succession. It can be considered that the National Games will be held in more and more provincial capital cities in the future, which will provide a full range of promotion for the construction and development of the entire city. As the National Games is the highest-level and largest-scale comprehensive sports event in the country, and the competition items are basically the same as the Olympic Games, so that it will strongly promote the construction of sports facilities in the host city every time it is launched.

As the largest comprehensive sports event in each province, the organization of the Provincial Games also bring in sports facilities, such as the Guangdong Games, known as the ‘Little National Games’, which has been successfully held for 14 sessions since 1956. Since the reform and opening up, the venue has been expanded from Guangzhou to Foshan, Jiangmen, Shenzhen, Zhanjiang and other cities. For example, in order to hold the 13th Guangdong Provincial Games in 2010, Huizhou invested about 1 billion in building 10 new venues, repairing, renovating, and

Comparison of the use demands among types of sports facilities Tab. 1-2

Type	Use demands	Example	Activity Scale	Activity Frequency
Competitive sports	Domestic and international large-scale comprehensive sports games	Olympic Games, Asian Games, Universiade, National Games, Provincial Games	Large	Low
	Single sports event	Individual championship	Large	Low
Mass sports	Mass exercise	Fitness, daily exercise	Small	High
	Other social services	Corporate activities, professional training	Middle/Small	Middle
School sports	Teaching and exercise of teacher and student	Daily exercise, teaching courses of the teachers and students	Small	High
Professional sports	Sports professional league	Super League, European Football, NFL, MLB, NBA, NHL and other leagues	Large/Middle	Middle
	Other sports games	Business competition	Large/Middle/Small	Middle/Low
Other functions	Entertainment	Festival celebrations, theatrical performances, concerts	Large	High/Middle
	Meetings and gatherings	Lectures, reports, opening ceremony	Large/Middle	Low
	Exhibition	Trade shows, consumer shows	Large/Middle	Middle
	Other business services	Supporting hotels, restaurants, shops	Large/Middle/Small	High

随着国民经济的发展，全国各省市综合实力的提升，无论是国际体育赛事、还是国内全运会、省运会，举办城市均由具有经济先发优势的城市主导，逐渐开放拓展至其他省市地区，从而带动更多地区体育设施的建设和完善。此外，城市运动会、民族运动会等综合性运动会，均给各地公共体育设施建设起到促进作用。

体育场馆需满足一定的体育活动要求，这使得体育场馆成为专业性较强的建筑类型。长期以来，不同类型的体育比赛活动发展出各自约定俗成的规则，进而对比赛场地和设施提出特定需求，落实到建筑层面成为单项体育组织协会根据运动项目需要制定的专业性规范和要求。

对于规模要求，大型运动会场馆建设规模大，例如，国际足联对于举办世界杯的体育场

expanding 7 venues; Zhanjiang invested about 2.25 billion in building the main venues and supporting facilities for holding the 14th Guangdong Provincial Games in 2015.

With the development of the national economy, and the improvement of the overall strength of all provinces and cities across the country, whether the international sports events, the National Games or the Provincial Games that the host cities are dominated by cities with economic advantages, are gradually opening up to other provinces and cities, thus promote the construction and improvement of sports facilities in more areas. In addition, the comprehensive games such as the City Games and the National Volk Games have all contributed to the construction of public sports facilities.

The sports venues need to meet certain sports requirements, which make them to be more professional buildings. For a long time, different types of sports competitions have developed their own conventions, which in turn place specific demands on the fields and facilities, and have been implemented to the architectural level as a professional specification and requirements for individual sports associations.

For the scale requirements, venues for large-scale sports events are built on a huge scale. For example, the FIFA's minimum requirements for the capacity of stadiums to host the World Cup are more than 60,000 in the finals, and 40,000 in the semi-finals. The number of fixed seats in the 'Bird Nest' stadium, namely the Beijing Olympics main stadium, has reached 80,000, and in addition there are 20,000 temporary seats; the National Aquatics Center 'Water Cube' contains 4,000 fixed seats and 2000 temporary seats. Furthermore, sports facilities built for large-scale sports events such as the Universiade, National Games, and Provincial Games are often designed based on the upper limit of construction

观众容量最低要求为决赛场 6 万座以上，半决赛场 4 万座以上。北京奥运会主体育场“鸟巢”固定座席达到 8 万座，临时座席 2 万座；国家游泳中心“水立方”包含 4000 个固定座席和 2000 个临时座席。此外，为大运会、全运会、省运会等大型体育赛事兴建的体育设施也往往以建设标准的上限为设计依据。随着全国各省市综合实力的提升，越来越多城市举办大型运动会，并兴建大型体育中心。

standards. With the improvement of the comprehensive strength of all provinces and cities across the country, more and more cities would be able to host the large-scale games and build large sports centers.

对于专业性要求，大型运动会场馆对工艺专业性要求高，体育工艺也通常以较高标准进行配置。然而大型运动会举办频率低，一般综合性或单项体育运动会时间为几天到十几天，如北京奥运会为 16 天，广州亚运会为 15 天，深圳大运会为 11 天。对于国内大型运动会，举办次数也相对稳定在 3-4 年 1 次至 1 年 1-2 次，这导致了大量大中型体育设施面临长期闲置的问题。

For the professional requirements, larger-scale sports venues require higher degree of professionalism, and the sports techniques are usually configured with higher standards. However, large-scale sports events are normally held at low frequency. The general comprehensive or individual sports games only last for several days to a dozen days. For example, the Beijing Olympics lasted 16 days, the Guangzhou Asian Games 15 days, and the Shenzhen Universiade 11 days. The number of holdings large-scale sports games in China is relatively stable from 1 time per 3-4 years to 1-2 times per year, resulting in a large number of large and medium-sized sports facilities to be in long-term idle.

与大规模、高专业配置的大量场馆建设相矛盾的是，目前国内体育产业尚不发达，造成体育场馆赛时赛后产生突出的使用需求落差。目前大型运动会场馆赛后运营已经成为全国范围内的普遍性问题。因此，对于这类体育场馆，在立项阶段进行总体定位、规模设定、工艺配置时，即应进行更加周全的考虑，务实谨慎地考虑场馆未来使用的可能。

Contradictory to the construction of large number of venues with large scale, high professional deployment, the current domestic sports industry is still underdeveloped, resulting in a serious gap in the utilization of sports venues after games. At present, the post-game operation of venues of large-scale sports games has become a universal problem across the country. Therefore, for such sports venues, more comprehensive consideration in the overall positioning, scale setting, and process configuration during the project phase, and the pragmatic and prudent consideration to the possibility of future utilization should be taken.

2. 群众体育

2015 年我国体育产业占 GDP 比重的约 0.7%，相比于美国的 3% 还有很大差距。按国际通行规律，当人均 GDP 达到 5000 美元时，体育产业会进入快速发展状态，目前我国人均 GDP 已超过 8000 美元。北上广深及部分沿海省份人均 GDP 已达中等发达国家水平，从“小康”进入“富裕”阶段。这一阶段中，居民的

2. The mass sports

In 2015, China’s sports industry accounted for about 0.7% of the total GDP, which is still a large gap compared to the 3% in the USA. According to the international laws, when the per capita GDP reaches 5,000 US dollars, the sports industry will go into a state of rapid development. At present, China’s per capita GDP has exceeded 8,000 US dollars. The per capita GDP of the first-

消费重点由食品等生活必需品支出，部分地转移到休闲娱乐、文化旅游、教育和健康等领域。其中居民体育消费能力的持续增长，为大众体育产业发展提供坚实的需求动力，为体育健身服务市场和竞技表演市场发展提供坚实的需求基础（图 1-13）。

由于生活水平和教育水平提高、闲暇时间增多、以及价值观和生活方式转变，居民对体育活动类型也产生多样化、个性化的需求，并具有娱乐性、参与性的特点。这在一方面促使体育项目产生多元化的发展趋势，而另一方面对体育场馆和设施要求相对较低，这体现在大众体育对场馆规模要求小，设备配置和体育工艺需求低。重视体育场馆向大众开放是体育设施公益性的体现，群众体育的健康发展也有利于提高场馆的使用率。如广东奥林匹克体育中

图 1-13　形式多样的大众体育

Fig.1-13　Mass sports of various forms

tier cities including Beijing, Shanghai, Guangzhou and Shenzhen, and some of the coastal provinces has reached the level of a medium-developed country, and has gone from the 'well-off' to the 'rich' stage. In this stage, the residents' consumption will be partly transferred to the fields of recreation, cultural tourism, education and health from the main consumption by food and other necessities. The continuous growth of residents' sports consumption capacity provides strong demand power for the promotion of the mass sports industry, and provides a solid base for the development of the sports fitness service, as well as the competitive performance market (Fig. 1-13).

As the increase of the living standards, education levels and leisure time, and the change of the values and lifestyles, residents become to have diverse and personalized demands for sports activities, which are characterized by entertainment and participation. On the one hand, this has motivated a diversified development trend in sports projects, and on the other hand, it requires relatively low requirements for sports venues and facilities, which could be seen in the small scale demand of fields and facilities for sports facilities and the low demand for the professional deployment. Paying attention to the opening of venues to the general public is a manifestation of the public welfare of the sports facilities, and the healthy development of mass sports in turn helps to increase the utilization of the facilities. For example, the natatorium in the Guangdong Olympic Sports Center is the swimming and diving hall during the Guangzhou Asian Games, and open to the public after the game. Due to the use of water of constant temperature, it can be operated for the whole year, so that become one of the most popular swimming venues in Guangzhou.

Therefore, the newly built venues for large-scale comprehensive sports games and professional leagues can also be considered as mass sports facilities after the games in

心游泳馆，赛时为广州亚运会游泳跳水馆，赛后对大众开放，由于采用恒温水，可常年运营，成为最受市民喜爱的游泳场地之一。

因此，对于因大型综合运动会和职业联赛而新建的体育场馆，在策划和设计中也可兼顾赛后作为大众体育服务设施，以实现体育场馆在社会效益和经济效益的共赢。随着大众体育需求的持续发展，体育场馆建设将会越来越多的趋向居民参与的健身服务、娱乐休闲和竞技表演等类型。

3. 学校体育

学校体育，与竞技体育、群众体育一起，被视为现代体育的三大组成部分。中华人民共和国成立后，我国的高校建设快速发展，1952年第一次院系调整，极大改变了原有的教育版图，建立起新中国的高等教育体系，扩大了高校规模。1978年—1998年，高等教育进入恢复和改革阶段，高校布局摆脱原苏联规划模式，开始探索适合中国国情的大学校园设计，以沿海地区高等学校为代表的一些校园开始学校体育场馆的建设。1998年由浙江大学开始的大学合并，带动第二次院系调整，以及随后1999年开始的大学扩招，高等教育向大众教育转移，也促成了校园建设的空前高峰，这一过程中，学校体育场馆，与图书馆等建筑一起，成为校园标志性建筑，因而也是各高校建设的重点对象，学校体育场馆在数量和质量上都得到快速发展。

一方面，我国学校目前普遍面临着体育场馆设施资源不足且在新老校区之间存在配置失衡问题。尽管教学和师生活动保证了学校体育设施一定的利用率，但总体使用效率不高，部分设施老化陈旧，功能单一，不能满足新型、多样化活动项目需求。学校体育教学的加强和普通高等学校建立高水平运动队的改革试验将会促使更多高校体育场馆的建设。

the planning and design stages, in order to achieve a win-win situation for the venues' social and economic benefits. With the continuous development of the demand for mass sports, more and more types of sports venues will tend to involve residents in fitness services, entertainment and athletic performances, etc.

3. The school sports

School sports, together with competitive sports and mass sports, are regarded as one of the three major components of modern sports. After the founding of the People's Republic of China, the construction of universities in our country developed rapidly. The first departments reorganization in 1952 drastically changed the original territory of education, established a new Chinese higher education system, and expanded the scale of colleges and universities; From 1978 to 1998, the higher education entered the stage of recovery and reform. In that period, the layout of universities was free from the Soviet Union planning mode, and the university campus design suited to the China's national conditions began to be explored. Some campuses, represented by colleges and universities in coastal areas, started the construction of school venues; in 1998, starting from the combination of Zhejiang University, the second department reorganization and the enrollment expansion of universities that began in 1999 were brought along. The transfer of higher education to mass education has also contributed to an unprecedented peak in the campus construction. In this process, school venues, together with libraries and other buildings has been set as the landmark of the campuses, and also the key target for the construction of colleges and universities.

另一方面，我国越来越重视学校体育场馆对社会公众开放，但目前面临着学校体育设施的教学属性和群众体育要求、师生使用时间和社会开放时间、公益性和盈利性之间的矛盾，以及责权区分上的困难。但学校场馆在保障本校师生体育教学和日常活动需求的前提下，在课余时间和节假日向学生和社会群众开放，是有效缓解群众体育场馆供给不足、提高学校体育场馆利用效率的有效手段。

2008年北京奥运会，在驻京的6所大学校园内，新建北京大学、中国农业大学、北京工业大学和北京科技大学的4座体育馆，改建北京航空航天大学和北京理工大学的2座体育馆。赛后作为各高校教学场地，并为社区居民提供健身场所。广州亚运会和深圳大运会均结合大学城、高校场馆作为比赛场地，对节约建设成本和解决场馆赛后利用问题进行了积极尝试（图1-14）。

4. 职业体育

国际上大型运动会和公益性的体育场馆大多依赖政府投资建设，而在商业化运营较为成功的领域则可以吸引民营资本，如欧洲的足球、美国的篮球、橄榄球、棒球和冰球等。职业体育有利于提高体育场馆的使用率，为解决大型运动会场馆赛后利用提供有效方式。例如2012年伦敦奥运会之后，位于东伦敦斯特莱福德的奥运主场赛后成为西汉姆联俱乐部的主场；1996年亚特兰大奥运会之后，一些比赛场馆改造成棒球职业比赛专用场；1984年洛杉矶奥运会主要场馆赛后均为职业体育俱乐部使用，其中最为著名的洛杉矶体育场成为洛杉矶职业橄榄球俱乐部袭击者队和南加利福尼亚大学橄榄球队的主场。

The school venues have been rapidly developed in both quantity and quality.

On the one hand, schools in our country are currently facing the lack of sports facilities and the allocation imbalance between old and new campuses. Although teaching, activities of the teachers and students can ensure a certain utilization rate of school sports facilities, the overall use efficiency is not high. Some facilities are ageing and have only single function that cannot meet the needs of the new and diversified activity projects. The strengthening of physical education in schools and the attempt of setting up high-level sports teams in ordinary colleges and universities will promote the construction of more sports venues in the campuses.

On the other hand, China has paid more and more attention to the opening of school venues to the public. However, it is currently confronted with the contradiction between the teaching attributes of school sports facilities and the demands of mass sports, between the usage time of teachers and students and the social opening time, and between the public welfare and the profitability, and also the difficulty in distinguishing the responsibilities and powers. Nevertheless, on the premise of guaranteeing the teachers and students' physical education and daily activities, the opening of school venues to the students and public during spare time and holidays, is an effective means to effectively alleviate the shortage of sports venues for the masses and to improve the utilization efficiency of the school sports venues.

At the 2008 Beijing Olympic Games, four gymnasiums of Peking University, China Agricultural University, Beijing University of Technology and Beijing University of Science and Technology were built, and two gymnasiums of Beijing University of Aeronautics and Astronautics and Beijing Institute of Technology were reconstructed in the corresponding 6 campuses in Beijing.

图 1-14　北京奥运中的 6 个高校体育场馆
Fig.1-14　6 venues in the campus of the Beijing 2008 Olympic Games

职业化体育是体育产业化的产物，职业化体育运动的发展及其市场体系的成熟，为民间融资进行商业化运营创造基础。以美国体育产业为例，以体育赛事为核心，以赞助商、体育资产、场馆、媒体、特许商品公司、营销及经纪公司为六大主体，形成完整的产业链，拥有国家橄榄球联盟（NFL）、美国职业棒球大联盟（MLB）、美国篮球职业联赛（NBA）和国家冰球联盟（NHL）等体量庞大的职业体育团体。这些职业队伍对训练和比赛的场地需求，为体育场馆的高效使用提供消费动力。

After the Olympic Games, they have been transformed into the teaching venues for colleges and universities, and provided fitness for community residents. Similarly, both the Guangzhou Asian Games and the Shenzhen Universiade have made use of the school gymnasiums as venues for competitions in the university towns and colleges, which made active attempts to save the construction costs and solve the problem of post-game utilization (Fig. 1-14).

4. The professional sports

Internationally, most of the sports games and public welfare sports venues rely on investment and construction by the government. However, those areas where commercial operations are relatively successful can attract private capital, such as European football, American basketball, rugby, baseball, and ice hockey, and so on. Professional sports are conducive to improving the utilization rate of the sports venues and provide an effective way to solve the post-game use of the large-scale sports venues. For example, after the 2012 London Olympics, the Olympic main stadium in East London's Streford became the home venue of the West Ham United Club; After the 1996 Atlanta Olympics, some venues were remodeled into the professional baseball games; In 1984, the main venues of the Los Angeles Olympic Games were totally used by some professional sports clubs after the match, in which the Los Angeles Stadium, the most famous one, became the home stadium of the Los Angeles Professional Football Club Raiders and the Southern California Football Team.

Professional sports are the products of sports industrialization. The development of professional sports and the maturity of their market system provide the basis for the commercial operation of private financing. Taking the American sports industry as an example, it forms a complete industrial chain with the sports events

体育场馆赛后必须保证较高的使用强度，才能维持场馆收支平衡乃至实现盈利。如亚特兰大奥林匹克体育场赛后每年使用80次左右，基本没有空置期。欧美体育场馆建设运营经验表明，一个万人馆需要作为至少两支球队的主场使用才可以实现正常运转，这大概可保证每年开放100场以上。一种常见的模式是篮球队和冰球队共用一个主场，两者赛季错开，再加上一定的文艺演出和展览活动，可保证运营的资金来源。

反观我国的职业体育，一方面，相关体育产业发展进程慢，以足球为例，自1957年开始国内联赛，经历了20年曲折的发展历程，到1978年恢复稳定的全国联赛，直到1994年才建立起职业足球联赛；同样地，我国乒乓球、篮球、排球等职业联赛尚未成熟，未来产业前景不明晰。

另一方面，当前国内职业体育场馆建设也存在诸多问题，如足球职业联赛比赛场以综合性的田径体育场为主，缺乏足球专用体育场的开发，配备设施落后，无法满足职业联赛的需求；部分体育场为单体开发，模式单一，远远不及国外超级综合体的水准；此外，还存在配套交通设施不足等问题。由于职业体育的群众基础尚未稳定，体育场馆观众人数存在巨大差异；这些职业体育团体通常不共用场馆，如东莞现有的四家职业篮球俱乐部都拥有各自的主场馆，而每个场馆仅有约10%的时间能得到有效利用，无论从场馆使用率的角度，还是观众聚集的角度，这样资源分散的形势对于长期经营而言十分不利。

as the core, and sponsors, sports assets, venues, media, licensed merchandise companies, marketing and brokerage companies as the six main bodies, and owns the National Football League (NFL). Major professional sports groups such as the Major League Baseball (MLB), the American Basketball Professional League (NBA) and the National Hockey League (NHL). The demands for training and competition from these professional teams provide consumption power for the efficient utilization of the venues.

After the match, the venues must ensure a high utilization intensity in order to maintain the balance and even get profitability. For example, the Atlanta Olympic Stadium operates about 80 times every year after the match, which basically has no vacant period. The experience of the construction and operation of sports venues in Europe and the United States shows that a gymnasium with a capacity of 10,000 needs to be used as a home gymnasium for at least two teams to achieve normal operation, which may guarantee the opening for more than 100 games per year. A common mode is that a basketball team and an ice hockey team share a home gymnasium, of which the two competition seasons are staggered. Together with certain theatrical performances and exhibition activities, the source of funding for operations can be guaranteed.

On the other hand, professional sports related industry has a slow development process in China. Take football as an example. Since 1957, the domestic league began, but after 20 years of twists and turns, it has been a stable national league until 1978, and the professional football league has been established until 1994. Similarly, the professional competitions such as table tennis, basketball, volleyball in our country has not yet become mature, and the future industry prospects are still not clear.

On the other hand, there are also many problems in the construction of professional

我国职业化体育正处在起步阶段，未来尚有很大发展空间，可以预见职业体育必然会推动更多专业性场馆的出现。但结合我国国情，目前以及未来很长时间内，专业化体育场仍无法取代综合性体育场馆的作用。因此，在项目中应兼顾综合性体育场馆比赛场地、观众席位、配套用房的灵活可变以适应不同体育活动项目的需求，整合和有效利用现有体育场馆资源，实现体育场馆可持续发展的目标。

东莞体育馆的前期研究中，我们提出改变根据城市人口计算容量的办法，充分考虑东莞拥有国内最为集中的职业篮球队伍的现状，尝试多个球队共享主场的国际做法。并附建20万平方米的“篮球城”，以满足群众的体育活动需求（图1-15）。

图1-15 东莞体育馆
Fig.1-15 Dongguan Gymnasium

sports venues in the country. For example, the field for football league is dominated by comprehensive athletics stadiums and there is a lack of development of professional football stadiums; The facilities are backward, which cannot meet the needs of professional leagues; Some stadiums are monomer development by a single mode, which is far from the level of some foreign super complexes. In addition, there are also problems such as insufficient supporting transportation facilities and other issues. As the mass base of professional sports is not yet stable, there are huge differences in the number of audiences; The professional sports groups usually do not share venues. For example, each of the existing four professional basketball clubs in Dongguan has its own main venues, and each venue can be effectively used for about 10% of the time. Whether from the perspective of the venue utilization rate or the audience gathering, the situation of decentralized resources is very unfavorable for the long-term business.

China's professional sports are in their infancy and there is still much room for development in the future. It can be predicted that professional sports will inevitably promote the emergence of more professional venues. However, considering the China's national conditions, at present and for a long time in the future, professional venues still cannot replace the role of comprehensive venues. Therefore, in the project, it should take into account the flexibility of the complex venues, audience seats, and ancillary rooms to meet the needs of different sports activities, integrate and effectively use the existing resources, and achieve the aims of sustainable development of sports venues.

During the preliminary study of Dongguan Gymnasium, we proposed to overlook the calculation of capacity based on urban population approach and gave full consideration to the fact that Dongguan had the most professional basketball teams of China. Consequently, we suggested that several teams shall share a home venue in the project. And 200,000m^2 of the 'Basketball Town' was added to meet the needs of the mass sports activities (Fig. 1-15).

5. 其他功能

作为大空间建筑，体育场馆除满足各种体育运动外，还具有更为多元的功能，可以通过空间灵活布置，开展文娱、庆典、会议、集会及展览等各种活动。

对于文艺演出和体育场馆的结合，场馆的体育活动区为文艺演出活动提供足够大的表演场地，可用于举办不同类型的节目，而场馆的看台设施也为文艺活动提供观演基础。有时文艺演出的使用频率可大大超出体育活动，例如香港红磡体育馆，自 1983 年启用以来，有数以千计的本地和国际文娱、体育节目在那里举行，每年举办的活动中体育节目比例不及 3%，而文艺演出节目约占 80%，这使得红磡体育馆整体使用率高达 96.7%。内地体育场馆使用率虽不及红磡体育馆的水平，但在文娱市场较为发达的城市，文艺活动在这类体育场馆中所占比重同样较为重要。这给体育场馆观演设计提出新的要求，为保证在单侧布置舞台时多数座席的视觉质量，一些体育场馆采用不对称的看台布局方式，如常州奥林匹克体育中心体育场、梅州市梅县区文体中心、华润深圳湾体育中心。此外，满足体育活动，同时又能兼顾文艺演出活动的场地设置、设备系统、舞台设施等要求，成为适宜性设计和建设的新课题。

对于展览活动和体育场馆的结合，主要存在两种形式，一种可以称作时间上的组合，在无体育活动的时间段开展展览活动。体育场馆所能提供的大跨度空间，可以满足各类展览活动的灵活布置，国内多数体育场馆在赛后运营活动中，展览均贡献了较为重要的比重。另一

5. Other functions

In addition to meeting a variety of sports activities, sports venues as a large space building, have more diverse functions. They can be arranged in various ways, such as entertainment, celebrations, meetings, gatherings and exhibitions.

For combination of theatrical performances and the sports venues, the sports area of the venues provides enough space for the theatrical performances, which can be used to hold different types of programs, and the grandstand facilities of the venues also provide a basis for watching the activities. Sometimes, the frequency of theatrical performances may exceed that of the sports activities. For example, in the Hung Hom Sports Centre in Hong Kong, there have been thousands of local and international cultural and sports programs held since its inception in 1983. The annual proportion of sports programs held there is less than 3%, while the theatrical performances accounted for about 80%, which makes the overall utilization rate of Hung Hom Stadium as high as 96.7%. Although the utilization rate of the venues in the Mainland is not as high as that of the Hung Hom Sports Centre, the proportion of cultural activities in such sports venues is also important in the entertainment market of the more developed cities, which puts forward new requirements for the 'watch and perform' design of the venues. In order to ensure the visual quality of most seats when arranging the performance stage on one side, some venues use asymmetrical layout of the grandstand, such as the Olympic Sports Center Stadium of Changzhou, the Meixian District Sports Center in Meizhou, and the Shenzhen Bay Sports Center. In addition, meeting the

种可以称作空间上组合，在项目选址上将会展建筑和体育建筑就近布置，形成体育会展中心。如南通体育会展中心，将总建筑面积为 8.9 万平方米的三个体育场馆和 3.2 万平方米的会展馆集中布置；常州奥林匹克体育中心，将总座席数近 5 万的体育场馆和 1000 个标准展位的会展中心组合布置；江门滨江体育中心，项目总建筑面积约为 20.3 万平方米，其中安排了 7.8 万平方米会展中心，具有 2050 个展位。相比之下，前者展览活动规模和级别均远不如后者。

然而，体育场馆作为文艺演出和展览功能使用，面临我国相关市场尚不发达的问题，以及相应类型建筑，如影剧院、会展建筑的竞争。在场馆建设中，如何在战略上把握好市场份额和服务对象，在战术上兼顾好体育活动和非体育活动的使用需求，以适应场馆运营的现实情况和未来趋势，在未来的研究和实践中有待探索。

广州大学城体育场馆建设项目及投资　　表 1-3

学校名称	项目名称	建筑面积 (m^2)	座位数（个）	概算投资（万元）	设计单位
华南理工大学	体育馆	12377	5000	8100	华南理工大学建筑设计研究院
	体育场	7113	3700	3000	
广东药学院	主体育馆	9786	5014	5871	华南理工大学建筑设计研究院
广东工业大学	体育馆	14050	5504	6940	清华大学建筑设计研究院
	体育场	7116	5140	2823	
广州中医药大学	体育馆	6503	2028	3600	广东省建筑设计研究院
	体育场	3755	2735	2300	
广东外语外贸大学	体育馆	9973	4415	5829	哈尔滨工业大学建筑设计研究院
	体育场	2702	3231	1927	
华南师范大学	体育馆	11105	4988	6663	中南建筑设计院
	游泳馆	8101	2000	4737	
	体育场	4845	3000	2500	
中山大学	体育馆	11639	4923	6800	中国建筑设计研究院
	体育场	3405	3009	1900	
广州大学	体育馆	8160	4000	5040	中建国际（深圳）设计顾问有限公司
	体育场	10000	5000	3500	
中心体育场	体育场	49791	40000	50864	广东省高教建筑规划设计院
		180421		122394	

资料来源：李传义 . 广州大学城体育场馆规划与建设回顾 [J]. 城市建筑，2007(11):21-24

needs of sports activities and at the same time taking into account the venue setting, equipment system, and stage facilities of theatrical performances, has become a new topic for the design and construction of suitability.

For combination of exhibition activities and the sports venues, there are mainly two forms. One can be called a combination of time, to carry out exhibition activities during the time period without sports activities. The large-span space provided by sports venues can meet the flexible arrangement of various exhibition activities. In most of the domestic sports venues, the exhibitions have contributed an important proportion in the post-game operations. The other can be called a combination of space, to arrange the exhibition buildings and sports buildings nearby in the site selection stage of the project, so as to form a sports and exhibition center. For example, in the Nantong Sports Convention and Exhibition Center, three sports venues with a total construction area of 89,000m^2 and exhibition halls of 32,000m^2 are centrally arranged; Changzhou Olympic Sports Center, will have venues with a total around 50,000 seats and an exhibition center with 1000 standard booths arranged in combination; Jiangmen Sports Center has a total construction area of 203,000m^2, of which an exhibition center of 78,000m^2 is arranged with 2,050 standard booths. By comparison, the scale and level of the exhibition activities of the combination of time are far less than that of the combination of the space.

In general, the use of sports venues as a theatrical performance and exhibition function faces some problems, such as the underdeveloped market in our country and the competition of similar types of buildings like the theaters and exhibition buildings. In the venues construction, how to strategically grasp the market share and service target, and tactically take into account the good use of sports activities and non-sports activities to be able to adapt to the actual situation and future trends of the venues operations, still wait to be studied in the future research and practice.

我国体育建筑的现存问题与原因

1. 场馆重复建设与总体发展失衡的矛盾

中华人民共和国成立初期至1990年代初期，受政府财力限制，集中在北京、重庆、广州、长春和上海等地建设大型竞技类体育场馆，其中1987年第六届全运会在广州举办，天河体育中心是国内首次集中兴建大型体育中心，也开创了“一场两馆”的模式。这一时期群众体育和学校体育虽然也得到一定建设，但相比之下投入较小，标准也较为简陋。1990年代以后，尤其是近二十年来，我国综合实力大幅提高，在奥运会、亚运会、全运会乃至省市运会推动下，大型竞技类体育场馆建设高潮迭起，对我

Construction and investment projects of sports venues at Guangzhou Higher Education Mega Center Tab. 1-3

University	Project name	Construction area (m^2)	Number of seats (piece)	Investment budget (ten-thousand Yuan)	Design agency
South China University of Technology	Gymnasium	12377	5000	8100	Architectural Design & Research Institute of SCUT
	Stadium	7113	3700	3000	
Guangdong College of Pharmacy	Main Gymnasium	9786	5014	5871	Architectural Design & Research Institute of SCUT
Guangdong University of Technology	Gymnasium	14050	5504	6940	Architectural Design and Research Institute of THU Co., Ltd
	Stadium	7116	5140	2823	
Guangzhou University of Traditional Chinese Medicine	Gymnasium	6503	2028	3600	Architectural Design and Research Institute of Guangdong Province
	Stadium	3755	2735	2300	
Guangdong University of Foreign Studies	Gymnasium	9973	4415	5829	Architecture Design and Research Institute of HIT
	Stadium	2702	3231	1927	
South China Normal University	Gymnasium	11105	4988	6663	Central South Architectural Design Institute Co., Ltd
	Natatorium	8101	2000	4737	
	Stadium	4845	3000	2500	
Sun Yat-sen University	Gymnasium	11639	4923	6800	China Architecture Design & Research Group
	Stadium	3405	3009	1900	
Guangzhou University	Gymnasium	8160	4000	5040	China Construction Design International (Shenzhen)
	Stadium	10000	5000	3500	
Center Stadium	Stadium	49791	40000	50864	Higher Education Architecture Planning and Design Institute of Guangdong Province
		180421		122394	

Source: Li Chuanyi. Review of the Planning and Construction of the Stadium of Guangzhou University Town [J]. Urbanism and Architecture, 2007(11):21-24.

Existing Problems in Sports Facilities and Causes

1. Contradiction between the partial repeated construction and the overall unbalanced development

From the early days of the founding of the country to the early 1990s, limited by the government's financial resources, large-scale sports venues were mainly constructed in Beijing, Chongqing, Guangzhou, Changchun, and Shanghai, among which the 6th National Games was held in Guangzhou in 1987 and the Tianhe Sports Center was for the first time to build a large-scale sports center and also create the 'one stadium and two halls' mode. Although the mass sports and school sports have also got certain construction during this period, relatively the investment is small and the construction standards are simple. After the 1990s, especially in the past two decades, the overall strength of China has increased dramatically. Under the impetus of the Olympic Games, Asian Games, the National Games, and even the provincial and municipal games, the construction of large-scale competitive sports venues has experienced an upsurge and has played a role in the development of China's sports venues. However, there are still many problems in the construction.

On the one hand, there is an imbalance in the spatial distribution of the sport facilities. Although the total number of sports facilities in our country is generally insufficient, there are serious repeated constructions in some cities or regions, among which the problem of competitive sports facilities is particularly prominent. Taking Guangzhou as an example, in the mid-1990s, in order to hold the 9th National Games in 2001, a number of venues were built to host the

国体育建筑发展起到极大推动作用。然而，体育场馆建设仍存在诸多问题。

一方面，场馆建设在空间分布上存在不均衡。我国体育设施总量总体不足，但在局部城市或地区却存在严重的重复建设现象，竞技类体育设施的问题尤为凸显。以广州为例，在1990年代中期，为迎接2001年九运会，新建一批可以承办高水平世界性单项和全国性综合大型运动会的场馆，包括广东奥林匹克体育中心体育场和分布在天河、黄埔、芳村和花都等区的8座场馆，并改造包括天河体育中心等15个场馆。其中主体育场投资12.3亿元，可容纳观众8万，但赛后运营情况极为不佳，常年月均开放不及1场。为承接2007年第八届全国大学生运动会，广州又在大学城内密集地新建1座体育场，可容纳4万观众，以及10座综合体育馆、10个标准田径运动场和1座游泳池，使得大学城体育设施建设强度过高，成为赛后维护和运营的负担。为举办2010年广州亚运会，广州更是动用70个比赛场馆，其中12个为新建，亚运赛后这些场馆出现冷热不均问题，不少场馆出现再利用困境，这有选址偏远造成可达性差的原因，也有由于重复建设、定位雷同造成场馆之间缺乏足够客流等因素（表1-3）。

另一方面，场馆建设在类型分布上存在不协调。竞技体育、群众体育和学校体育一起构成现代体育的三大组成部分，目前我国存在重竞技体育场馆，而轻群众和学校体育设施的现象，三者在发展上存在较为严重的不均衡。梅季魁教授曾指出“我国几座大城市同纽约、巴黎、东京等发达国家的大城市相比，竞技体育设施的数量和质量可能不在其下，但社区和中小学的体育设施则相差甚远。这种不协调发展应该引起各方面的重视，摆上议事日程”。场馆建设在类型分布上的不均衡，会削弱体育运动的群众基础，并阻碍体育场馆的可持续运

high-level, single- and multi-national sports games, including the Guangdong Olympic Sports Center Stadium and the 8 venues distributed in Tianhe, Huangpu, Fangcun and Huadu districts, and 15 venues, including the Tianhe Sports Center, were retrofitted. Among them, the main stadium was invested with 1.23 billion yuan and could accommodate 80,000 spectators. However, the post-game operation is extremely bad that the opening is less than one time per month. In order to undertake the 8th National University Games in 2007, Guangzhou intensively built a new stadium that can accommodate 40,000 spectators, as well as 10 integrated gymnasiums, 10 standard athletics fields and a swimming pool in the University Town, making the sports facilities construction strength of the University Town to be excessively high, which become a burden on maintenance and operation after the game. In order to host the 2010 Asian Games, Guangzhou again put 70 competition venues to use, of which 12 venues were newly built. After the Asian Games, these venues suffered from uneven operation condition, and many venues faced the re-use difficulties. The reasons for such situation include factors such as poor accessibility caused by remote location, and lack of passenger flow due to the repeated construction and the similar positioning among venues (Tab. 1-3).

On the other hand, there is an inconsistency in the distribution of venue types. Competitive sports, mass sports and school sports together constitute the three major components of the modern sports. However, the developments of the three are serious imbalance. At present, the sports venues for competitive sports get far more attention than the facilities for mass sports and school sports in China. Prof. MEI Jikui once pointed out that ‘The number and quality of competitive sports facilities in several big cities in China may not be lower than that of the big cities in developed countries such as New York, Paris, and Tokyo, but the sports facilities in the community, primary

营。这些问题产生的背后，有着急功近利地追逐政绩和脱离理性的主观决策造成对项目建设目标的误导，使得很长时间内国内体育场馆建设违背其服务体育和公众的核心价值。

and secondary schools are very insufficient. This kind of uncoordinated development should cause attention from all aspects and start to be discussed.' The uneven distribution of the construction will weaken the masses foundation of sports and hinder the sustainable operation of the venues. The emergence of these problems is caused by misleading the goal of project due to the pursuit of political achievements and subjective decision-making that is out of rationality, which has made the construction of domestic sports venues to violate the core values of serving sports and the public for a long time.

2. 建设与运营的矛盾

国内北京、广州、深圳等一线城市，借助举办奥运会、亚运会和大运会等大型体育赛事，完善自身的体育设施建设。随着全运会、省运会等举办权利走出一线城市局限，许多省会城市乃至二三线城市也陆续争夺举办国内重要赛事的机会，并借此兴建大批大型体育中心和体育场馆（图 1-16）。

在这个过程中突显出许多决策层面的问题，尽管在建设前期都强调节俭建设以及重视赛后的场馆利用，但实际情况是许多场馆赛后

2. Contradiction between the construction and operation

The first-tier cities such as Beijing, Guangzhou, and Shenzhen in China have improved their sports facilities by taking the advantage of hosting the major sports events such as the Olympic Games, Asian Games and Universiade. With the rights of hosting the National Games, the Provincial Games go out of the limits of the first-tier cities, many provincial capital cities and even second and third-tier cities are also competing for the opportunity to host major national events, and to build a large number of large-scale sports centers and venues (Fig. 1-16).

In this process, many problems at the decision-making level can be seen. Despite the strong emphasis on the frugal construction and the post-game utilization of venues in the early construction period, the actual situation is that many venues do not operate well after the games. For example, the 'Bird Nest' of the Beijing Olympic Games was demonstrated during construction, and the concession period of 30 years after the games was handed over to Beijing CITIC Stadium Sports Operations Co., Ltd. for operation and maintenance. However, the 'Bird Nest' was always at a loss during the year

图 1-16　各城市体育中心热潮

Fig.1-16　Sports center wave of different cities

运营不佳。如北京奥运会的“鸟巢”在建设时经过论证，将场馆赛后30年特许经营期交予北京中信联合体体育场运营有限公司，负责运营维护工作，然而奥运会结束后一年，“鸟巢”始终处于亏损状态，使得中信联合体即放弃经营权，这也说明项目的前期策划和论证未达到预期效果。类似地，广州亚运会和深圳大运会提出节俭建设，但实则大兴土木，兴建大量大中型体育场馆，许多场馆赛后无人问津，运营压力巨大乃至亏损，广州亚运会后一些性质雷同的场馆之间为争夺客源而展开恶性竞争，深圳大运会后各场馆为争夺作为职业联赛主场而进行长时间的角逐。

after the Olympic Games. As a result, the CITIC consortium gave up its right to operate, which also showed that the preliminary planning and demonstration of the project did not achieve the desired results. Similarly, frugal construction was proposed in the Guangzhou Asian Games and the Shenzhen Universiade. However, a great number of large and medium-sized sports venues were built, of which many venues were unattended after the match, and the operation was stressful and even loss. Competition between some similar venues after the Guangzhou Asian Games was in contention and even vicious to struggle for customers. After the Shenzhen Universiade, the venues competed for a long time for being the home venues of the professional league.

我国目前体育产业尚不发达，大型体育场馆赛后综合利用困难，脱离国情和城市实际需求，急功近利地建设“大跃进”，盲目追求高标准，缺乏科学性决策，造成形势不容乐观。由此可见，体育场馆建设需要决策者摒弃主观意念，更正价值观，让体育场馆回归其社会责任，在决策阶段采纳体育部门、技术设计相关人员的积极反馈，回到科学决策的轨道来。

At present, China’s sports industry is still underdeveloped. Large-scale sports venues are difficult to get comprehensive utilization after the games. The situation is not optimistic due to the out of the national conditions and the actual needs of the city, the urgently construction ‘great leap forward’, blindly pursuing high standards, and the lack of scientific decision making. It can be seen that the construction of venues requires policy makers to abandon their subjective ideas, correct their values, return sports facilities to their social responsibilities, and adopt positive feedback from relevant personnel of sports departments and technical design in the decision-making stage, and return to the path of scientific decision-making.

3. 唯我独尊与融入城市的矛盾

改革开放以后，国民经济高速增长，城市化进程尤其在近三十年来发展迅速，这一时期也正值体育建筑建设高潮。用地庞大的体育中心通常被布置在城市边缘，作为新城开发的助推器，成为许多城市新区发展的建设策略。大型体育活动作为城市中大事件对城市发展具有重要影响，国内体育场馆建设目标也由早期为

3. Contradiction between the self-emphasis and the urban-integration

Since the reform and opening up, the national economy has grown at a rapid rate, and the process of urbanization has developed rapidly especially in the last three decades, which is also the boom period of the construction of sports facilities. The huge sports center is usually located on the edge of the city, as a booster for the development

满足举办赛事，发展到作为标志性、启动性项目参与到城市的发展进程中。广州天河体育中心即是以 1987 年第六届全运会为契机，在政策支持和引导下成功带动城区发展的典型案例。体育场馆依存于城市环境，并对城市可持续发展起到重要作用，然而目前国内由于规划等方面失策，譬如选址不当，缺乏交通可达性，布局模式单一，使得一些建成的体育场馆与城市环境之间无法形成有机互动关系（图 1-17）。

一方面，未能准确地把握城区发展，使得对体育场馆建设在时间和空间上把握不当，将新场馆选址在过于远离市区的地带，不便于城市居民的到达，使得场馆长期使用率低下，而市区群众则缺乏体育设施。梅季魁教授曾指出："不能在市区内获得足够用地的情况下，是建集中的体育中心远离市民，还是分散几处建中小型体育中心亲近市民并为提高利用率创造条件，已是优化设计的重大课题。"

of the new districts, which has become a strategy for the development of many new urban areas. Large-scale sports events, as a major event of the city, have significant impacts on the urban development. The goal of building sports venues in our country has also evolved from the early stage of meeting the sports events themselves, to participate in the development of the city as iconic and start-up projects. Guangzhou Tianhe Sports Center is a typical case of the 6th National Games in 1987 as an opportunity to successfully promote the urban development under the policy support and guidance. Sports venues depend on the urban environment and play an important role in the sustainable development of the city. However, due to the lack of planning and other aspects in the country, such as improper site selection, lack of accessibility to traffic, and single layout mode, the venues built are in no organic interaction with the urban environment (Fig. 1-17).

On the one hand, the failure to accurately predict the development of the urban areas makes it difficult to grasp the time and space for the construction of sports facilities. New venues are set too far from the urban areas that are inconvenient for the arrival of urban residents, which make the utilization rate low in the long-term use of the venues, while the urban residents lack sports facilities at the same time. Prof. MEI Jikui once pointed out: 'In the case of unable to obtain adequate land use in urban areas, to centrally build a sports center that is far from the public, or dispersedly build some small and medium-sized sports centers that are close to the citizens to create conditions for increasing the utilization rate, have already become issues for the design optimization.'

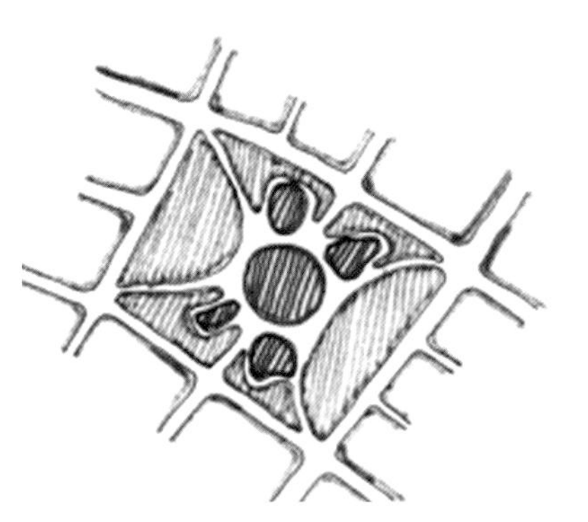

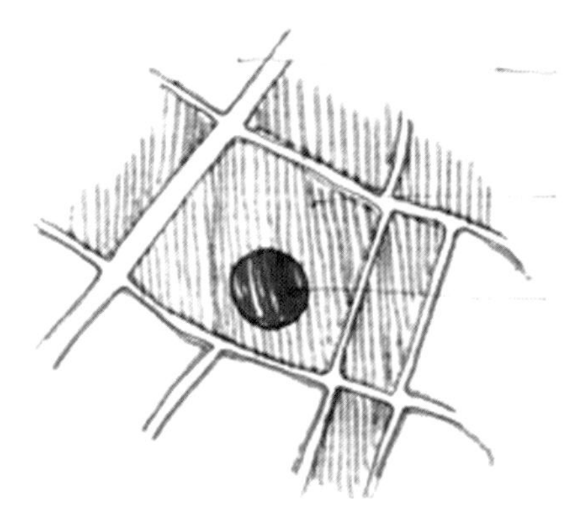

图 1-17　体育中心与城市相融与隔离

Fig.1-17　The integration and isolation of sports center with city

On the other hand, the layout pattern within the sports lands is single, which inherit the inherent thinking, and cannot face the future development of urban space and sports industry itself. For many years, the 'one stadium and two halls' and 'self-emphasis'

另一方面，体育用地内的布局模式单一，采取固有思维布置，不能面向未来城市空间和体育产业自身的发展。许多年来体育中心“一盅两件”“唯我独尊”的布局模式被延续下来，将体育场馆和室外体育设施布置在用地中央，外部环绕以停车场或大片绿地，这与决策者和设计人员将体育场馆作为标志性建筑，突出宏伟和纪念性的惯性思维有关。这一模式一方面从空间形态上缺乏“街墙”，与环境格格不入，另一方面也违背运营规律，许多这类体育中心在周围城市环境成熟后，需要在地块沿街道的用地边界处增建配备的功能设施，这也反映出原有体育场馆布局方式的局限性。而外部开敞空间则通常被视为附属空间，往往孤立于城市整体环境。粗糙地建立大广场、缺乏室外活动功能设置、缺少步行交通联系，忽视不同类型的活动与停留空间细部设计，使得许多体育用地的外部空间未能有机融入城市公共空间，成为不受市民欢迎的空旷冷清地带。

我国体育建筑建设与养护矛盾突出的主要原因在于定位混淆，存在前期公共投入控制不严、后续支持力度不够等问题；其次是我国体育场馆建设的盲目性。由于许多城市将大型体育场馆作为标志性工程建设，项目决策由主观肇始，建设初始缺乏科学论证，建筑标准定位不当，导致建设主体内容不准确、规模确定随意、项目策划不科学，重复建设、恶性竞争严重。

layout mode of sports center have been continued, where the venues and outdoor sports facilities are located in the center of the site, and surrounded with large parking lots or green areas. This is decided by the inertial thinking of creating the sports center as a landmark, and highlighting the grandeur and memorial image of the venues by the decision makers and designers. On the one hand, this mode lacks a ‘street wall’ and is incompatible with the urban environment. On the other hand, it also violates the law of operation. Many of these sports centers need to make up functional facilities at the site boundary along the street after the surrounding urban environment is mature, which also shows the limitations of the original layout. The open space within the site is usually regarded as an ancillary space and is often isolated from the exterior urban environment. Rough establishment of large squares, lack of function settings for outdoor activities, insufficient pedestrian transport links, and neglect of the detailed design of different types of activities and accommodation spaces make the open space of many sports center unable to integrate organically into urban public spaces and finally become a deserted area that is not welcomed by the public.

In China, the obvious contradiction between the construction and the maintenance of sports facilities can be primarily imputed to confused targeting, negligent preliminary control of public input, inadequate follow-up support, etc. The secondary cause is the aimlessness of design and construction. In constructing large sports venues as landmark projects, subjective decision-making, insufficient justification and improper criterion-targeting result in ambiguous subject-matter, inconsiderate scale, irrational project planning, repetitive construction and vicious competition prevail.

4. 功能定位单一与灵活多样使用的矛盾

当前我国体育场馆普遍面临严峻运营压力，我国体育产业尚处起步阶段，文艺演出市场尚不发达，竞技活动和文艺观演活动也缺乏群众基础。我国体育场馆平均使用时间仅有5%-10%，且观众上座率不足30%，保证场馆的多功能性以使得在赛后可以容纳更多非体育活动的策略被逐渐接受。

但实际情况是，由于匆忙应对大型赛事而忽略科学精细的前期决策，以及缺乏多功能灵活适应性设计的方法指导，许多场馆失去在赛后通过合理改造满足多样化用途的可能性，造成运营的困难。例如为迎接1987年六运会而新建的场馆之一广州市黄埔区旧体育馆，内部平面布置即缺乏灵活性，场地设置上采用固定看台环绕单个篮球场，配套设施陈旧且不足，导致赛后利用率低下，由于无法满足2010年广州亚运会比赛要求而被提前拆除。同样地，沈阳五里河体育场，始建于1988年，由于硬件条件无法满足2008年奥运会分赛场的要求而在2007年拆除（图1-18）

图1-18 沈阳五里河体育场爆破

Fig.1-18 The blasting of the Shenyang Wulihe Stadium

4. Contradiction between the single function and the diverse usage

At present, the sports facilities in China are faced with severe operational pressures. The sports industry in our country is still in its infancy. The market for theatrical performances is still underdeveloped, at the same time the sports activities and theatrical performances are also lack of widespread mass base. The average utilization rate of sports venues in our country is only 5%-10%, and the audience attendance rate is less than 30%. The strategy to ensure the versatility of the venues so that they can accommodate more non-sports activities after the game is gradually accepted.

However, the reality is that, due to the hurry arrangement to deal with the large-scale competitions, the ignorance of scientific and advanced decision-making, and lack of flexible and adaptive design guidance, many venues have lost the possibility of adapting to diversified usages after the game, and are facing the difficulty of operation. For example, the old Huangpu Gymnasium in Guangzhou, one of the new venues for the 6th National Games in 1987, was lack of flexibility in its interior layout, which used fixed stands to surround a basketball court and was equipped with inadequate facilities. These resulted in the low post-game utilization rate, and it was demolished ahead of schedule due to the inability to meet the requirements of the 2010 Guangzhou Asian Games. Similarly, the Shenyang Wulihe Stadium built in 1988 was demolished in 2007 since it could't meet the requirements of the 2008 Olympic Games (Fig. 1-18).

5. 资金技术投入高与经济节能效益低的矛盾

新中国成立初期，体育建筑受到政府重视，在北京、重庆、广州、长春和上海等城市集中财力兴建大型体育场馆，在经济条件有限的情况下，很长一段时间内尊重“实用、经济、美观”的建设原则。这一原则直至1990年北京亚运会场馆中依然清晰可见，彼时的体育建筑多功能研究，结构技术创新方面展现出那时的

5. Contradiction between the high capital investment and the low economic and energy efficiency

In the early years after the founding of the People's Republic of China, sports construction was highly valued by the government and large-scale venues were centrally built in Beijing, Chongqing, Guangzhou, Changchun, and Shanghai. ‘Practical, economic, and beautiful’ was

研究水平。近二三十年以来，随着国家经济水平提高和大量体育赛事的成功申办，体育场馆获得前所未有的资金投入，各地兴建体育场馆的造价节节攀升。

2004年关于北京奥运会“鸟巢”的去顶和瘦身引发大众热议，也折射出早期决策阶段对项目可行性研究的缺乏，例如招投标中要求的40亿造价成本，即便按照9万个设计座席进行计算，每个座席也要超过4万元。北京凯迪拉克中心（原五棵松体育馆）中标方案采用将商业活动用房布置在体育比赛馆上方，这一“大空间在下，小空间在上”的空间布局，违背了体育建筑设计的基本原则。两年后将商业活动用房取消，而赛后又投入1.3亿元进行改造，也反映出前期决策的失误和对运营考虑的不足（图1-19）。

究其背后原因，更多的投入并没有用于对建筑功能的提升和后续运营的支持上，而是在各级政府行政意志下催生对标志性的追逐。体育建筑该有的研究并未跟进，相关设计的非理性趋势反而上涨。体育场馆的规模和配置标准得到大幅提高，然而多功能性和灵活适应性

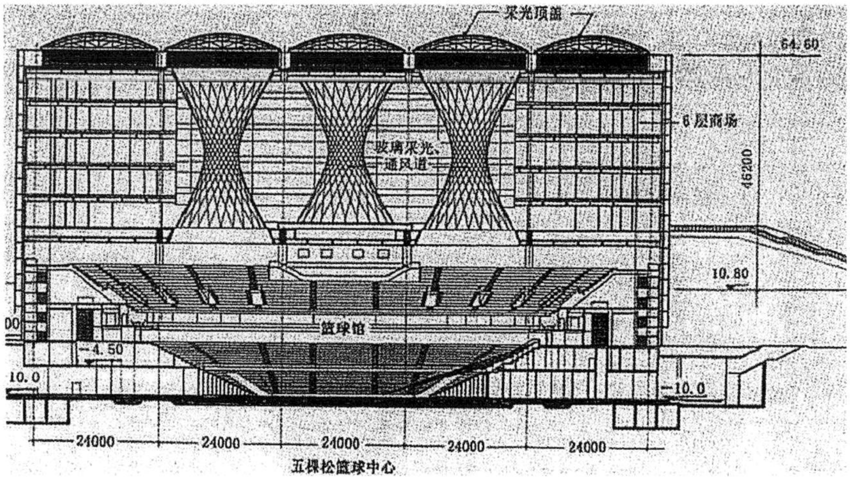

图1-19　北京五棵松体育中心中标方案
Fig.1-19　The winning bid for Wukesong Sports Center

respect as the construction principles for a long period of time under the limited economic conditions. This principle was still clearly visible in the venues of the Beijing Asian Games in 1990. At that time, the multi-functional research on sports venues, and the structural and technological innovation demonstrated the research level at that time. In the past two or three decades, with the improvement of the national economy and the successful organization for a large number of sports events, the sports venues have got unprecedented capital investment, and the cost of construction all over China has been climbing.

In 2004, the top-down and slimming of the 'Bird Nest' of the Beijing Olympics sparked the public debate, which also reflected the lack of feasibility studies in the early stage of decision-making, such as the cost of 4 billion Yuan required in the bidding. Even with 90,000 design seats for calculating, each seat had to exceed 40,000 Yuan. The winning bid scheme of the Songkai Cadillac Center (formerly Wukesong Gymnasium) arranged a commercial activity room at the top of the sports hall. Such spatial layout of the 'large space below and the small space above' violated the basic principles of sports building design. Two years later, the commercial housing was canceled, and after the game, 130 million yuan was added for reconstruction, which also reflected the mistakes of previous decisions and the lack of consideration to the post-game operations (Fig. 1-19).

The reason behind this is that more investment has not been used to support the promotion of building functions and follow-up operations. Instead, it has gone to the iconic pursuits at all levels of the government administration. The research on sports facilities do not follow up, but the irrational trend of related design rise instead. The scale and configuration standards of the venues have been greatly improved, but there is no general improvement in the versatility and flexibility. Some venues

却没有普遍提升。一些场馆投入高昂的费用采购进口设备和高科技技术，却忽视考虑实际应用需求和技术设备的适宜性，以及采用自然采光、自然通风的被动式节能降耗策略，采用复杂的浮夸的结构形式，导致结构用钢量远远超出实际需求。在我国体育场馆运营环境尚不成熟的现实条件下，过高的造价投入到怪异的形态和浮华的表皮中，必然造成国家和地方政府乃至社会的沉重负担。

决策者攀比心理和好大喜功促使更多奢华昂贵的体育场馆诞生。每每听到我市的‘鸟巢’、我县的‘水立方’等报道见诸媒体，都不禁让人担忧。相对应的是，体育建筑造价不断攀升，用钢量、单方造价等经济指标屡遭忽视，公共建筑投资成本加大，维护运营的预算却仍然不足，赛后使用困难重重。大型盛会赛前关于设施经营的美好许诺，赛后荡然无存，随之而来的是新场馆改造的不断进行、后续投入的不断增加，其根本原因还是在于项目规划建设的科学性有待加强。

作为大空间公共建筑，体育建筑节能潜力巨大，其能耗大小直接关系体育建筑自身的可持续性，因而一直是研究的重点课题。“绿色奥运”“绿色亚运”“绿色大运”等理念在各大体育赛事中得到推行，并指导体育场馆建设，在许多项目中推行了绿色技术的示范性实践。

have been invested with high costs in the procurement of importing equipment and high-tech technologies, however been neglected to consider the practical application requirements and the suitability of technical equipment, as well as the passive energy-saving and consumption-reduction strategies of the natural lighting and ventilation, and the use of complex and exaggerated structural forms resulted that the structural steel far exceeded their actual demand. Under the realistic conditions in which the operation environment of our sports venues is not yet mature, the excessively high costs being thrown into those weird shapes and flashy skins, would inevitably cause a heavy burden on the national and local governments, and even the society.

The comparison psychology and craving for greatness of the decision-makers promote the creation of more and more expensive sports venues. Reports about ‘another Bird Nest’ and ‘another Water Cube’ in the media are severely disconcerting. Meanwhile, the construction costs of sports facilities are on the rise, the quantity of steel, the construction cost per square meter and other economic measurements are ignored from time to time. The cost and investment of public construction have increased, while the operation maintenance budget is still inadequate, putting post-game uses in awkward situation. The peaches and cream promises of satisfactory post-game operation of facilities made before large events result in nothing afterwards through repetitive transformation and investments. The fundamental reason is the poor scientificity of project planning and construction.

As a large-scale public building, the potential for energy saving in sports facilities is huge, and the energy consumption is directly related to their sustainability. Therefore, it has always been a key research topic. Concepts such as ‘Green Olympics’, ‘Green Asian Games’

但与此同时，在一些案例中，牵强附会的添加一系列“先进”的节能技术，通过绿色建筑评价获得星级认证，有时候成为舆论宣传的噱头。已有的建设经验有待长期的跟踪研究，是否具有普及推广的价值有待考验。过高的技术造价有时得不偿失，并且建筑作为一个整体，更需要的是系统的可持续理念、技术和设计作为支撑。

and ‘Green Universiade’ have been promoted in major sports events, which have guided the construction of many venues, and the demonstration practice of green technologies has been implemented in many projects.

At the same time, in some cases, there is also phenomenon to add a series of far-fetched ‘advanced’ energy-saving technologies, obtain star certification through green building evaluation, which sometimes become the publicity gimmicks. The existing construction experience needs to be followed up for a long time to test whether it has the value of popularization and promotion. Too high technology costs sometimes outweigh the gain and the sport facilities as an integrated whole needs more support from systematic sustainable technologies and the design theory.

我国体育建筑在各个环节均存在问题，缺乏基于可持续的决策和设计研究。

在决策环节上，各级政府在政绩工程的驱动下，将行政意志施加于项目建设，催生许多违背体育建筑基本规律的项目，脱离社会发展实际和体育产业现状，随意定位和确定规模，追随形式上的标新立异而忽视内在的功能设置。整个建设决策机制中，长期以来项目实际参与者，包括投资、设计、建设和使用各方难以形成统一连贯的建设目标和决策思路，致使很多项目需求调研缺乏、定位不明确、建设重复、建成后使用运营艰难的困境。

在设计环节上，一方面在总体布局上迎合领导意志，片面强调体育场馆的标志性，不能与城市环境有机融合，忽视体育建筑的服务群众、服务体育的核心价值。在功能上定位单一，缺乏灵活适应性设计，使得场馆无法应对未来体育活动项目和其他非体育类活动项目的多

There are problems in all aspects of the construction of sports facilities in China, where there is a lack of research on sustainable decision-making and design.

In the decision-making process, government at all levels is imposing administrative will on project construction under the drive of political performance projects, and has spawned many projects that violate the basic laws of sports construction, which are free from the actual conditions of the social development and the status of the sports industry, freely positioning and scale determining, following the formal maverick, and neglecting the internal function settings. In the entire construction decision-making mechanism, long-term project participants, including investment, design, construction, and operation are difficult to form a consistent and continuous construction goal and decision-making ideas, resulting in a lack of investigation of many project needs, unclear positioning, repeated construction and operational difficulties after the game.

功能使用。在技术应用上缺乏系统思维，忽略建筑自身通过体积控制、自然采光和通风等方法实现节能降耗的可能，部分地片面追求昂贵的高科技技术和设备，堆砌绿色建筑技术手段，造成过高的技术成本。此外，现有的分工体系使得可持续设计难以在可行性研究、规划和建筑设计、建造以及使用后评价的各个环节充分贯彻。

In the design process, on the one hand, it caters to the will of decision-makers in the overall layout, emphasizes the iconic of the venues, cannot integrate organically with the urban environment, and ignores the core values of serving the public and sports activities. In terms of functional positioning, the lack of flexibility and adaptability design, the venues could not cope with the multifunctional use of future sports activities and other non-sports activities. There is a lack of systematic thinking in the application of technology, ignoring the possibility of energy saving through building volume control, natural lighting and ventilation. The partial pursuing of expensive high-tech technologies and equipment, piling up green building technologies, causes excessively high technical cost. In addition, the existing labor division system makes it difficult for sustainable design to be fully implemented in all aspects of feasibility studies, planning and architectural design, construction, and the post occupancy evaluation.

二、走向精明营建

体育建筑可持续研究的立足点

1. 基于国情

我国国情构成体育建筑可持续研究的外在条件，体育建筑可持续决策与设计研究必须充分关注我国体育建筑建设情况和体育事业发展状况。

体育建筑建设现状。近二三十年来，我国大中型体育场馆建设高潮迭起，然而以人均体育设施作为指标，我国仍存在体育设施总体不足的问题，但在局部城市或地区却存在严重的重复建设现象，其中竞技类体育场馆的问题尤为突出，目前我国存在重竞技体育场馆，而轻群众和学校体育设施的不均衡发展现象。大多数为承办体育比赛而兴建的大中型场馆赛后运营情况不佳，虽然场馆造价日益攀升，但更多的投入并没有用在对建筑功能的提升和后续运营的支持上，而是在各级行政意志下催生对标志性的追逐。体育建筑该有的研究并未跟进，相关建设的非理性趋势反而上涨。体育场馆的规模和配置标准得到大幅提高，然而面向多功能使用的灵活适应性却没有普遍提升。一些场馆投入高昂的费用采购进口设备和高科技技术，却忽视考虑实际运营需求和技术设备的适宜性。由此可见，体育场馆建设需要摒弃主观意念，避免非理性价值观，让体育场馆回归其服务于体育的社会责任，回到科学决策的轨道来。

II. MOVE TOWARDS SMART PLANNING AND BUILDING

Standpoint of Sustainability Research

1. Based on the conditions of China

The national conditions of China constitute the external conditions for the sustainable studies of sports facilities. The research on sustainable decision-making and design of sports facilities must pay full attention to the construction status and the development of sports industry in China.

As for the construction status of sports facilities, in the past two or three decades, the construction of large and medium sports venues in China has experienced an upsurge. But with the per capita sports facilities as indicators, the overall sports facilities are still insufficient. However, in some cities or regions there are serious problems of repeated construction, in which the phenomenon of the competitive venues is particularly prominent. At present, there are serious unbalanced development between the competitive venues and the facilities for masses and school sports, in which the former has always got more emphasis than the latter. Most of the large and medium-sized venues built for hosting sports games have poor post-game operations. Although the cost of venues has been increasing, more investment has not been spent on the improvement of building functions and the support for the subsequent operations. Instead they are given to the iconic chase driven by the administrative will at all levels. The necessary related research on sports construction does not follow up, but the irrational trend of construction rise instead. The scale and configuration standards of the venues have been greatly improved, but the flexibility for multi-functional usage has not been generally enhanced. Some venues have been invested with high costs in the procurement of importing equipment and high-tech technologies, but neglected to consider the actual operational requirements and the suitability of technical equipment. It can thus be seen that the construction of sports venues needs to abandon those subjective ideas, avoid the irrational values, go back

体育事业发展现状。目前我国体育场馆主要服务于举办国内外大型综合运动会、单项竞技赛事、职业联赛等商业比赛，以及服务于群众锻炼的大众体育、服务于师生日常教学和锻炼的学校体育，此外还包括文艺演出及展览等非体育活动，几大主要使用需求对于场馆建设及其可持续运营具有决定性影响。然而，我国体育产业尚处起步阶段，文艺演出市场尚不发达，竞技活动和文艺观演活动也缺乏群众基础。这使得当前我国体育场馆普遍面临严峻运营压力。但从另一方面也应看到，我国体育事业的高速发展将会促使设施经营环境逐步改善，投资和运营模式的多元化趋向也为体育建筑建设和运营提供多样化的解决方案，保证场馆的多功能性以使得场馆可以容纳更多类型活动的策略被逐渐接受。这些发展现状和趋势将会给未来体育场馆建设带来新的机遇和挑战。

to the path of scientific decision-making, and return the facilities to their social responsibilities for serving sports.

As for the development status of the sports industry, at present, our sports venues mainly serve the holding of large-scale domestic or international comprehensive sports games, individual competitions, professional leagues and other business competitions, as well as mass sports for the public, school sports for teachers and students for daily teaching and exercise, and also those non-sports activities such as theatrical performances and exhibitions. The major use requirements have a decisive influence on the construction of venues and their sustainable operation. However, the sports industry of China is still in its infancy, the market for theatrical performances is still underdeveloped, and there are also lack of audience bases for competitive activities and literary performances. All these factors make the current sports venues in China generally face severe operational pressure. However, on the other hand, it should also be noted that the rapid development of China's sports industry will lead to a gradual improvement in the operating environment of the facilities. Diversification of investment and operating modes will also provide diversified solutions for the construction and operation of sports facilities, and the strategies of ensuring the venue's versatility is gradually accepted to accommodate more types of activities. These status and trends will brıng new opportunities and also challenges to the construction of sports facilities in the future.

2. 基于体育

体育运动自身特点及其对建筑的需求构成体育建筑可持续研究的内在条件，体育建筑可持续决策与设计研究必须充分关注体育建筑自身的核心价值和特点。

2. Based on sports

The characteristics of sports and their requirements for the facilities constitute the inherent conditions for the sustainable development of sports facilities. The sustainability research of decision-making and design on sports facilities must pay full attention to the core values and characteristics of sports facilities themselves.

体育建筑核心价值的遵循。体育建筑为竞技比赛、体育锻炼、体育教育和体育娱乐等活动提供合适的空间场所。全民健身作为体育设施的最基本功能，必须予以尊重，这也是体育建筑公益性的重要体现。职业体育和文艺演出及展览的使用需求，在我国产业现状背景下值得特别关注。服务于民众的体育活动是体育建筑的价值所在，这使得体育建筑很大程度上具有公共性特点。体育建筑应遵循和发挥其核心价值，而这也构成其自身可持续运营的基础条件。

体育建筑自身特点的尊重。这些特点表现在投资运营、空间结构、设备系统等各个方面。作为城市大型公共投资项目，体育建筑占地面积大、消耗资金多，对城市功能和空间整体发展承担着重要角色，同样由于大量的资金投入，体育建筑的运营关系到城市和社会的经济运行，因此体育建筑的规划选址、规模定位和建设标准需要特别关注。作为大型公共建筑，体育建筑主空间体量和跨度大，投入的设备和系统复杂，需要更高的建设造价和建造技术投入，而建成后的改造和拆除也更为困难，长期的运营过程中能源消耗、维护和更新成本也更为高昂，因此体育建筑功能和空间的灵活适应性、设备和系统的集约适宜性尤其重要。这些特点赋予体育建筑特殊性，也正是体育建筑可持续发展需要解决的关键问题（图 2-1）。

体育建筑可持续研究立足于实现其可持续性目标，关注其定位和规模、建设和运营、功能和空间，为可行性研究、规划和建筑设计、建造以及使用后评价的全过程提供研究基础和指引。

Follow the core values of sports facilities. Sports facilities provide suitable space for sports competitions, physical exercises, physical education and sports entertainment. The construction of sports facilities mainly serves the holding of competitive sports such as comprehensive sports games, individual competitive events and professional leagues, and mass sports that serve the public, and school sports that serve teachers and students in daily teaching and exercise. Among them, the exercise activities for the general public are the major function of sports facilities that must be fully respected, which is also an important manifestation of the public welfare of sports facilities. The use of professional sports, theatrical performances and exhibitions deserves special attention in the context of China’s industry. Serving the people’s sports activities are the value of sports facilities, which makes sports buildings to have public characteristics to a large extent. Sports construction should follow and play its core values, which also constitute the basic conditions for its own sustainable operation.

Respect for the identities of sports facilities. These features identities are reflected in various aspects such as the investment and operation, space and structure, equipment and systems. As a large-scale public investment project in the city, sport facilities occupy a large arca and consume great amounts of money, so that they play an important role in the overall urban functions and space. Similarly, due to a large amount of capital investment, the operation of sports facilities is related to the economic operation of the cities and even the society. Therefore, the location planning, scale positioning, and construction standards of the sports facilities require special attention. As a large-scale public building, sports buildings have large space and span, and the equipment and systems involved are complicated. They require higher construction investment, and the reconstruction and demolition after the games are even more difficult. In the long-term operation process, the energy

图 2-1 赛后得到很好运营的中国农业大学体育馆

Fig. 2-1 The China Agricultural University Gymnasium is being successful operated after game

从城市的可持续性出发，结合城市设计理念，从功能整合和空间整合两方面研究基于城市整体环境的体育建筑决策与设计方法，探讨体育建筑对城市区域的影响及其与城市环境的协调发展策略。从功能的可持续性出发，结合体育产业发展趋势，探讨建筑功能灵活适应性设计方法，包括建设规模、功能配置、空间利用策略，为实现体育建筑功能可持续运营提供决策与设计基础。从技术的可持续性出发，探讨结构选型、设备系统、自然通风、天然采光、容积控制对室内光、热、声环境、能耗的影响和相应的设计方法，研究适宜技术条件下实现体育建筑节地、节能、节水、节材和环境保护目标的设计策略。

consumption, maintenance, and renewal costs are high as well, so the flexibility and adaptability of the functions and spaces of the sports facilities, the intensive suitability of equipment and systems is extraordinary worthy of attention. These characteristics give the sports facilities their identities, and are also the key issue that needs to be solved for their sustainable development (Fig. 2-1).

The sustainability research of sports facilities is for achieving its sustainable goals, focusing on its positioning and scale, construction and operations, functions and space, and providing research foundations and guidelines for the whole process of feasibility studies, planning and architectural design, construction, and post occupancy evaluation.

From the perspective of the city sustainability, combining the urban design concepts, the studies summarize the decision-making and design methods of sports facilities based on the overall urban environment from the aspects of functional integration and spatial integration, and discuss the impact of sports facilities on urban areas and the coordinated development strategy with urban environment. From the perspective of the functions sustainability, combining with the development trend of the sports industry, the studies investigate the flexibility and adaptabiltiy design methods for building functions, including construction scale, function allocation, and space utilization strategies, which provides decision-making and design foundations for the sustainable operation of the building functions. From the perspective of technology sustainability, the studies discuss the structural selection, equipment systems, natural ventilation, natural lighting, the impact of volume control on the indoor optical, thermal, acoustic environment and energy consumption, and sum up corresponding design methods and strategies for saving the land, energy, water and material, and protecting the environment.

体育建筑可持续研究的目标

1. 回归理性

作为大型公共建筑和城市重大公共投资项目，体育建筑不仅功能复杂、建造技术要求高，而且承担着许多社会责任，关系到城市的可持续发展，其建设迫切需要回归科学定位、理性决策、复合逻辑的轨道中来。体育建筑可持续决策与设计研究关注体育建筑建设决策的科学理性、规模定位的务实合理、功能空间的灵活适应、设备系统的集约适宜、场馆运营的永续发展，为体育建筑回归理性的科学决策和设计提供基础论证和指引。

2. 回归城市

体育建筑是城市功能和空间的重要组成部分，体育建筑由于其自身功能和形态的特殊性，对城市整体功能和空间有重要影响。体育建筑的健康运转是城市持续发展的有机组成部分，城市整体也为实现体育场馆的可持续发展提供外部条件的保证。体育建筑设计应关注城市的整体性，突破传统由内而外的单向思维，摒弃以自我为中心的单体建筑设计思维。体育建筑可持续决策与设计研究关注更多地从城市整体功能和空间需求出发由外而内的进行设计，促进体育建筑与城市的功能和空间相协调发展（图 2-2）。

图 2-2　尊重城市空间的一个体育建筑案例

Fig. 2-2　A sport facility case with respect to the urban environment

Objectives of Sustainability Research

1. Return to rationality

As large-scale public buildings and city major public investment projects, sports facilities not only have complex functions and high technical requirements for construction, but also bear many social responsibilities and are related to the sustainable development of the cities. The construction of the sports facilities urgently needs to return to the track of scientific positioning, rational decision-making, and complex logic. Sustainable research of decision-making and design of sports facilities focuses on the scientific rationality of sports facilities construction decision-making, pragmatic and rationalization of the scale positioning, flexibility and adaptability of the function and space, intensivism and suitability of the equipment and systems, sustainable development of the stadiums and gymnasiums, and provide basic demonstrations and guidelines to the scientific decision-making and design for the return of sports facilities to rationality .

2. Return to city

Sports facility is an important part of urban function and space, and has an important influence on the overall function and space of the city because of its particular function and form. The healthy operation of sports facilities is an organic part of the city's sustainable development, and the city as a whole in turn provides external conditions for the sustainable development of sports venues. The design of sports facilities should pay attention to the integrity of the city, break through the traditional thinking that carry out only from the inside out, abandon the single thought of self-centered architectural design. The research on sustainable decision-making and design of sports facilities pays more attention to designing from the outside to inside based on the overall function and space requirements of the city, and promotes the coordinated development between the sports facilities and the cities (Fig. 2-2).

3. 回归功能

体育建筑为竞技比赛、体育锻炼、体育教育和体育娱乐等活动提供合适的空间场所。体育是根据人体生长发育、技能形成和技能提高的规律，通过身体和智力活动的手段，达到改善身体素质，增强运动能力，提高生活质量的一种社会活动，体育事业的发展水平已经成为国家和社会发展进步的一个标志。体育建筑可持续决策与设计研究关注其服务于民众体育活动的核心功能，也通过这一价值的实现，促进体育建筑自身的可持续运营（图 2-3）。

4. 回归技术

体育建筑建设技术难度大，使其往往成为某一时期最高建筑技术的代表之一。建筑技术属于工程实用技术，是将相关科学知识或技术发展的研究成果应用于建造实践，以达到建设目的的手段和方法，具有实用可行、经济合理的特点。建筑技术一方面需要避免配备不足而影响场馆的正常使用，另一方面也要避免盲目超标准地投入而造成浪费。体育建筑可持续决策与设计研究关注通过适宜性建筑技术，优化结构选型、改善建筑室内环境，提高能源利用效率，实现建筑可持续发展。

3. Return to function

Sports facilities provide suitable space for sports competitions, physical exercises, physical education and sports entertainment. Following the laws of human growth and development, skill formation, and skill improvement, sport is a kind of social activity that improves physical fitness, enhances athletic ability, and improves quality of life through physical and intellectual activities. The development level of sports has become a sign of the national and social progress. The research on the sustainable decision-making and design of sports facilities focuses on the core functions of serving the people's sports activities, and also promotes the sustainable operation of sport construction itself through the realization of this value (Fig. 2-3).

4. Return to technology

The large construction technical difficulty of sports facilities makes it generally become one of the highest building technology representatives in a certain period. Building technology belongs to practical engineering technology that is a method for applying research results of related scientific knowledge or technology development to the construction practice to achieve certain purposes of construction, which is practical, feasible, and economically reasonable. On the one hand, construction technology needs to avoid the lack of equipment that affects the normal use of the venues. On the other hand, it must avoid blindly over-standard investment that causes waste. Research on sustainable decision-making and design of sports buildings focuses on the sustainable development through appropriate building technologies, optimizing the structures selection, improving the buildings' indoor environment and the energy efficiency.

图 2-3 赛时和赛后可能在体育建筑内举行多种活动

Fig. 2-3 Variety of activities may be held in a sport facility during and after game

“精明营建”的理念

1. 基于可持续性，注重“全过程”

建筑可持续的研究在我国逐渐受到关注，但目前大多数关注点集中在建筑节能的应用技术层面，关注设备和系统、材料和结构等建造方式。在整个建设环节上看，这些工作处于末端，对既定的方案进行性能评价和优化工作，能收到一定的成效，但问题在于，这些研究成果无法提前反馈到项目前期阶段，为决策者提供决策参考和建议，甚至也没有反馈到设计前期阶段，为规划师和建筑师提供方案设计依据和指导，因此从根源上限制了可持续发展目标的实现。

在项目建设前期的体育建筑决策，明确建设目标，制定建设计划，涉及到定位、规模、投资、运营、功能和技术的各个方面，这一过程有利于确保项目的建设过程和良好效益，决策的成果不仅面向规划管理，而且面向建筑设计和运营。一方面，作为城市大型公共建筑和公共投资项目，体育建筑决策涉及到体育设施选址、用地、交通和城市公共空间体系，也关系到政府财政和社会资金的投资回报效益问题，对城市空间格局、社会经济效益有着重要影响。另一方面,决策环节的场馆规模定位、建设标准、运营策划等研究，作为规划和建筑设计的依据，对建筑场馆建设有着直接把控，影响甚至决定了场馆建造和运营的可持续性。

体育建筑建设特别需要从“全过程”研究可持续性，从决策阶段充分考虑场馆建设和运营，在设计阶段进行精明设计，才能在合理控制场馆建设成本的情况下，保证长期的可持续性运营。而这其中，建设初始环节决策的科学性，在一定程度上决定了可持续目标是否能够达成，因而成为可持续设计策略研究的关键。强调全过程，重视早期科学决策，突破传统研究局限，提高体育设施规划布局的灵活机动性；结合国情，从控制初始成本、维护成本和

The Concept of ‘Smart Planning and Building’

1. Based on sustainability and focus on the overall process

The study of building sustainability has gradually attracted attention in our country. However, most of the current concerns are focused on the application technology of building energy conservation, focusing on the equipment and systems, materials and structures. Judging from the entire construction process, these works are at the end, in which the performance evaluation and optimization of a given program can achieve certain results, but the problem is that these results cannot be fed back to the early stage of the project and provide references and recommendations for decision-making to the decision makers, and even cannot feed back into the early stages of design, to provide design basis and guidance for planning and design to the planners and architects, thus limiting the achievement of sustainable development goals from the root.

The decision-making of sports facilities construction in the early stage of the projects, related to determining the construction goals and formulating the construction plan, which involve all aspects of positioning, scale, investment, operation, function and technology. This link helps to ensure the construction process and good benefits of the project. The results are not only for the planning management but also for the building design and operations. On the one hand, as a large-scale urban public building and public investment project, the decision-making of sports construction involves the location of sports facilities, land use, transportation, and urban public space system, as well as the issue of the investment return of government finances and social funds, and have an important influence on the urban space, social and economic benefits. On the other hand, the research on the positioning of venues, construction standards, and operation planning in the decision-making process,

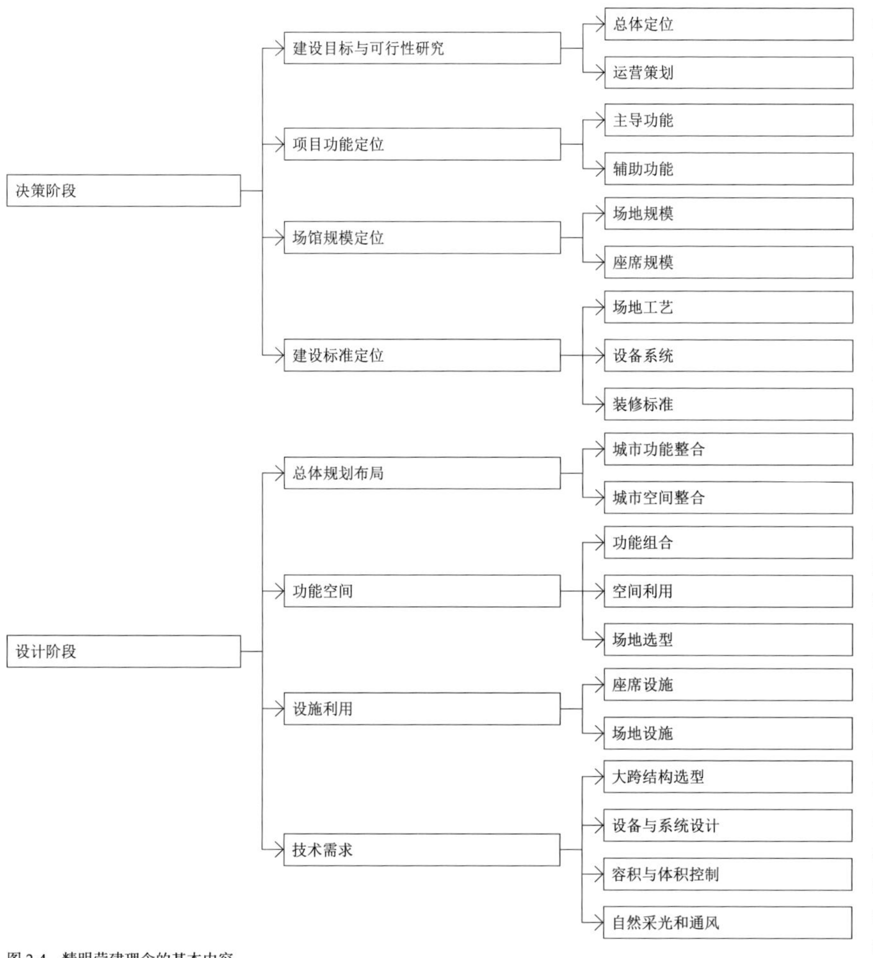

图 2-4　精明营建理念的基本内容

更新成本三方面研究降低全寿命周期成本设计策略，以集约建设为原则，研究符合国情的低成本、低损耗的体育建筑设计策略（图 2-4）。

as a basis for the planning and architectural design, has direct control over the construction of building venues, influences and even determines the sustainability of the construction and operation of the sports facilities.

In summary, the key is to focus on the overall process, to emphasis on decisive moments at early stage, conquer conventional limits, optimize operational flexibility; to control construction, maintenance and update funding for a reduced total input. The ultimate goal is to develop low-expense, low consumption sports facilities suitable for domestic conditions. The construction of sports facilities particularly requires sustainability study from the 'whole process', sufficient consideration to the construction and operation of venues from the decision-making stage, and smart planning and design in the design stage so as to ensure the long-term sustainable operation under the reasonable control of construction cost of the venues. Among these, the scientificity of the decision-making process at the initial stage determines to a certain extent whether the sustainable goals can be achieved, and thus becomes the key to the research of sustainable design strategy (Fig. 2-4).

2. 基于“整体环境”，强调“城市理性”

1999 年，我们从城市的可持续性出发，在国内首先提出“基于城市的体育建筑设计”理念，探讨体育建筑对城市区域与街区的影响，强调体育设施应注意与城市空间环境的协调性和匹配度，尝试建立与室外环境影响相关联的科学分析程序，建立与建设强度、交通模式相关联的科学分析与设计方法。从功能整合和空间整合两方面优化体育建筑设计方法策略。所谓“整体环境”与“城市理性”，是基于将城市作为整体考虑，城市建筑在整体格局下担当

2. Based on the general environment and srengthen the urban rationality

Bearing urban sustainability in mind, we invented the concept of city-based sports facilities design in 1999 to explore the impact of sports facilities on urban districts and neighborhoods, in which we emphasized the coordination between sports facilities and urban spatial environment, and tried to establish scientific analysis procedures pertinent to outdoor environmental impact and design methods pertinent to construction intensity and traffic patterns. The design methods and strategies

Fig. 2-4　Basic content of smart planning and building

不同的角色作用，即：作为日常的建筑和作为丰碑的庆典、殿堂建筑。对于体育建筑而言，仅仅作为丰碑式建筑的单一思维，导致了体育建筑在中国成为充满遗憾的殿堂，远离了城市日常生活，城市理性趋于消失。

体育场馆由于其自身形态和功能的特殊性，对城市整体环境的空间体系、功能布局和交通组织等都有重要影响。体育建筑的合理投资和运营是城市持续发展的有机组成部分，其定位和规模、功能和空间对城市形态产生极大影响。城市整体经营条件也为实现体育场馆的可持续发展提供基础，城市整体发展为体育场馆构建空间框架，城市居民活动为体育场馆运营提供需求动力。体育建筑可持续设计策略

for sports facilities should be optimized through functional integration and spatial integration. By general environment and urban rationality, we meant to consider the city as a whole and to assign different roles for different urban buildings under the big picture, namely, the common buildings for daily use, and monumental buildings for special occasions. The simplistic thought that sports facilities nothing more than monumental reduced sports facilities in China a pity isolated from the daily urban life. Rational development of urbanism is on the brink of disappearing.

Because of the unique shape and function, sports venues have an important influence on the spatial system, function distribution, and traffic organization of the city's overall environment. As a large-scale public building and urban large public investment project, the rational investment and operation of sports facilities is an organic part of sustainable development of the whole city, and their positioning and scale, function and space have a great influence on the urban form. The overall operating conditions of the city also provide the external environment for the sustainable development of sports venues. The overall development of the city creates a spatial framework for sports venues, and meanwhile the activities of urban residents provide the driving force for the operation of sports venues. In the process of decision-making, design, construction and operation, the sustainable design strategies of sports facilities focus on the spatial integration and functional integration of sports venues with the overall environment of the city, including the mutual promotion and development of construction site selection and urban development, organic integration of the venue layout with the urban texture, space public system, the mutual benefits between the sports activities and non-sports activities of the venues and the demands of urban residents, etc.

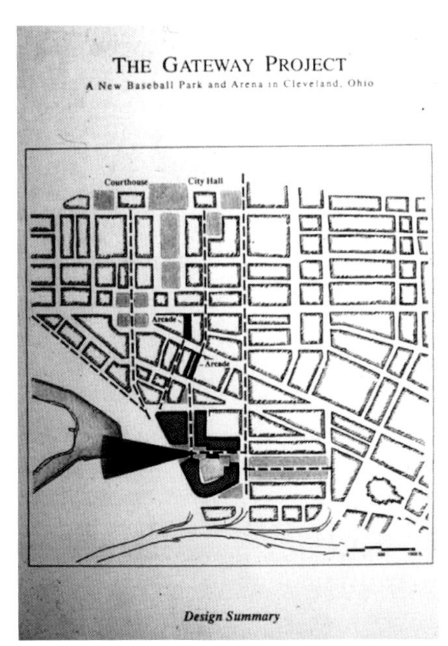

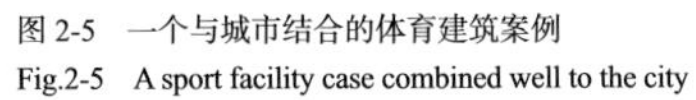

图 2-5　一个与城市结合的体育建筑案例
Fig.2-5　A sport facility case combined well to the city

在决策、设计、建造和使用过程中，关注体育场馆对城市整体环境的空间整合和功能整合，包括体育场馆选址与城市发展的互促发展、场馆布局与城市肌理和空间公共空间体系的有机融合、场馆体育活动和非体育活动功能与城市居民使用需求之间的互惠共赢等。

我们无法否认体育建筑在城市中的纪念性意义，作为殿堂的体育建筑有着极其重要的精神作用，而伴随着庆典般的赛事，其精神意义更加显著。然而，现代体育是植根于日常生活的，过分强调体育建筑的纪念性，甚至因此排斥体育日常性需求的结果将导致城市公共生活场所的缺失。当今中国的体育建筑主要是由政府主导公共投资建设，即便是缘起于大型运动会的殿堂性活动，赛后依然要面对日常需求与运行，这不仅仅是体育设施利用率的问题，本质上是公共建设的价值观问题。在设计中巧妙区分体育建筑的不同角色、身份，将直接影响设施的使用、居民的生活、公共利益与公共投入的公平性。只有将体育建筑的建设决策，纳入“城市理性”的范畴，才会更加符合城市公共利益（图 2-5）。

We cannot deny the monumental significance in sports facilities in a city. Sports facilities play an extremely important spiritual role, particularly during celebrative events. However, modern sport is rooted in daily life. Too much emphasis on the monumental significance in sports facilities in repulsion of daily sports needs will eventually result in the sacrifice of public urban space. The sports facilities in China today are mostly public investment and construction projects guided by the government. Yet even venues where major games have taken place are facing daily needs and operation after the games. This is not just an issue of sports facility utilization. Essentially, it is an issue of values of public building. Appropriately identifying the difference roles of sports facilities in its design will directly affect the use of facilities, the lifestyle of residents and the fairness of public interest and public investment. Only by including the decisions on the construction of sports facilities into the category of urban rationality can we better serve the urban public interest (Fig. 2-5).

3. 基于灵活、适应性，坚持“功能理性”

体育建筑的多功能使用研究不是新的话题，但建筑发展今天的问题是功能研究的被忽略，特别是功能性强的公共投资项目。在体育场馆全寿命周期内，其使用需求必然会随着国民经济和体育产业发展而不断动态变化，为了在全寿命周期内提高场馆的利用率，可多功能利用的策略是许多场馆实际运营的必然选择，这使得体育设施使用需求上具有动态、不确定性的特点。与其他类型的建筑相比，体育场馆的大跨度结构和大空间体量使其自身的技术要求和建设成本高，并且建成之后的拆除和改造难度大。

因此体育建筑建设应从决策阶段考虑场馆的多功能利用，在设计阶段做出充分的灵活适应性设计，从而在全寿命周期内提高场馆的使用率，实现场馆的可持续性。所谓“灵活性”，强调应对动态需求的方法，所谓“适应性”，强调体育场馆具有与环境、外部需求等条件相适合的能力。“灵活适应性”与“原则性”存在着一种辩证关系，是在原则性规律、法规和制度下的灵活处理问题的方法，灵活适

图 2-6 可灵活进行功能转换的体育建筑

Fig.2-6 A sport facility that can change function neatly

3. Based on flexibility and daptability and adhere to the functional rationality

The study of multi-purpose of sports facilities is not a new topic, but the problem today in architecture development is the ignorance of function study, especially for that od public investment projects with high functionality. During the entire life cycle of the sports venues, the demands for them will inevitably change dynamically with the development of the national economy and the sports industry. In order to increase the utilization rate of the venues throughout the life cycle, the multi-functional utilization strategy is the inevitable choice for actual operation of many venues, which makes the operation of sports facilities with dynamic and uncertain features. Compared with other types of buildings, the large-span structure and large space volume of the sports venues raise their own technical requirements and construction costs, and make it difficult to be removed and reconstructed after construction.

Therefore, the construction of sports facilities should take into consideration the multifunctional use of venues from the decision-making stage. In the design stage, flexibility and adaptability design should be made to increase the utilization rate and realize the sustainability of the venues throughout the life cycle. The so-called ‘flexibility’ emphasizes the method of responding to the dynamic demands. The so-called ‘adaptability’ emphasizes that sports facilities have the ability to adapt to the conditions such as the environment and external demands. There is a dialectical relationship between ‘flexibility and adaptability’ and ‘principality’, and it is a method of flexibly handling problems under principled laws, regulations, and systems, while the reasonable requirements for flexibility and adaptability can also provide feedback to the formulation of principled rules and systems. The greater the flexibility of sports venues, the lower the risks and losses due to the inability to adapt to changes in functional requirements during

应性的合理要求又可以对原则性规则和制度的制定提供反馈。体育场馆灵活适应性越大，后期运营过程中因无法适应功能需求变化的风险和损失就越小，两者呈反比例关系。

the latter operation, which means the two are inversely proportional.

面对体育建筑功能需求的不确定性，在前期策划和设计阶段做好弹性应变准备，为场馆功能、空间、设备系统和外部环境等要素进行合理定位，预留采用最小成本进行优化、重组、转换和更新的可能性，使得场馆具备付出少量成本满足多功能利用的条件。而明确坚持“功能理性”也是避免公共投资变相流失的重要环节（图 2-6）。

Faced with the uncertainty of the functional requirements of the sports facilities, preparing the elasticity in the early planning and design stages, rationalizing the functions of the venues, space, equipment systems, and the external environment, and reserving for the optimization and reorganization possibility of conversion and update at the minimum cost, allows the venue to meet the conditions for a small amount of costs to satisfy the multi-functional utilization. Explicit adherence to functional rationality is an important measure against disguised erosion of public investments (Fig. 2-6).

4. 基于集约、适宜性，满足“技术理性”

目前我国建筑能耗在全国能耗总量中的比重达到 35% ~ 40%，其中公共建筑能耗巨大，建筑节能直接关系到国家的可持续发展。体育建筑与其他类型的建筑相比，大量混凝土和钢材的消耗、大型结构和复杂设备系统的投入，也使得其往往成为一个城市最为昂贵的建设工程之一，而建成后体育场馆的运营成本通常还更高于初始建设成本。从决策阶段充分考虑场馆建设和运营成本控制，采用适宜性技术，对于提高能源利用效率，降低能源消耗具有很大潜力。

4. Based on intensivism and suitability to meet the technical rationality

At present, the proportion of building energy consumption in the total energy consumption in China is 35%-40%, of which the weight of public buildings is huge. The building energy conservation is directly related to the country's sustainable development. Compared with the other types of buildings, the consumption of large amounts of concrete and steel, the large-scale structure, and the investment in complex equipment and systems make it usually one of the most expensive construction projects in the city, and the operating costs of the venues after games are normally higher than the initial construction cost. The adequate consideration of venue construction and operation cost control from the decision-making stage and the adoption of appropriate technologies have great potential for improving the energy efficiency and reducing the energy consumption.

我国体育产业发展尚处于起步阶段，场馆运营条件尚未成熟，因而体育建筑的建设应集约节约，采用适宜性技术。“集约化”强调提高资源利用效率，“适宜性”强调采用合适的技术手段。集约适宜的“技术理性”关注决策和建设阶段的精明策划和设计，关注大跨度选型、设备与系统设置、容积与体积控制、自然采光和自然通风的可持续设计策略。结合地域条件，综合被动式和主动式技术，采取科学实用的结构技术和体育工艺，既要避免盲目迷信昂贵的高技术和设备，又要避免技术配备不足而影响场馆的正常使用和可持续运营。

The development of China's sports industry is still in its infancy, and the operating conditions of the venues are not yet mature. Therefore, the construction of sports facilities should be intensive and economical, and suitable technologies should be adopted. 'Intensivism' emphasizes the improvement of resource utilization

“技术理性”一方面要求在决策环节立足于场馆建设作为城市公共服务设施的基本需求，尽量集约节俭地建设，进行科学合理的规模和建设标准定位，既要避免过度追求高标准和标志性，也不能为压低初始建设成本而造成功能的不足和运营成本的提高。另一方面要求在技术手段上应结合我国国情，优先采用适宜性技术，进行结构、设备、体型、采光和通风的精明设计。将集约适宜的原则应贯穿到体育建筑立项、可行性研究分析与评估、决策、设计、施工到运营使用的整个过程中，在全寿命周期控制体育建筑建设投入，改善建筑室内环境，降低能源消耗，提高体育建筑的可持续性。

efficiency, and ‘suitability’ emphasizes the use of appropriate technological means. Intensive and suitable ‘technical rationality’ emphasizes smart planning and design at the decision-making and construction stages, focusing on sustainable design strategies for long-span structure selection, equipment and systems arrangement, capacity and volume control, natural lighting and ventilation. Combining regional conditions, the technical rationality integrates passive and active technologies, adopts scientific and practical structural technologies and sports techniques, so as to avoid superstitiously consuming expensive high-tech and equipment, and the lacking of technical equipment that might affect the normal use and sustainable operation of the sport facilities.

‘Technical rationality’ requires, on the one hand, that the decision-making process be based on the construction of venues as the basic needs of urban public service facilities, intensive construction as far as possible, scientific and reasonable scale and positioning of construction standards. It is necessary to avoid excessive pursuit of high standards and landmarks, and also the functional deficiencies and increase operating costs caused by irrational initial construction cost. On the other hand, it is required that technological measures should be combined with China's national conditions, priority should be given to the use of appropriate technologies, and smart design of structures, equipment, volume, lighting and ventilation should be carried out. The principle of intensivism and suitability should be carried throughout the entire process of the projects, including the analysis and evaluation of feasibility studies, decision-making, design, construction and operation, and control the overall investment in sports construction throughout the life cycle, improve the indoor environment of the buildings and reduce energy consumption, so as to improve the sustainability of sports facilities.

三、城市理性

III. URBAN RATIONALITY

人类聚集生活产生了城市，城市是人类向往舒适生活而创造的聚居形式，随着聚集度的提高，密度的增加，城市机能趋向复杂，城市本身却给人类带来了双重的可能。曾经人们对城市的乐观理解，在20世纪各种城市缺陷泛滥的情形下破灭。面对所谓的城市病，人们认识到，城市并非一定让生活更美好，只有好的城市才能让生活更加美好。在建设方面，为达到好的城市需要许多工作，最为重要的就是遵守公共规则。复杂的城市构成当中如何让人类生活和工作保持便利、有效是城市建设的根本。因此，好的城市不是没有制约的城市，而是公共利益明确、公共空间完整，公共生活丰富多彩。

体育建筑是城市功能和空间的重要组成部分，由于其自身功能和形态的特殊性，对城市整体功能和空间形态有重要影响。体育建筑的健康运转是城市可持续发展的有机组成部分，城市整体也为实现体育场馆的可持续发展提供外部条件的保证（图3-1）。

自古以来体育建筑在城市中的作用：提供公共活动场所，不是形态的主角，而是空间的载体。体育建筑的近代发展与城市的关系一直是一种动态变化的过程。在奥林匹克运动初期，体育建筑与城市的关系最为密切（图3-2）。

图3-1　罗马斗兽场
Fig.3-1　The Roman Colosseum

图3-2　雅典大理石体育场
Fig.3-2　Panathenaic Stadium

City takes form when human gather up to hive. It is a form of inhabitation created in quest of comfort. As inhabitation and density advance, the functionality of city gets complex while the city itself has presented dual possibilities to mankind. People had an optimistic impression about city, but that busted when various urban flaws became rampant in the 20th century. Faced with the so-called urban disease, they get to realize that city does not necessarily make life better, except when it is a good one. In terms of construction, it takes a lot of work to build a good city, and the most important of all to comply with public rules. It is the root of urban construction that how a city of complex composition makes our life and work easy and effective. Therefore, a good city is not a city without constraints, instead, it shall provide clear public welfare, integral public space and colorful public life.

The role of sports facilities among the rest is providing public venues, as a space carrier instead of a form leader. Since the very beginning, the relationship between the modern development of sports facilities and the city has been a dynamic process of change. The relationship between sports facilities and the city started in the most intimate way when Olympic Games first appeared (Fig. 3-1).

Sports facility is an important part of urban function and space, and has a significant influence on the overall functional and spatial influence on the city because of its particular function and form. The well working of sports facilities is an organic part of the city's sustainable development, and the city as a whole also provides external conditions for the sustainable development of sports facilities (Fig. 3-2).

To present a magnificent cluster in the 1936 Olympic Games, Germany tried to make an concentrated arrangement of sports centers and venues by paying attention to the massing relationship, axial relationship, outdoor environments and sculptures. The preparation for Olympic Games in Rome was parallel with the preparation for the World Expo and the planning and construction of the new district of Rome (Fig. 3-3).

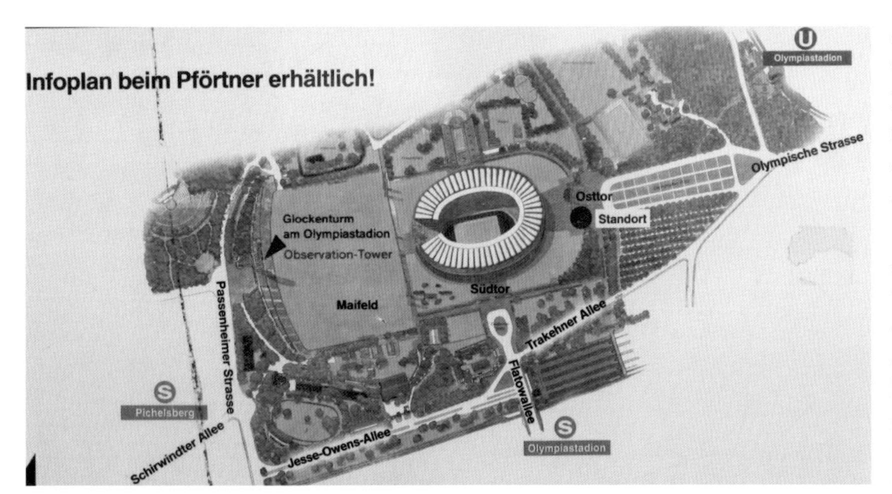

图 3-3　1936 年奥运会体育中心和实景图

Fig.3-3　The sports centers and live scene of the 1936 Olympic Games

1936 年奥运会为了形成集中壮丽的场景，德国尝试集中建设体育中心，体育场馆集中布置，注意了体量关系、轴线关系、室外环境、雕塑小品。同时筹备的罗马奥运会甚至是伴随筹办世界博览会、建设罗马新区一并规划建设的（图 3-3）。

随着现代体育建筑的功能日益复杂，容量大、集聚快的功能带来了对大跨度建筑结构选型的复杂需求。而现代建筑思想要求“建筑设计从功能出发，建筑形式是内部功能的忠实反映”，这一曾经被奉为天条的现代建筑运动基本原则，对体育建筑而言似乎一直是天经地义。回顾体育建筑的发展史，我们可以发现，在理性主义的设计理念之下，曾涌现众多的大师与杰作。奈尔维的罗马小体育馆、丹下的东京代代木体育馆都是令人折服的隽永之作（图 3-4）。

As the functionality of modern sports facilities gets increasingly complex, large capacity and fast clustering effect entail complex needs for structures selection for large-span venues. The modern architectural ideal of ‘architectural design starts from functionality, and architectural form is a faithful reflection of internal functionality’ have been regarded as a heavenly instruction of modern architecture movement, appears to be proper for sports facilities. A review of the history of the development of sports facilities reveals that design concept of rationalism has inspired a large number of masters and masterpieces. To name a few, the admirable small Roman Gymnasium by Nervi, and Tokyo Yoyogi Gymnasium by Tange (Fig. 3-4).

图 3-4　罗马小体育馆

Fig.3-4　The Palazzetto Dello Sports of Roma

体育场馆现状布局模式与反思

体育建筑作为特殊的建筑类型，由于体型庞大和对人员的聚集效应，在形式和功能上均与其他建筑形成鲜明对比，构成城市空间节点的标志物。体育建筑承载着满足人们体育运动需求的功能，由于运动场地和观演功能对大空间的需求，体育建筑往往成为同一时代最高建筑技术的代表类型之一，在很长一段时间内，

The Status of venues Layout and the Reflections

As a special type of building, sports facilities constitute a landmark of urban space nodes due to its large size and the effect of gathering people, and the stark contrast with other buildings in terms of form and function. Sports facilities carry functions that meet the sports demands of the residents. Due to the demand for large

由于技术局限，体育建筑和其他背景建筑之间保持相对和谐的体量比例关系，因此长期以来与城市紧密结合，协调发展。到了近现代，由于新建筑材料的出现以及结构技术、设计和施工方法的进步，体育建筑突破原有空间跨度和造型局限，获得前所未有的体量膨胀和造型表现的机会，并因此摆脱和周围建筑原有的比例关系。在各种建筑思潮和决策者意志影响下，体育建筑创作走上了强调自我地位，表现造型，炫耀新材料、新结构和新技术的道路。

早期兴建的体育建筑用地大都在建成区，除了个别利用公园兴建外，大多较为局促 (图 3-5) 。改革开放以后，随着我国国民经济高速增长，城市化进程尤其在近二十年来发展迅速，这一时期也正值体育建筑建设高潮。用地庞大的体育中心布置在城市郊外作为新城开发的助推器，成为许多城市新区发展的建设策略。1987 年建成的天河体育中心是第一个跳出城市建成区，完整兴建的大型体育中心，得到了广泛的好评。项目获得了体育建筑的国际

spaces of sports activities, sports venues are often one of the representatives of the highest building technologies at the same time. For a long time in history, due to the technical limitations, Sports facilities maintain a relatively harmonious proportion relationship with the other background buildings, and therefore have long been closely integrated and developed in coordination with the urban environment. In modern times, due to the appearance of new construction materials and advances in structural techniques, design, and construction methods, sports construction has broken through the original space span and styling limitations and has gained unprecedented opportunities for volume expansion and styling expression, which made it get rid of the original proportion with the surrounding buildings. Under the influence of various architectural trends and the will of decision-makers, the construction of sports facilities has embarked on the road of emphasizing self-position, expressing the styling, and showing off the new materials, new structures and new technologies.

Most of the early sports facilities were sited in developed areas, hence their layout was compact, except for those sited in parks (Fig. 3-5). After the reform and opening up, with the rapid growth of China's national economy, the process of urbanization has developed rapidly especially in the past two decades. This period also coincided with the upsurge of the construction of sports facilities. The large sports center located in the outskirts of the city as a booster for the development of new districts has become a strategy for the development of many new urban areas. For example, the Tianhe Sports Center, built in 1987, was the first large-scale, complete sports center erected outside the urban area, which had widely received positive reviews. The project was the second domestically original sports facilities that received international recognition since the Beijing Asian Games Center. Henceforth, the Tianhe Sports Center has become a mode for the 'one stadium and two halls' design

图 3-5　建于 1932 年的广东省人民体育场

Fig. 3-5　The Guangdong Provincial People's Stadium built in 1932

图 3-6　建于 1987 年的广州天河体育中心

Fig. 3-6　The Guangzhou Tianhe Sports Center built in 1987

奖励，成为继北京亚运中心之后的第二个获得国际认可的中国原创体育建筑。之后，天河体育中心几乎成了“一场两馆”体育中心的样板，深深影响了各个类型城市的体育中心建设。而天河体育中心经过近30年的发展，空间形态也在发生着本质的变化（图3-6）。广州天河体育中心直接带动了天河区的整体发展，并成为城市新中轴线上的重要节点。当前的体育中心建设还通过与其他核心功能的并置一起构成功能复合的城市中心区。例如，将体育场馆与会展中心并置，形成体育会展城市中心区，如南通体育会展中心和常州奥林匹克体育中心；将体育场馆与休闲娱乐、金融商业、旅游购物、文艺演出等结合为大型多功能综合城市中心区，如江门滨江体育中心。

standard and influenced the construction of various sports centers. Essential changes have also occurred in the spatial form of the Tianhe Center over nearly 30 years of development (Fig. 3-6). Guangzhou Tianhe Sports Center has directly driven the overall development of the Tianhe District and has become an important node on the new central axis of the city. The current construction of the sports center also forms a functional complex urban center area by juxtaposition with other core functions, for example, juxtaposing the sports venues with exhibition centers to form the center of sports and exhibition central areas, such as the Nantong Sports Exhibition Center and Changzhou Olympic Sports Center; combining sports venues with leisure and entertainment, finance, commerce, tourism and shopping, and theatrical performances to form the comprehensive urban center area with large-scale functions, such as Jiangmen Binjiang Sports Center.

对于这类作为城市新区中心的体育建筑，其规划建设效果有待城市的发展来检验，但目前已经暴露出来许多问题需要引起关注。例如在许多城市规划中，规划方案不断大幅更改，声东击西，使得体育建筑面临的外部条件存在巨大的不确定性。规划文件中提倡的功能复合、集聚发展、公共交通、以人为本、营造活力场所等理念并未充分贯彻，取而代之的是孤立于城市、远离居民的体育中心，以及超尺度、门庭冷落的城市空间。究其原因，当下体育建筑规划布局存在以下问题：

For such a sports facility as the center of the new urban district, its planning and construction effects need to be examined with the city's development. However, many problems have been exposed, which should draw our attention. For example, in many urban planning, the planning scheme is continuously changed to a large extent, which sells the dummy and makes a huge uncertainty to the external conditions faced by sports facilities. The concepts of complex function, agglomeration development, public transportation, people-centeredness, and creation of vitality sites advocated in the planning documents have not been fully implemented. Instead, they are over-scaled and uncivilized spaces isolated from cities and far away from residents' sports centers. The reason for this includes the following problems in the layout planning of the sports facilities:

1. 固化的布局模式与变化的城市生活需求之间的矛盾

总体布局模式固化，缺乏对多样化城市生活需求的适应性。许多年来体育中心一场两馆“品”字形或者“一”字形的布局模式被延续下来，将体育场馆和室外体育设施布置在用地中央，外部环绕以停车场或大片绿地，这与决策者和设计者将体育场馆作为标志性建筑，突出宏伟和纪念性的惯性思维有关，这一固化的布局方式，不能适应未来多样化的城市生活需求。这一模式一方面从空间形态上缺乏“街墙”，与环境格格不入；另一方面也违背运营规律，缺乏适应性和灵活性，不利于满足未来城市生活和体育活动自身的发展，许多这类体育中心在周围城市环境成熟后，需要在地块内沿街道的用地边界处增建配套功能设施，这也反映出原有体育场馆布局方式的局限性。从未来多样化的城市生活需求出发，探讨多元化的体育设施布局模式，对体育建筑可持续发展至关重要。

1. Contradiction between the curing layout modes and the changing needs of urban life

The solidified overall layout mode lacks adaptability to the diverse demands of the urban life. For many years, the layout mode of ‘delta-shape’ or ‘line-shape’ for the ‘one stadium and two hall’ of the sports centers has been continued. The venues and outdoor sports facilities have been placed in the center of the site and surrounded by large parking lots or green areas. This is in line with inertial thinking of decision makers and designers that regard sports venues as a symbolic building, highlighting grand and commemorative. This solidified layout cannot adapt to future diversified urban life. On the one hand, this mode lacks a ‘street wall’ from the perspective of the spatial form and is incompatible with the environment. On the other hand, it also violates the law of operation, lacks adaptability and flexibility, and is not conducive to the future development of urban life and sports activities. In many of these cases, when the surrounding urban environment is mature, the sports center needs to build additional supporting facilities at the land boundary along the street, which also shows the limitations of the original layout mode. From the perspective of the diversified demands of urban life in the future, the exploration of the diversified sports facilities layout mode will be crucial to the sustainable development of sports facilities.

2. 商业配套功能与体育建筑公益形象之间的矛盾

配套功能设置上缺乏与体育建筑和城市整体特征的统一，造成与体育建筑公益形象的矛盾。许多的体育建筑配备了商业配套设施，但仅仅服务于体育场馆自身的经营，而不是出于城市整体的功能需求，这种商业配套功能注

2. Conflict between the commercial supporting functions and the public image of sport facilities

The lack of integration of supporting functions with the overall identities of the sports facilities and cities has caused conflicts with the public interest image of sports facilities. Many sports venues are equipped with commercial facilities, but only serve the sports venue itself, not for the overall functional needs of the city. This kind of commercial supporting

定了其成为体育建筑的附属功能，也容易因缺乏城市需求动力而造成难以运营。另一种情况则反之，在缺乏统一规划下，商业功能的营利性经济目标压倒体育建筑的公益性社会目标，造成体育建筑整体形象的损害。

functions is destined to become an auxiliary of sports facilities, and is also easy to be difficult to operate due to the lack of cities demand power. In the other case, on the contrary, in the absence of a unified plan, the profit-making economic objectives of the commercial function have overwhelmed the social goals of the public sports construction, and have caused damage to the public interest image of sports facilities.

3. 突出体育建筑标志性与回归城市建筑群体之间的矛盾

形体处理手法上以突出自我为目标，缺乏城市建筑群体视角的考虑。当前大量体育建筑设计延续着自我为中心的传统思维，这与体育建筑回归城市群体，与城市共同实现可持续发展的理念存在矛盾。在一定的城市环境中，突出体育场馆作为区域标志物，有利于统领全局,形成有场所感的空间秩序。但许多案例中，体育场馆未能尊重城市环境的整体性和延续性，由于庞大体量造成对周边建筑的压迫，并导致城市街道空间界面的破损和城市肌理的破碎。把握体育建筑一定的标志性和与周围城市建筑群体的协调性，是体育建筑形体设计的关键。

3. Contradiction between highlighting the iconicity of the sports facilities and the return to urban building group

The form treatment aims at highlighting oneself and lacks the consideration to the urban buildings. At present, a large number of sports venues designs still stay in the traditional thinking self-centered, which is in contradiction with the concept that sports facilities return to the urban environment and achieve sustainable development together with the cities. In a certain urban environment, highlighting the sports venues as the regional landmarks is conducive to organizing the overall situation and forming a spatial order with a sense of place. However, in many cases, the sports venues have failed to respect the integrity and continuity of the urban environment, that due to its large size, it has caused oppression of the surrounding buildings and caused damage to the city's street interface and fragmentation to the city's texture. It is the key to the form design of the sports facilities by balancing the iconicity of sports facilities and the coordination with the surrounding urban building groups.

4. 体育建筑附属空间与城市公共空间之间的矛盾

室外空间处理消极，缺乏与城市公共空间体系的有机整合。由于固化的总体布局模式，将体育场馆布置在用地中央，周围环绕以大片停车场、绿地和广场。室外开敞用地通常被视为附属空间，并且通过围墙和围栏与用地外的城市街道空间相隔，使其孤立于城市整体环境。

4. Contradiction between the space attached to sports facilities and the urban public space

The outdoor space is negatively treated, which lacks organic integration with the urban public space system. Due to the inherent overall layout mode, the venues are located in the center of the site and surrounded by large parking lots, green spaces and squares, while the outdoor open

事实上，体育场馆对于室外附属空间的需求有限，主要为满足后勤使用。体育用地内的大部分开敞空间，均可以从城市公共空间的角度进行系统性定位和设计，采用积极主动的策略为市民营造有品质和场所空间，以发挥其作为城市大型公共空间的作用。然而，现实中粗糙地建立大广场，缺少步行交通联系，缺乏室外活动功能设置，忽视不同类型的活动与停留空间细节的设计屡见不鲜，使得许多体育建筑地块内的开敞空间未能有机融入城市公共空间，成为不受市民欢迎的空旷冷清地带。

体育建筑逐渐缺乏乃至失去城市理性，背后有着多层面的原因：

在规划和决策层面，体育建筑向来受到政府关注，在政绩工程驱动下，规划和决策方案容易受行政意志影响，催生许多非理性产物。长期以来项目实际参与者，包括投资、决策、设计、建设和使用各方难以形成统一连贯的建设目标和决策思路，致使很多项目缺乏明确目标和科学定位，也使得进一步设计工作缺乏依据。大型体育赛事和体育建筑成为政治、国家和民族炫耀资本的做法可以追溯到 1936 年的德国柏林奥运会，在国内新中国成立早期作为社会主义建设成就和近年来城市名片的做法也屡见不鲜。其操作思路都是将体育建筑尤其是大型公共体育场馆作为城市肌理中的特别对象，并着力突出与其他城市建筑群体的不同，强调其自身的宏伟和标志性，最终造成体育建筑布局、体量、形态种种的“脱离群众”。

以广州天河体育中心为例，天河体育中心是为迎接 1987 年第六届全运会，在国内首次集中兴建的大型体育中心，也开创了一场两馆和“品”字形布局的模式。建设之初，天河体育中心位于城市郊区，经过多年发展，如今城市环境大幅变化，天河商圈已经成为广州最为繁华的城市中心区之一，其土地价值和开发

space is often considered as an ancillary space and is isolated from the overall urban environment and separated by fences from the urban street. In fact, the venues have a limited demand for outdoor ancillary space, which mainly includes the logistical use. Most of the open space within the sports lands can be systematically positioned and designed from the perspective of urban public space. Active strategies can be used to create space of good quality for citizens as a large public space. However, in reality, the rough construction of large squares, the lack of pedestrian transport links and functional settings for outdoor activities, and the neglect of details design for different types of moving and staying spaces are commonly occurred, making the open space within many sites unable to integrate into the cities and become a deserted area that is not popular with the public.

There are many reasons behind the gradual lack of sports facilities and even the loss of urban rationality:

At the level of planning and decision-making, sports facilities have always been concerned by the government. Under the drive of the achievement project, planning and decision-making programs are easily affected by the administrative will, and many irrational products are spawned. For a long time, the actual participants of the project, including investment, decision-making, design, construction and operation, are difficult to form a consistent and successive construction goal and decision-making ideas, resulting in lack of clear goals and scientific positioning for many projects, which also makes the lack of basis for further design work. The practice of making large-scale sports events and sports constructions to be the political, national and ethnic showings of capital can be traced back to the Berlin Olympic Games in Germany in 1936. The achievements of socialist construction in the early days of the founding of China and the practice of city cards in recent years also commonly

强度与建设之初有数量级上的提高，这促使天河体育中心为适应土地价值提高而做出反应。天河体育中心布局也随之陆续做了许多改变，最为突出的一点是，围绕东西两边入口处增加了许多体育设施，面向市民开放，增加了一些沿街商铺，使得原来强调自身完整性的“品”字形布局逐渐被城市肌理同质化。这一过程中天河体育中心原有功能和空间形态都需要应对实际发展需求做出调整，也反映出原有的体育建筑规划布局方式的局限性（图 3-7）。

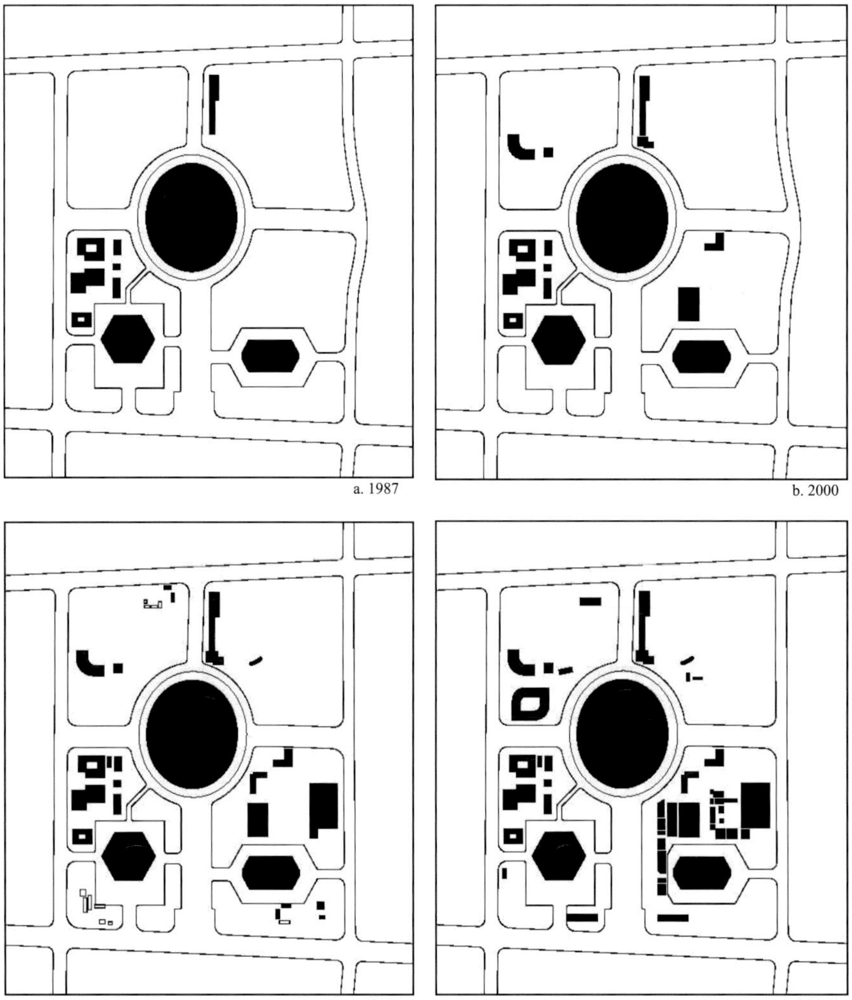

occur. Among them, the operation idea is to regard sports facilities, especially the large-scale public sports venues, as special objects in urban textures, and to highlight the differences with other urban building groups, emphasize its own grandiose and iconic, and ultimately result in 'break away from the masses' of the sports facilities, in terms of the layout, volume, form, and so on.

Taking the Tianhe Sports Center in Guangzhou as an example, the Tianhe Sports Center is a large-scale sports center that was the first time centrally built in the country to welcome the 6th National Games in 1987. It also created a 'delta shape' layout mode composed of one stadium and two halls. In the beginning of construction, Tianhe Sports Center was located in the outskirts of the city. After years of development, the urban environment has changed dramatically, and the Tianhe shopping district has become one of the most prosperous urban centers in Guangzhou. The land value and development intensity have an increase in the order of magnitude compared with the beginning, which has prompted Tianhe Sports Center to respond to the change. For that reason, the Tianhe Sports Center layout has also made many changes in succession, of which the most prominent point is that many sports facilities opening up to the public have been added around the entrance to the east and west of the site, and some shops added along the street, so gradually the original layout that emphasizes its integrity has been gradually homogenized by the urban fabrics. In this process, the original functions and spatial forms of the Tianhe Sports Center need to be adjusted in response to actual development, which also reflect the limitations of the original planning and layout of sports center (Fig. 3-7).

a. 1987

b. 2000

c. 2004

d. 2010

图 3-7 天河体育中心发展历程的图底分析

Fig.3-7 Figure-ground analysis of the evolution of Tianhe Sports Center

At the level of urban design, the traditional labor division mode is to carry out the plan from top to bottom as a control requirement throughout the architectural design process.

在城市设计层面，传统的分工模式是自上而下地将规划方案作为控制要求贯穿到建筑设计过程中，由于城市设计处在城市规划和建筑设计的过渡位置，本应为两者互动沟通的良好渠道，但恰恰是这一环节的工作不足，使得规划与建筑设计脱节，建筑设计也缺乏对规划条件进行自下而上反馈的有效渠道。实践证明，脱离城市的体育建筑在造成城市功能和空间尴尬的同时，其自身也面临着运营和发展的压力，城市设计的方法为应对这些问题提供有效途径。

Since urban design is in the transitional position of urban planning and architectural design, it should be a good interaction channel between the two. However, it is the lack of work in this link, making the planning and architectural design out of line, meanwhile the architectural design also lacks an effective channel for bottom-up feedback to the planning conditions. Practice has proved that while the sports facilities out of the city have caused urban function and space defects, they themselves face the pressure of operation and development. The methods of urban design provide an effective way to deal with these problems.

在建筑设计层面，一方面，迎合行政意志，追求建筑物的自我表现，将规划和决策层面的一些非理性意愿实现到体育场馆的设计中。另一方面，由于建筑材料、结构技术、设计和施工方法的进步，体育建筑突破原有空间跨度和造型局限，获得前所未有的体量和造型表现机会，在设计师缺乏城市系统性思维的情况下，体育建筑创作走上了表现造型，炫耀新材料、新结构和新技术的道路。此外，由于体育建筑设计长期以来由内而外的单向设计思路，造成建筑物内部功能与城市外部功能需求的脱节，使得许多用房需要通过改造和转换才能适应实际运营需求。

At the level of architectural design, on the one hand, it appeals to the administrative will, pursues the self-expression of the building, and executes some irrational intentions in planning and decision-making to the design of the venues. On the other hand, due to the advances in building materials, structural techniques, design, and construction methods, sports venues have broken through the original span and structural limitations, gaining unprecedented opportunities for the expression of the mass and style. In the absence of urban systemic thinking by designers, the creation of sports facilities embark on the road of expressing the form, showing off the new materials, new structures and new technologies. In addition, because of the one-way design ideas of sports facilities from the inside out for a long time, the internal functions of the building are seriously out of step with the external functional requirements of the city, so that many venues needs to be transformed and converted to adapt to the actual operational demand.

美国克利夫兰市体育建筑案例及其启示

在欧美，随着战后经济的蓬勃发展，职业化的体育赛事丰富而频繁，体育产业的强大更加突出强调体育建筑的舒适与便利使用。而所谓舒适与便利大多围绕服务于小汽车的大型

The Case of the Sports Construction in Cleveland, USA and its Enlightenment

In Europe, frequent and colorful professional sporting events come after the vigorous development of postwar economy. The prosperous sports industry places

停车设施、大规模人流积聚与安全疏散、多样性满足家庭成员需求等综合功能要求。大型体育场馆与公益性的社区体育设施的发展逐步分离，大型设施开始远离都市。1960、70 年代，美国许多大型职业化比赛的场馆都被停车场包围孤立。而同时，美国城市遭遇郊区化漫延式发展的影响，老城区日益衰败（图 3-8）。

由于美国体育产业巨大的市场号召力和经久不衰的人气，衰败的城市开始努力挽回大型体育场馆。作为美国第一个宣布破产的城市，克利夫兰是做较早尝试的城市，其基本思想是：利用大型文化体育娱乐设施的建设，让消费人口重新返回都市，即便是短暂的消费，也将有助于城市的经济复兴。

这就要求选址具有前瞻性，交通便捷，便于大量居住在郊区的中产阶级驾车前往。内容丰富，有助于观赛之外各种活动的展开。克利夫兰拥有棒球、橄榄球和篮球三大职业赛事的明星队伍。

图 3-8　被停车场包围孤立的比赛场馆

Fig.3-8　Competition stadium isolated in the middle of parking lots

prominent emphasis on the comfort and convenience of sports facilities. The so-called comfort and convenience, in most cases, are centered around other integrated functional requirements such as large parking facilities, massive foot traffic and safe evacuation, and diversity that meet the needs of family members. It becomes common that the development of large sports venues gradually drifts apart from community sports facilities that are provided as a public welfare. Large facilities begin to keep away from cities. In 1960s and 1970s, many large professional competition venues in the USA. were isolated among parking lots. Meanwhile, old cities were gradually on the wane as American cities yielded to spreading suburbanization (Fig. 3-8).

Considering the huge market appeal and enduring popularity of the US sports industry, declining cities started to win over large venues. Cleveland, the first American city declaring bankruptcy, was a forerunner in the game. It followed a basic idea of building large cultural, sports and entertainment facilities to bring the consumer population back to the city, in the hope that even short-term consumption would also contribute to the revival of its economic recovery.

This required insightful siting that provides convenient accessibility, so that the facilities were within the driving distance of a large number of middle-class population from the suburbs. Rich functionality was also part of the requirement for holding various activities in addition to sports events. Another factor in the equation is that Cleveland has three star teams for baseball, football and basketball.

The key was to adjust the elevation of sports facilities to conveniently coordinate between public spaces and urban streets; to structure rich spaces for intimate interaction between private and public; to introduce mixed use of land to meet the needs for competition and consumption; to develop the site plan

重点在于：调整体育建筑场地的标高，使公共空间与城市街道便捷联系。构建丰富的物质空间，引导个人和公众的紧密联系。提供土地的混合使用功能，满足比赛和消费要求。用城市设计引领地段内部的建筑设计，为他们的建筑边界、体型、功能甚至内部空间的确定建立严格而又能够容纳相当灵活性的控制导则。市中心的商业拱廊延伸过来，通过规划中的新的开敞空间，沿着新的街道，众多的出入口被安排在靠近城市中心的方向，赛后拥挤的人流自然涌向毗邻市中心的餐馆、酒吧、商店。

项目 1990 年开工，1994 年建成，棒球场由 HOK 设计，体育馆由 EB 公司完成，Sasaki 公司负责城市设计和公共空间的景观设计（图 3-9、图 3-10）。

upon the urban design with exquisite yet flexible massing and function. In the masterplan, the downtown commercial arcade extended through the new open space along planned streets, where many entrances and exits were arranged towards the city center so that the crowd after the games could naturally flow into the restaurants, bars, and shops downtown.

The project started in 1990 and completed in 1994, with contribution from HOK, which designed the baseball field; EB, which completed the gymnasium; and Sasaki, which handled the urban design and the landscaping (Fig. 3-9, Fig. 3-10).

图 3-9　美国克利夫兰市体育建筑案例总体布局

Fig. 3-9　The overall layout of the Case of Cleveland sports architecture

基于城市设计理念的思考与尝试

尽管城市设计的思想由来已久，但狭义上的“现代城市设计”理念源于对现代功能主义规划和建筑理论以及在其指导下的城市消极现状的反思。当城市功能不再和谐与混合，单一的功能分区导致了人性空间的失落，大量缺乏关联性的城市空间，不仅无助于城市公共生活，

Thinking and Attempt Based on Urban Design Concept

Although the idea of urban design has a long history, the concept of ‘modern urban design’ in the narrow sense is originated from the examination on the planning and architectural theory of modern functionalism the negative status of the the city under its guidance. When harmony and hybridization are taken away from urban functionality, monotonic division of functional areas will lead to the loss of humanistic space, which results in a substantial number of irrelevant urban spaces worthless of urban public life and accountable for urban isolation. Therefore, a distinct value has come with modern urban design from its date of birth. The modern urban design theory emphasizes the people-centeredness that observes and thinks about urban issues from the user's point of view, advocating the multiple and diverse development of urban functions, giving play to the role of the public space as a catalyst, enhancing the intensity of land development, compactness of the space and identifiability of the image, and encouraging public and walking traffics. Through the process, functional diversification and

图 3-10　美国克利夫兰市体育建筑案例总体鸟瞰

Fig.3-10　The general aerial view of the Case of Cleveland sports architecture

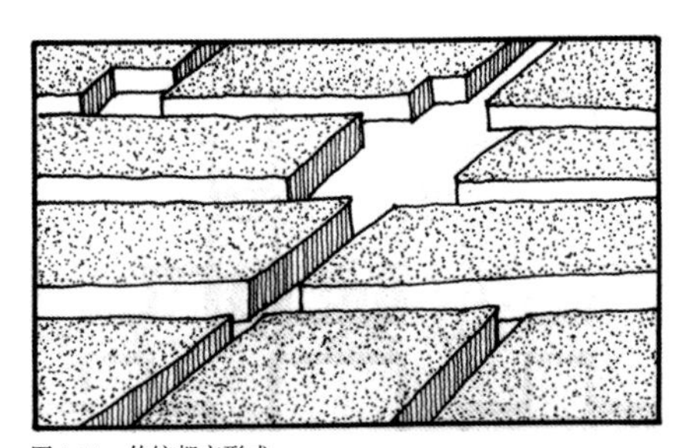

图 3-11 传统都市形式
Fig. 3-11 The traditional urban form

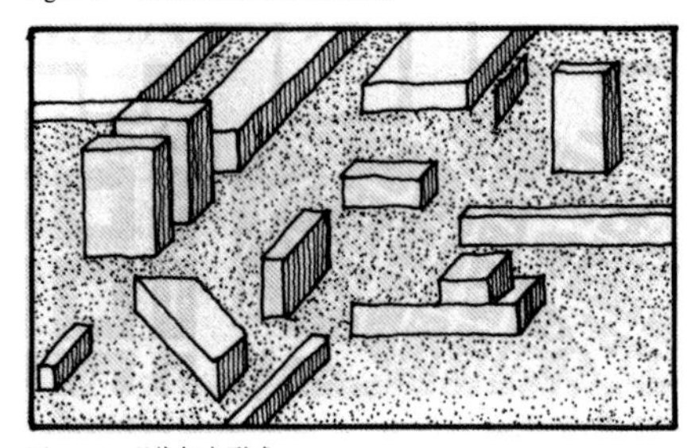

图 3-12 现代都市形式
Fig.3-12 The modern urban form

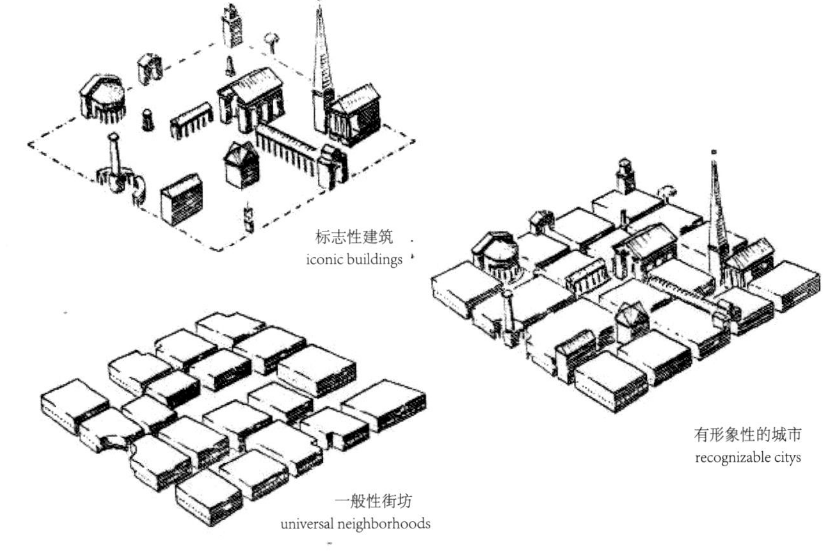

图 3-13 克莱尔的理想城市模型
Fig. 3-13 The ideal city model of Claire

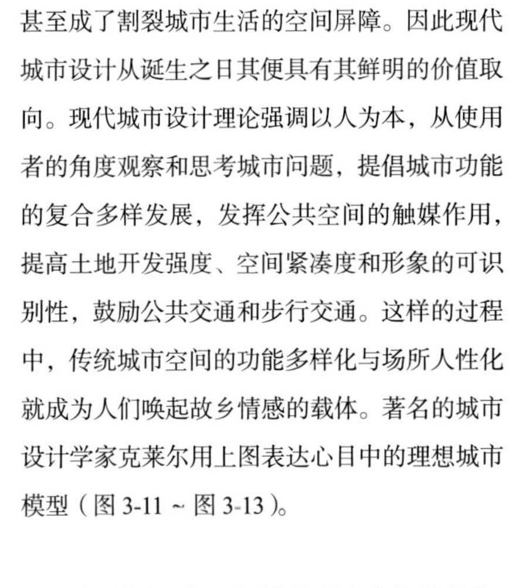

甚至成了割裂城市生活的空间屏障。因此现代城市设计从诞生之日其便具有其鲜明的价值取向。现代城市设计理论强调以人为本，从使用者的角度观察和思考城市问题，提倡城市功能的复合多样发展，发挥公共空间的触媒作用，提高土地开发强度、空间紧凑度和形象的可识别性，鼓励公共交通和步行交通。这样的过程中，传统城市空间的功能多样化与场所人性化就成为人们唤起故乡情感的载体。著名的城市设计学家克莱尔用上图表达心目中的理想城市模型（图 3-11 ~ 图 3-13）。

这里我们可以看到传统城市和谐的景象：城市大量的日常生活是由普遍性的街区和街坊构成的。街坊组成的街廓构建并体现了城市的尺度、生活的内容，街坊既是市民家的构成也是日常劳作的场所。同时，以公共活动为主要内容的建筑，在城市日常生活的网络中则起到了标志性的牵引，公共建筑及其所属的公共空间为城市创造了日常之外的精彩（图 3-14）。

humanization of traditional urban space would become vessels for nostalgia. The famous urban designer, Claire, explained his ideal city model in the figure above (Fig. 3-11 ～ Fig. 3-13).

From the figures we see a harmonious scene of traditional cities: the majority of daily life in city takes place in universal neighborhoods; where the profile of neighborhoods represents the dimension of a city and the content of life. Neighborhoods are collections of homes of citizens and places for practices. Meanwhile, buildings mainly intended for public events serve guidance in the network of daily urban life. Public buildings and their spaces provide activities higher than common life (Fig. 3-14).

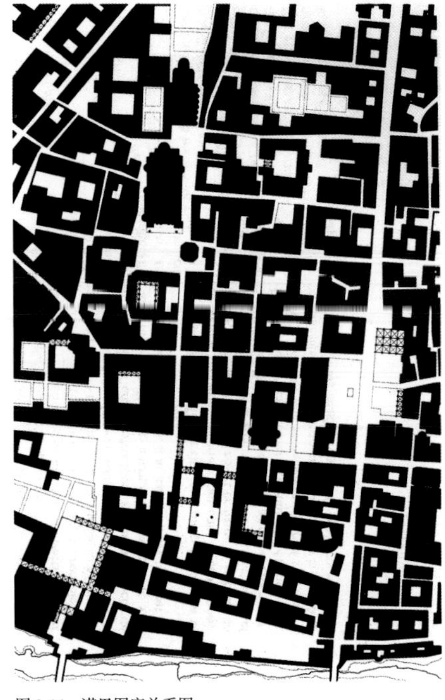

图 3-14 诺里图底关系图
Fig.3-14 Figure-Ground relation proposed by Nolli

图 3-15 中国城市照片
Fig.3-15 A photo of a Chinese city

这样的图景并非西方城市专属，中国城市同样具有相似的类型构成（图 3-15）。

对照现代城市设计理念肇始的一些发达国家，体育设施的建设不仅与大型体育赛事有关，而且与城市的日常生活相呼应，体育设施所形成的公共空间、公共活动对城市日常生活的影响与互动关系，得到了应有的重视。经过审慎的城市设计研究，越来越多的体育场馆避免在远离市民的郊外集中新建，而是合理分散地融入市区功能，亲近市民，成为城市公共活动场所的有机组成。

1996 年现代奥运会诞生 100 周年，第 26 届奥林匹克运动会在美国亚特兰大召开，当时新建的主体育场的设计特别注重与街道的关系，通过合理布置辅助用房，让建筑满足奥运会开闭幕式要求和赛时田径比赛的要求，赛后则迅速转换、改造成为一座标准的棒球场，达到了赛前投资商的要求。建筑形式不再是简单的结构与功能的表达，更加希望形成传统街道的空间尺度。赛后改造缩小建筑体量后还形成了具有纪念意义的奥林匹克广场。而与其一路之隔的相邻旧体育场，结构表达清晰，几何关系明确，反映了两种完全不同的建筑观念。

The ideal model is not exclusive in Western cities, similar compositions can be found in Chinese counterparts (Fig. 3-15).

Developed countries planned with modern urban design concept since their outset usually nurture sports facilities built not only for major sports events, but also in correspondence to daily urban life. The interaction between public space and public activities associated with sports facilities and daily urban life has received the attention it deserves. After deliberative study of urban design, more and more sports venues are avoiding being constructed in the suburbs which are far away from the citizens. Instead, they are integrated to the functions of the urban areas in a rational and decentralized manner, get close to the citizens, and become an organic component of urban public activities.

For the 26th Olympic Games held in Atlanta in 1996, the 100th anniversary of the modern Olympic Games, the designer of the new main stadium actuated on the stadium's relationship with streets by rational distribution of auxiliary rooms to meet the requirements of the opening and closing ceremonies and the track and field events. After the Games, the main stadium was quickly converted and transformed into a standard baseball field to meet the requirements proposed by the investors before the Games. The architectural forms was no longer a simple expression of structure, but a proportional adaption to street scale. A commemorative Olympic Plaza was reshaped after the transformation and reduction of building masses after the Games. In comparison, the old stadium across the street of articulated, expressive structure and clear geometry, embodies a completely different architectural concept. Afterwards, the demolishment of the old stadium marked a new transitional phase of urban transformation (Fig. 3-16).

Another example is the Sacramento Kings New Arena and the 'Railway Plaza'

随后旧体育场拆除，城市空间进入崭新的转换时代（图 3-16）。

另一个例子是美国加州萨克拉门托国王队新球馆和“铁路广场”项目，通过公共体育场馆的兴建，成为城市中心再发展的重要推力。政府计划在该区建设区域联合运输中心，在前期发展策划和评估中，认为应该给这一区域投入公共基础配套设施。因此在“铁路广场”北部设置萨克拉门托体育娱乐区，其中包含可容纳 17000 ~ 18000 位观众的国王队球馆作为核心设施，以及综合办公商务区、铁路科技博物馆、零售 / 娱乐 / 餐饮区、滨水广场和停车区等，此外还在体育娱乐区北部和东部规划了居住区和办公区。该项目策划及概念规划运用了体育娱乐区来促进城市中心的再发展，将体育建筑作为城市发展的一个重要部分，并采用错位互补、最大化混合功能的模式，促进城市用地的高密度利用、多业态的聚集和繁荣，为体育建筑和周边城市可持续发展建立了良好的基础（图 3-17）。

project in the United States. Through the construction of public sports venues, it has become an important thrust for the redevelopment of the city center. The government planed to build a regional joint transportation center in the area. In the preliminary development planning and evaluation, it was believed that this area should be invested in public infrastructure facilities. Therefore, the Sacramento Sports and Entertainment District was schemed in the northern part of the ‘Railway Plaza’, which included the King’s Team Arena that could accommodate 17,000-18,000 spectators as a core facility, as well as a comprehensive office business district, a railway technology museum, a retail/entertainment/catering district, a waterfront plaza and the parking areas. In addition, the residential areas and office areas were planned in the north and east of the sports and entertainment district. The project and concept planning made use of the sports and entertainment districts to promote the redevelopment of the urban centers, took sports facilities as an important part of the urban development, and adopt the maximum the complementary mixed functions to promote the high-density utilization of urban land, the gathering

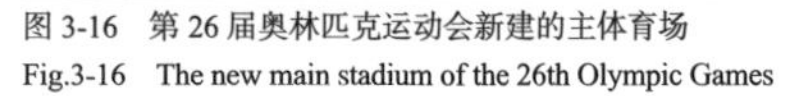

图 3-16　第 26 届奥林匹克运动会新建的主体育场

Fig.3-16　The new main stadium of the 26th Olympic Games

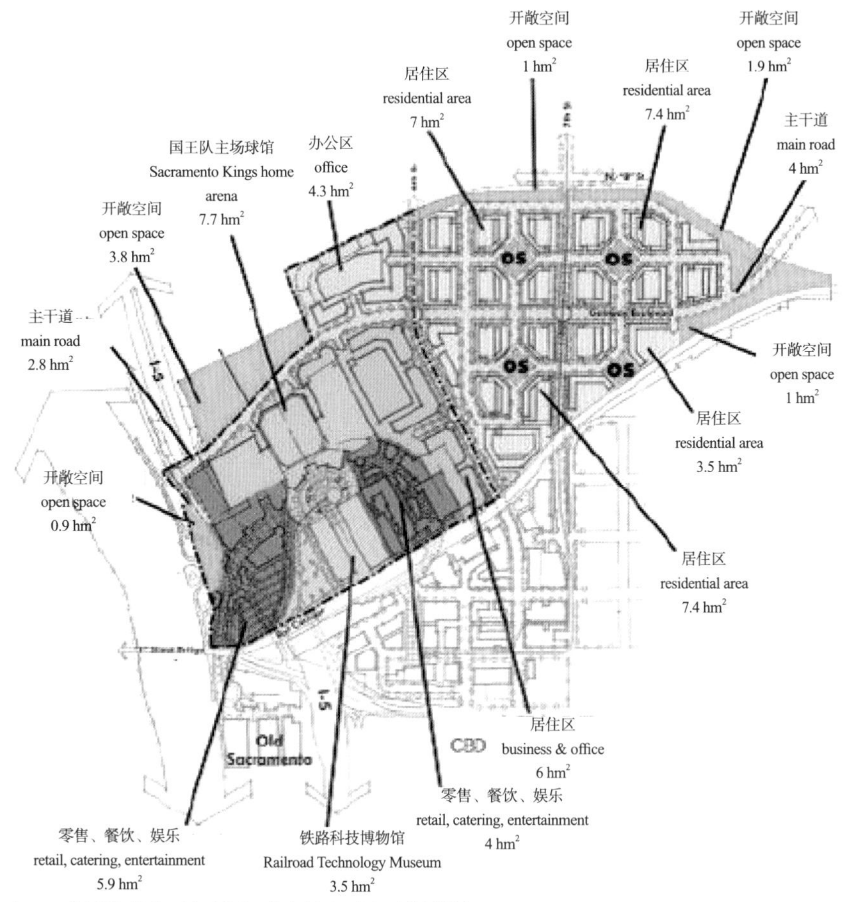

图 3-17 萨克拉门托国王队新球馆和"铁路广场"项目土地使用计划

Fig.3-17 The land-use plan for the new gymnasium of Sacramento Kings and 'Railway Square' project

体育建筑设计应突破传统由内而外的单向思维，更多地从城市整体功能和空间需求，由外而内地进行设计。尽管萨克拉门托项目是关于城市中心区再发展的问题，但其城市设计理念和做法可以延伸到新区建设中来。其中基于区域城市需求分析，将体育设施和商业设施、居住办公功能等集聚建设、协同发展的思路，为体育建筑和城市的协同发展奠定了良好的基础。反之如果体育场馆和其他功能各自独立地分散于各个角落，则一方面不利于发挥对城市更新或者新区开发的推动作用，另一方面也造成场馆运营失去外部条件。应对体育场馆运营问题不仅需要在内部灵活适应性方面作好安排，而且应该对外部条件作充分研究。只有在时间和空间上把握好城市发展，

and prosperity of the multiple business. It had established a good foundation for the sustainable development of sports facilities and the surrounding urban function and space (Fig. 3-17).

Sports facilities design should break through the traditional one-way thinking that only carry out from the inside to the outside, conversely should start from the overall function and space demands of the city to the design of the construction project. Although the Sacramento project is concerned with the redevelopment of urban center, its urban design concepts and practices can be extended to the construction of those new city districts. Based on the analysis of regional urban demands, the idea of agglomeration and collaborative development of sports facilities, commercial facilities and residential office functions has laid a good foundation for the coordinated development of sports facilities and the cities. On the contrary, if the sport facilities and other functions are scattered independently in various regions, on the one hand it is not conducive to exerting a role in promoting the urban renewal or development of new districts. On the other hand, the sport facilities lose their external conditions. For the operation of sport facilities, it requires not only the arrangements for internal flexibility and adaptability, but also the fully study on external conditions. Only by grasping the development of the city in the dimension of both time and space, and ensuring the coordinated development of the function and space between the venues and the cities can it be possible to realize the sustainable operation of the sport facilities.

Urban design plays a key role in the transition between the urban planning and architectural design. On the one hand, it can provide feedback and corrections to the urban planning process. On the other hand, it gives a certain 'rules' to the design of sports facilities. However, the urban design is what seriously lacking in the construction

保证体育场馆与城市的功能和空间协同发展，才有可能实现体育场馆的可持续运营。

城市设计作为规划和建筑设计之间的过渡环节，具有关键作用，一方面对规划环节可起到反馈和修正，另一方面为体育建筑设计起到一定的“规定”作用，而这正是现有建设程序和理念中严重欠缺的。这一环节对于体育场馆这类体量庞大的城市标志性建筑物而言具有特殊意义，可有效减小总体布局、体型设计、功能设置等方面的随意性和决策者的行政意识影响。因此设计方应跳出用地红线，将关注视角放大到更广的范围，关注城市的整体性，摒弃以自我为中心的单体建筑设计思维以城市设计的理念指导体育建筑的设计，促进体育建筑和城市整体的可持续发展。

procedures and ideas of the present. This is of special significance to the sports facilities as large-scale urban landmarks, which can effectively reduce the arbitrariness of the overall layout, form design, function setting, and the influence of administrative awareness from the decision makers. Therefore, the designer should jump out from the red line of the land, broaden the perspective to a larger scope, follow with interest of the city integrity, and reject the self-centered design thinking for single independent building. The concept of urban design guides the design of sports facilities and promotes the sustainable development of the facilities and the city as a whole.

基于城市设计理念的“都市理性”

1999年，借体育建筑专业委员会昆明大会和《建筑学报》，我们在国内首次提出“基于城市的体育建筑设计”的主张，试图为现代体育场馆设计提供了富有启发性的都市理性方法和策略。针对体育建筑发展中存在的问题，从如何善待城市环境，正确处理大尺度体验建筑与城市的关系提出了4个观点与策略：

1）尊重街廓，促进街道生活；
2）降低体量，寻求完美的多视点观赏；
3）双重个性，积极应对环境；
4）弹性应变，可持续协调发展。

这是基于国内体育建筑实践的特殊性而综合考虑的。目前，体育建筑在我国基本上是公共投资。官方与民间传统思想根深蒂固，新建筑的标志性、可视度是每一个城市和甲方都期盼的。因此，缺少了视觉独特性的体育建筑投标方案，就无法获得中标机会。最近20年，对体育建筑标志性的要求愈演愈烈，简单照搬西方都市设计理念是无法获得认同的。

'Urban Rationality' Based on Urban Design Concept

On the occasion of 1999 Kunming General Assembly of Sports Building Committee and Architectural Journal, we first proposed the city-based sports architecture design principle in attempt to provide an inspiring urban-rationality method and strategy for modern venues designs. Four opinions and strategies are presented to address the problems in the development of sports facilities regarding protection of urban environment and correct approach to the coordinate between large-scale experiential architecture and the city:

1) Respect blocks, energize street life;
2) Reduce mass, and seek perfect multi-perspective visual percaption;
3) Develop dual personality and positive response to the environment;
4) Seek elastic adaptability and sustainable, coordinated development.

These are based on the uniqueness of domestic practices adopted for sports facilities. At present, sports facilities in China basically rely on public investments.

我们在参与的体育建筑项目中，基于可持续发展理念，运用城市设计的方法进行体育建筑设计的探索。这些项目所面对的城市区位各不相同：有的位于已经发展成熟的社区或大学校园内，如北京工业大学体育馆、中国农业大学体育馆；有的位于城市公园中，如梅州市梅县区文体中心；有的则位于尚未开发的城市新区，如江门市滨江体育中心、江苏淮安体育中心、佛山世纪莲体育中心。应对不同的环境条件，我们坚持从城市设计理念出发，系统探讨都市理性框架下的体育建筑策略，从四个方面思考和践行都市理念下的体育建筑发展。

Subject to the traditional ideals of both the governors and the folks, the iconicity and visibility of a new building is expected by every city and every employer. Therefore, tender projects for sports architecture, if in lack of visual uniqueness, will not stand a chance. In the last 20 years, as the demand for iconic sports facilities grew stronger and stronger, simply copying Western urban design philosophy does not apply to the national condition. In our projects of sports construction, based on the concept of sustainable development, the urban design methods are used for the exploration of sports facilities design. The locations of these projects are different: some are located in mature communities or university campuses, such as the Beijing Industrial University Gymnasium and the China Agricultural University Gymnasium; some are distributed in the city parks, such as the Mei County District Sports Center in Meizhou City. Some are in the untapped urban areas such as the Jiangmen City Riverside Sports Center, Jiangsu Huaian Sports Center, and Foshan Century Lotus Sports Center. Facing different environmental conditions, we insist on starting from the urban design concept, systematically exploring the strategy of sports facilities under the framework of urban rationality, considering and practicing the development of sports facilities under the concept of urbanism from four aspects.

1. 内外结合，遵从城市肌理与都市涵构

体育建筑设计从内向外是长期以来的设计思想。复杂的内部功能、大跨度的结构技术决定了由内向外思想的合理性，也是现代主义建筑长期以来坚持的。然而，城市长期形成的物质形体环境具有内在的逻辑关系。所谓西方学者提出的“context”一词在大陆被译作“文脉”，导致了关于肌理、涵构内容的缺失。尊重逻辑关系，意味着建筑设计从内到外之外，需要在理解和分析城市涵构关系的基础上，补充实现从外到内的过程。

1. Internal and external integration, to comply with the urban fabric and urban culvert structure

It has been a long-standing axiom on sports facilities design to proceed from interior to exterior. Complicate internal functions, and large-span structure technologies justify the rationality of going from interior to exterior. The logic has been what the modernistic architects insist on since ever. However, the physical form of urban environment evolved over time has its inherent set of logic. The mistranslation of the word ‘context’ defined by Western scholars to

相较而言，我国体育建筑长期缺失“从外向内”，大部分设计思路局限于“用地红线”范围，导致了体育建筑与城市空间的扭曲关系，从思想上认识这一点更加紧迫。跳出用地红线范围，内外结合、从城市街区范围思考体育场馆布局，让体育建筑遵从都市公共空间体系的涵构关系，从初始阶段奠定体育场馆与城市协调发展的良好基础。

在新疆体育中心方案设计投标中，我们从城市的角度进行方案的构思，在更大范围内探讨建筑和城市空间的结合。基于对周边环境的分析，我们在设计过程中提出扩大用地规划范围的想法，将北部初步定位为居住区及高新技术区的现有空地也纳入到整体的规划中，并将其和体育建筑区两部分进行统一的空间规划，以保证空间上的完整性和连续性。在设计手法上，考虑到体育场馆和居住建筑在体量、形式上的明显区分，我们将居住建筑组团作为设计背景处理，而将体育建筑明确突出，作为地段城市空间的标志性节点，从而建立起城市空间的整体性和秩序感。该方案在比体育建筑场地

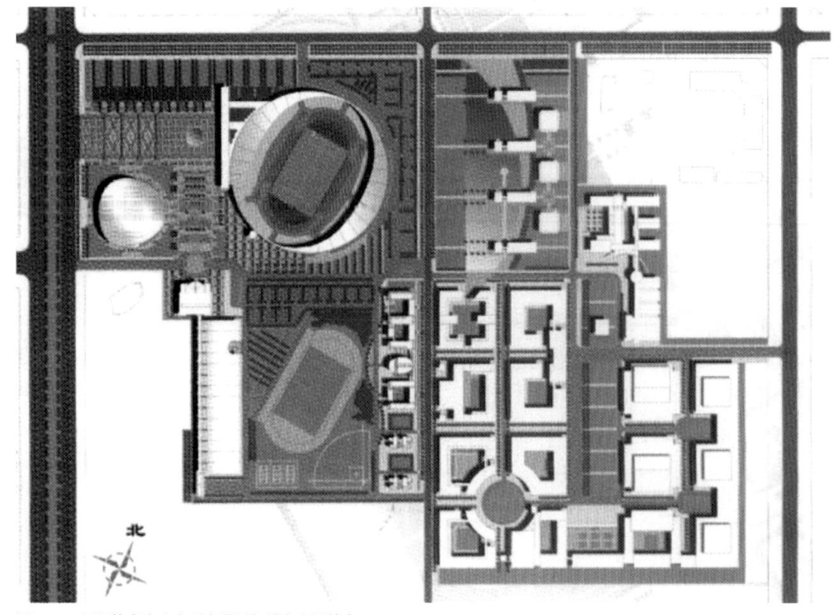

图 3-18　新疆体育中心在更大范围与城市空间结合

Fig.3-18　Xinjiang Sports Center fits in with the urban space in a broader picture

mainland Chinese as ‘grains’ led to the absence of the connotation of fabric and context. Respect for logical relationship implies that, in addition to the interior-to-exterior design process, architectural design requires us to, on the basis of understanding and analyzing the relationship of urban context, supplement with the exterior-to-interior process.

It is a more urgent task to ideologically understand the importance of urban context as in China, design for sports facilities suffers a long-term neglect of exterior-to-interior process. Instead, most of the design merely comes from the parcel boundary, resulting in a distorted relationship between sports facilities and the urban space. Supposedly, designers should break away from the constrains of planned boundaries, combine interior and exterior, put sports venues in the context of its urban neighborhoods, bring sports facilities in line with the urban public space system, so as to lay a good foundation for the correlated development between venues and the city at the beginning of design and construction.

In the design bid for the Xinjiang Sports Center, we began to conceive the plan from the perspective of the city and explored the combination of the building and the urban space in a wider context. Based on the analysis of the surrounding environment, we proposed the idea of expanding the scope of land use planning during the design process. The open space that was initially positioned as a residential area and a high-tech area would also be incorporated into the overall planning and the sports construction area. Unified spatial planning of the two parts was carried out to ensure the spatial integrity and continuity. As to the design techniques, taking into account the obvious distinction between sport venues and the residential buildings in terms of volume and form, we treated the residential building group as a design background, and highlighted the sports venues as an iconic node of the local urban space, thus

自身更大的用地范围内考虑整体布局的设计方法，从重视自我转向关注城市整体空间形态和效益的思路，为体育建筑总体规划提供一个有益的参考（图 3-18）。

2009 年，淮安体育中心设计竞赛，鉴于体育中心完全建设于新城区，控制性详细规划存在许多未落实的地方，我们向淮安市规划局提出了在体育中心竞赛成果中，包含对周边地区城市设计的建议。设计中我们将体育设施做了不同性质的区分。体育场突出强调其结构简洁、明细、完整，力图用理性内敛的方式体现出标志性。位置规划于数条道路与视觉交汇之处，成为城市区域的标志性建筑。体育馆、游泳馆则靠近城市街坊区域，与周边街区相互融合，以利于日常使用的便捷。为减少体育中心巨大街区形成交通拥堵的可能，规划了一条便于穿行的内部道路，以备未来城市发展的需要。在这种规划理念下，开敞空间集中有效了，街道空间得以展开了，标志性的需求也在低成本的条件下也得到了满足(图 3-19 ~图 3-21)。

以上案例表明，无论是位于发展成熟的城市中心还是尚未开发的城市新区，在一个更高的视角，从城市整体发展角度考虑体育建筑的可持续性设计，不仅有利于将体育建筑的建设

establishing from the overall integrity of the urban space and sense of order. The design method of the project considered the overall layout with a larger land area than the sports building site itself, changed the perspective from focusing on the building itself to the overall space and benefits of the city, which provided a useful reference for the overall planning of the sports facilities (Fig. 3-18).

In the design competition for Huaian Sports Center in 2009, considering that the sports center would be located in a new urban district where many uncertainties about the detailed planning existed, we suggested to Huai'an Urban Planning Bureau that the competition should include urban design for the adjacent areas. We rendered the stadium with iconicity in a rational and internalized way by highlighting its simple structure, details, and completeness. Its planned location, where several roads and visual lines intersected, made it a landmark of the urban area. The gymnasium and the natatorium area were close to the urban neighborhoods for them to be merged into the surroundings and for convenient daily use. To reduce possible traffic jams in the huge block near the sports center, we planned a convenient internal access to meet the need for future urban development. The plan united the open space, unfolded the inter-block spaces, and meet the requirement of iconicity at a low cost (Fig. 3-19 ~ Fig. 3-21).

The cases mentioned above show that, whether it is located in a mature urban center or an undeveloped urban new district, considering a sustainable design of the sports facilities from a higher perspective of the overall city development, would not only contribute to the integration of the construction of sports facilities to the overall development of the city, enhance the function and space of the city as a whole, but also be able to form a good feedback to the urban planning, which is conducive to the realization of multi-level and multi-directional sustainable development goals.

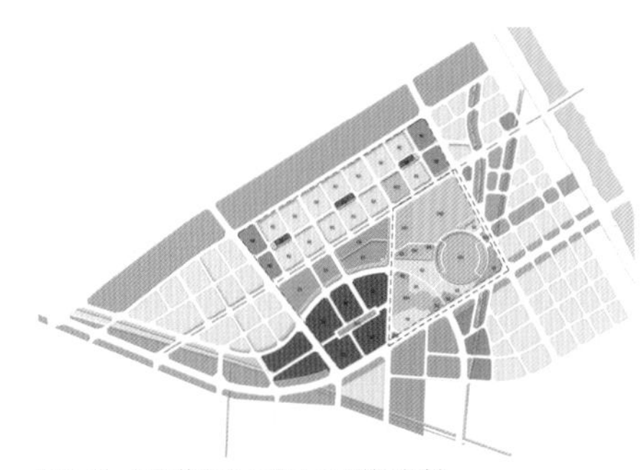
图 3-19　淮安体育中心片区土地利用规划
Fig.3-19　Land use planning of Huai'an Sports Center zone

图 3-20　淮安体育中心规划结构图
Fig.3-20　Planning structure of Huai'an Sports Center

图 3-21　淮安体育中心总体城市设计
Fig 3-21　Overall urban design of Huai'an Sports Center

融入到城市的整体发展中，对城市整体起到功能和空间的提升，而且可以对城市规划形成良好反馈，从而有利于多层次、多方位的可持续发展目标的实现。

2. 重视街廓，强化城市公共空间

城市街廓是组成城市重要的比尺，现代主义城市的发展首先摧毁的是传统城市完整亲切的街廓。随着街廓的消失，大量消极的城市空间出现，隔绝公共生活，阻断步行路径，使城市生活网络支离破碎。体育建筑与街廓看似背道而驰，实际不然。

今天，许多体育场馆都布局在新区，利用体育场馆等大型公共建筑项目作为启动，拉动城市新区的开发，这使得新区成了许多体育场馆所处的典型环境类型之一。在这种条件下，体育场馆建设面临一个空旷、缺乏人气、有着很大不确定性的城市格局，而这一环境的演变和成熟需要经历很漫长的过程。对于城市新区体育场馆建设，为与城市保持功能和空间上的协调发展，需要使场馆在总体布局上具有较强的适应性。但实际情况是由于缺乏外部条件的限制，强调自我、突出自身标志性，将体育场馆置于用地中央的传统总体布局模式更加容易被使用。随着周围城市新区的成长，一场两馆“品”字形和“一”字形的布局，忽略城市公共空间的设计，将不利于体育场馆在一个相对长的时期内适应城市空间与功能的变化。

城市设计的方法为应对外部环境的变化问题提供了很好的思路。一方面，城市设计的方法提倡城市的多功能混合和高密度开发，以及应对变化的灵活适应性，使得项目总体布局具有较大的应变能力，为体育建筑与城市功能的协调发展提供基础。另一方面，城市设计的方法强调体育场馆自身及其外部空间与城市公共空间的整体性和延续性，形成良好

2. Attach importance to street blocks and strengthen urban public space

Street blocks are an important scale reference for a city, but no matter how complete and friendly they used to be in a traditional city, they bear the attack when modernistic cities grow. The disappearance of street blocks has left a large number of urban space passive, public life and pedestrian paths isolated, and the urban life network broken. Sports facilities and street blocks are seemingly exclusive to each other while they are not.

At present, many sports venues are located in the new district. The large public building projects such as sports venues are regarded as a start-up to stimulate the development of new urban areas. This has made the new district one of the typical types of external environment in which many sports venues are located. In such conditions, the construction of sport venues faces an open and unpopular urban area with great uncertainties. The evolution and maturation of this environment requires a long process. For the construction of sports venues in the new urban areas, it is necessary to make the venues more adaptable to the overall layout in order to maintain the coordinated development of the functions and spaces with the whole city. However, the actual situation is that the traditional layout mode is more easily used, by emphasizing the building itself, highlighting its own landmark, and placing the stadiums in the center of the site, due to the lack of restrictions on external conditions. With the growth of the new urban areas around the city, the layout of the ‘delta shape’ and ‘line shape’ of the ‘one stadium and two halls’, ignoring the design of urban public spaces,

的空间框架，为体育建筑与城市空间的协调发展提供基础。因此，相比于传统的在用地范围内思考体育场馆自身的布局方法，运用城市设计的理念，从城市整体功能和空间环境出发，积极主动地将体育建筑与城市街道和公共空间节点有机结合，营造界面清晰的街廊、可达性好和层次丰富的公共空间。

2003年萨萨基（sasaki）公司中标实施的北京奥林匹克中心，良好的体现了其一贯的城市设计理念。通过调整街区，让轴线宏观层次上统领和延续布局城市空间的特色，中观层次上城市公共空间从亚运体育中心开始，序列整齐紧凑，而高密度的街区设计为奥运会前后的城市开发提供了优质的发展用地，密实的街廓空间烘托了体育建筑的标志性，满足国人百年奥运梦想的寄托。今天，回顾萨萨基公司的最初设计图，良好的城市设计概念确保了后续基本城市空间的完整（图3-22、图3-23）。

图3-22　北京奥林匹克中心原城市设计方案

Fig.3-22　The original design of Beijing Olympic Center

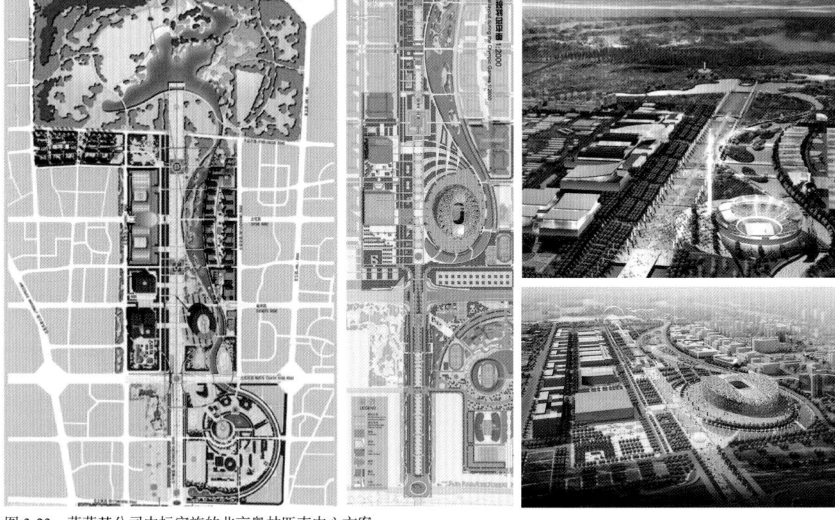

图3-23　萨萨基公司中标实施的北京奥林匹克中心方案

Fig.3-23　The design of Beijing Olympic Center, the bid won and implemented by Sasaki

would't be conducive to adapting the venues to the changing urban spaces and functions within a relatively long period of time.

The urban design method provides good ideas for dealing with the changes in the external environment. On the one hand, it promotes the city's multi-functional mixing and high-density development, as well as flexibility and adaptability responding to the uncertainties, making the overall layout of the project more resilient and providing the basis for the coordinated development of sports facilities and the city functions. On the other hand, the urban design method emphasizes the integrity and continuity of the facilities themselves and the external urban public space, and forms a favorable spatial framework, which provides the basis for the coordinated development of the sports facilities and urban space. Therefore, compared with the traditional way of thinking about the layout of venues within the scope of the red line, using the concept of urban design, starting from the overall function and the space environment of the city, actively integrate sports facilities with the urban streets and public space nodes, create a place with clear interface, good accessibility and rich spatial hierarchy.

The Beijing Olympic Center, whose bid won by Sasaki in 2003, is a good embodiment of Sasaki's consistent urban design theories. It adjusted the block on the macro-level by using an axis to lead and extend the feature of the urban space in issue; on the meso-level by densifying adjacent blocks in relation to their distance to the center so as to provide quality land for urban development before and after the Olympic Games. The dense blocks served as a background to the iconic sports architecture, which fulfilled the Olympic symbolization cherished by the people for a century. Today, when reviewing the original drawings of Sasaki's design, we can still identify how the concept of urban design ensures the completeness of basic urban structure (Fig. 3-22, Fig. 3-23).

图 3-23　萨萨基公司中标实施的北京奥林匹克中心方案（续）
Fig.3-23　The design of Beijing Olympic Center, the bid won and implemented by Sasaki (followed)

在 2002 年的佛山世纪莲体育中心国际竞赛中，我们设计团队同时参加城市设计和体育中心建筑设计的国际竞赛。设计中尝试突破常规的体育中心布局模式，将体育中心的规划与佛山新城市中心区的城市设计相结合。在总体布局上，延续作为城市公园的城市绿化轴，将体量庞大的体育中心靠用地西侧和南侧道路布置，在体育场东面预留宽裕的休憩空间，使得城市绿轴得以贯通南北。这一做法回避了将体育场作为纪念物置于场地中央的做法，但并不因此牺牲建筑个性，反而将体育场融入城市，共同积极地塑造城市街道空间，成为居民生活的一部分。体育中心外部环境进行明确化，满足各种功能上的需求，在西面和南面设置临街广场，配置商业和服务于体育场的功能用房，北面作为体育赛事的疏散空间，采用硬质广场铺地，也为居民日常体育锻炼提供宽敞的室外大空间。此外，我们在项目中引入体育公园的概念，为体育场馆创造空间类型丰富、尺度亲切、可达性好、具有灵活适应性的外部环境，避免将外部环境作为消极空间的处理方法，将体育场馆外部场地设计为有趣、便于市民休息游玩的园林式公共空间，使得体育中心内外部环境成为可以愉快轻松进行比赛和观赏的场所。

In the 2002 international competition of Foshan Century Lotus Sports Center, our design team participated in both the total urban design and the architectural design of the sports center. In the project, we tried to break through the conventional layout mode of the sports center, by combining the sports center’s planning with the urban design of the new downtown area of Foshan. In terms of overall layout, in order to continue the city green axis, which was the city park, the large-scale sports center would be laid close to the roads on the west and south sides, and ample open space would be reserved on the east side of the stadium so that the city’s green axis would be able to link north and south. This approach evaded the traditional way of placing the stadium as a monument in the center of the venue. However, it did not sacrifice the identity of the buildings, and to the contrary, it integrated the stadium into the city and actively shaped the urban street and space and became a part of the residents' lives. The sports center’s external environment was clarified to meet various functional demands. The frontage square was set up in the west and south, and the function rooms for commercial and sports stadiums were deployed. In the north, the evacuation space for sports events was adopted, with hard square for paving, and it also provided a spacious outdoor space for residents' daily physical exercise. In addition, we introduced the concept of

图 3-24 佛山体育中心实施布局演进
Fig.3-24 Evolution of the Implementation layout of Foshan Sports Center

最终，我们的城市设计方案获得采纳，建筑设计则采用了德国 GMP 的设计方案，而我们的团队作为 GMP 的合作单位。接下来的问题在于如何协调建筑师的体育中心布局与城市设计的矛盾。德国建筑师的体育中心之所以

sports parks in the project to create an external environment with plentiful space types, friendly scale, good accessibility, flexibility and adaptability for the sports venues. In this project, we avoided treating the external environment as a negative space. Instead, the external site was designed to be a garden-style public space that is fun and convenient for the public to rest and play, which made the sports center's internal and external environment a pleasant and leisure place for sporting and sightseeing.

Ultimately, our urban design and the architectural design of GMP, a German cooperator of ours, were adopted. The next problem was to coordinate the contradiction between the sports center and the urban design. The design of the sports center proposed by the German architect convinced the employers with its iconic magnificence, impressive axial symmetry and the tensile lotus shape. The German party quickly modified its site plan, which nonetheless remained symmetrical. Finally, the coordination issue was referred to the Planning Bureau, which issued indicative documents based on the urban planning and specified the distance of the stadium from the west and south roads. The German designer, in the spirit of professionalism, conservatively agreed to leave as much

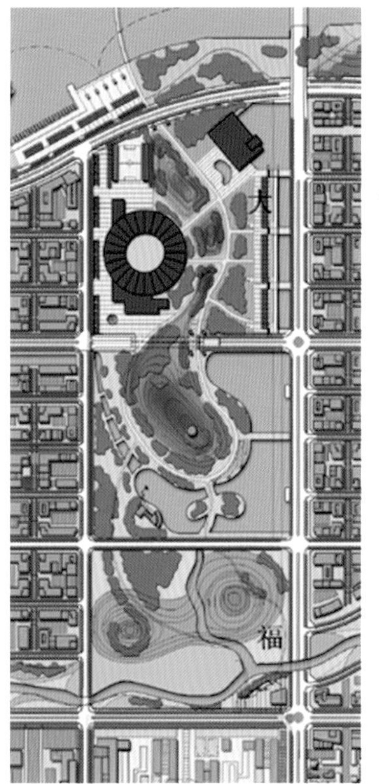

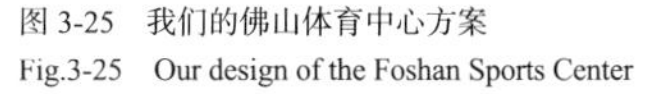

图 3-25 我们的佛山体育中心方案
Fig.3-25 Our design of the Foshan Sports Center

被最终采纳，得益于其雄伟的标志性设计理念，强烈的中轴对称和富于张力的莲花造型打动了业主。中标后德方迅速修改了总图，但依然是对称布局。最后不得不通过规划局协调。规划局根据城市设计出文，强行规定了体育场距西侧、南侧道路的距离，德方设计师基于良好的专业精神，保留意见的基础上同意在确保建筑设计效果的前提下，将体育场东侧留出尽可能多的公园空间，实现了城市设计的连续步行路径串连城市公共空间的理念。体育中心建成5年后，城市增建网球中心，也因为当年体育中心总体布局将东侧土地的空出，新的网球中心得以在东南角建设，从而印证了城市设计理念对可持续发展的重要（图3-24、图3-25）。

江门市滨江体育中心的总体布局设计中也运用了城市设计的方法，将片区城市总体空间的格局贯穿到场地内的建筑布局中。项目位于广东省江门市滨江新城核心地段，作为新城开发的启动区，是集体育运动、商贸会展、休闲娱乐、文艺演出为一体的大型综合性活动中心，其中包含体育馆、体育场、游泳馆和会展中心等重要单体建筑。在设计前期的概念方案设计中，被选中的一家国外公司采用传统的体育中心布局模式，将几个重要单体作为标志物布置在场地中心。2012年我们参与了体育中心和建筑设计的国际竞赛，打破原概念布局方案，尤其是对南区体育场和游泳馆的布局中，将两者整体向南侧移动，摒弃原概念方案中将建筑自身置于场地中央的方式。这让整个地块的核心区成为开敞的公共空间，水道和绿地贯穿其中，使得场地西侧天沙河滨江景观带和场地东侧城市中轴线上的碧莲湖公园之间的视廊得以贯通，更好地延续了地区概念规划中的城市空间总体格局。此外，将游泳馆和扩展经营用房紧凑组合，并向西侧移动至靠近城市道路，形成连贯而变化丰富的街道界面，这也为未来场馆和沿街商业的可持续运营准备良好的空间基础。这一基于城市环境的设计理念受

park space as possible provided the effect of architectural design remains, thus the concept of continuous walking paths connecting with urban public spaces was protected. Five years after the completion of the Sports Center, a tennis center was planned to be built at the south east corner reserved in the original site plan of the Center, which demonstrated the significance of urban design concept for sustainable development (Fig. 3-24, Fig. 3-25).

The overall design of the Jiangmen City Riverside Sports Center also adopted the method of urban design to permeate the overall urban space of the district into the architectural layout within the site. Located in the core area of Jiangmen Binjiang New City in Guangdong Province, the project was a large-scale comprehensive activity center integrating sports, trade fairs, entertainment, and artistic performances, which included important buildings such as stadiums, gymnasiums, natatorium, and exhibition center halls. In the conceptual design of the early stage, a selected foreign company adopted the traditional layout mode of sports center, which placed several important buildings as markers in the center of the venue. In 2012, we participated in the international competition of sports center and architectural design, and broke through the original conceptual layout. Especially for the layout of the stadiums and the natatorium in Southern District, we moved the whole of the two to the south, and abandoned the original conceptual plan that placed the venues in the center of the site, which made the core area of the entire site an open public space, with waterways and green spaces running through, allowing the viewing corridors between the Tiansha River Riverside landscape belt on the west of the site and the Bichain Park of the city's central axis on the east to be linked. The overall structure of the urban space in regional conceptual planning was better continued. In addition, a compact combination of the management rooms of the natatorium and the extended operating

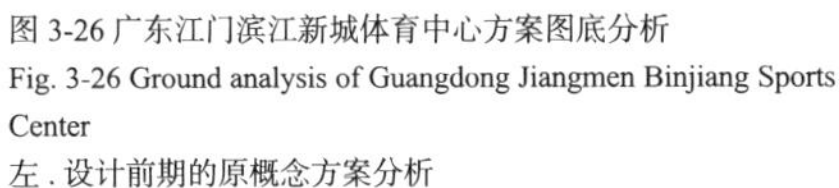

图 3-26 广东江门滨江新城体育中心方案图底分析
Fig. 3-26 Ground analysis of Guangdong Jiangmen Binjiang Sports Center
左 . 设计前期的原概念方案分析
Left. analysis of original pre-design conceptual scheme
右 . 优化方案分析
Right. analysis of optimized scheme

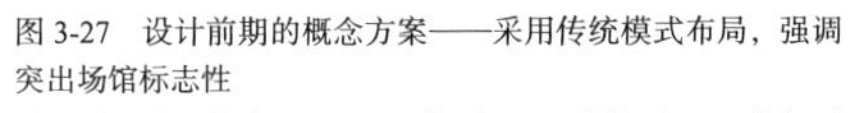

图 3-27　设计前期的概念方案——采用传统模式布局，强调突出场馆标志性
Fig.3-27　Pre-design conceptual scheme - following traditional layout pattern and highlighting the stadium iconicity

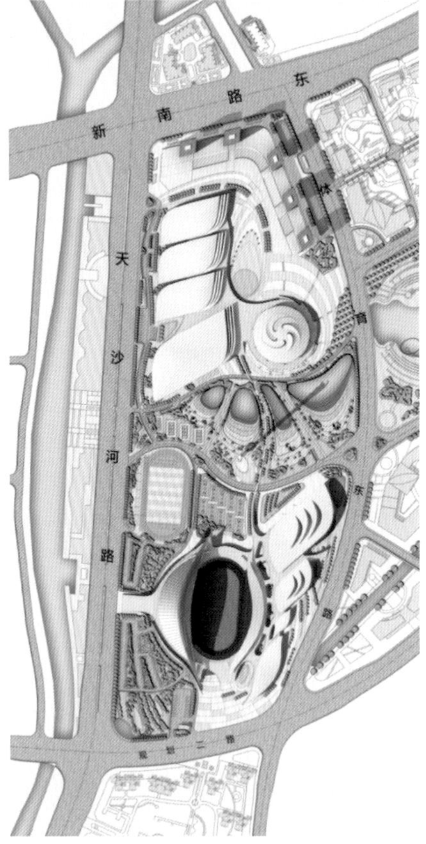

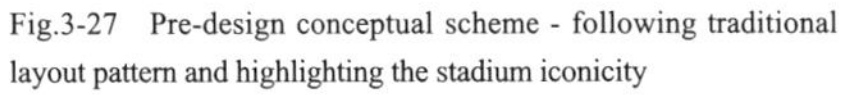

图 3-28　经过设计优化后的总体设计——重视街廓，强化城市空间架构
Fig. 3-28　Overall design after design optimization- emphasizing street blocks and strengthening urban space structure

到认可，我们的方案从众多的国际设计单位参赛方案中脱颖而出（图 3-26 ~ 图 3-28）。

3. 营造场所，提升城市节点活力

当体育场馆处于城市市区的校园或社区中时，周围环境相对发展成熟，用地局促、建筑密度高，这是总体设计需要面对的另一种典型城市环境。此时体育场馆总体布局需要与周围现有建筑和开敞空间相协调，延续和整合城市肌理，重视视廊的通达性，避免尺度上的强烈反差造成对周围建筑的压迫和城市肌理的破坏；且应注重交通上的可达性，营造体育场馆外部空间场所，发挥其对城市节点空间的整合和提升作用。

rooms was moved to the west side to get close to the city road, forming a coherent and highly varied street interface, which also provided a good spatial basis for the sustainable operation of future venues and street businesses. This design concept based on the urban environment was finally recognized, and our solution emerged victorious from the numerous international competitive bidding schemes (Fig. 3-26 ~ Fig. 3-28).

3. Creating places to enhance the vitality of city nodes

When the venues are located in a campus or a community in an urban area, the surrounding environment is relatively developed and matured, where there is

北京奥运摔跤馆所处的中国农业大学是一座典型的中国20世纪50年代规划、修建的大学校园，逻辑严整的校园建筑布局、亲切的毛主席塑像留住了那些并非久远却在快速逝去的历史。其中独具一格的是校园内几栋老旧的砖混结构教学建筑，虽然造型与50年代的教学建筑一样普通，材料也是质量一般的红砖，但建筑师将砖体按规律稍作嵌出，丰富了墙面质感。几栋主要教学建筑虽细节有所不同，但均采用了类似的手法，加上养护良好的树木绿化，顿时形成独特的校园氛围。

在如此独特的校园环境之下，我们对体育建筑的构思开始于对校园空间环境所进行的严谨的图底关系分析。为迎接奥运，校方确定了位于校园体育活动区内一个150m×200m的体育建设用地，在成熟的校园环境中，这是不可多得的核心地段。一方面，相比于用地周边的教学楼等校舍，体育馆在体量和功能方面的特殊性，必然成为这一区域最为突出的主体建筑，因而在形体和景观处理手法上应能适当突出主体建筑的标志性；另一方面，考虑到周边逻辑严整、分布紧凑的建筑布局，应保证建筑形式和体量上的自律，避免张扬，使得体育馆能与周边相对小尺度的建筑相协调。因此把握好体育馆作为主体建筑的标志性及其与校园环境之间的协调性，是本项目总体布局和形体设计的关键。

基于以上分析，我们从校园总体规划的角度考虑体育馆的布局，考虑到用地北侧存在三号学生公寓大楼和食品实验楼，将体育馆布置在用地中部或北侧都会使得整体空间显得局促。将其靠地块南侧布置，不仅有效避免对周围环境的压迫感，而且使得周边几处原本较为松散的开敞空间得到组织和界定，并由此突显体育馆作为该片区统领全局的作用。游泳训练

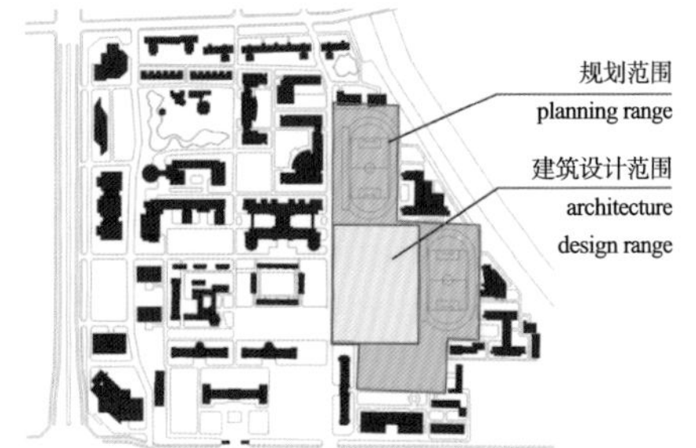

图3-29　中国农业大学体育馆基地环境图底分析
Fig. 3-29　Ground analysis of site environment of the China Agricultural University Gymnasium

cramped land and high building density, which is another typical urban environment that the overall design needs to face. For this situation, the overall layout of the venues should be coordinated with the surrounding existing buildings and open space, continues and integrates the urban fabric, attaches importance to the accessibility of the visual corridors, and avoids the strong contrast on the scale caused by the oppression of the surrounding buildings and the destruction of the urban texture; In addition, attention should be paid to the traffic accessibility, creating the outdoor space of the venues, and playing its role in the integration and promotion to the urban node.

China Agricultural University, where the Beijing Olympic Wrestling Hall is located, is a typical Chinese campus planned and built in the 1950s. The rigorous layout of the university and the amiable statue of Chairman Mao retain the transient sense of recent history. A few distinct brick-concrete structure academic buildings inside the campus, although as ordinary as any other campus building in style and material, have a special wall texture of pixelated bumpiness in red bricks. Although different in details, these academic buildings follow similar approaches, composing a scenery in concert with the well-maintained vegetation, instantly creating a unique campus atmosphere.

Given such a unique campus environment, we initiated the design a rigorous figure-ground relationship analysis of the campus space environment. In order to hold the Beijing Olympics, the school had established a 150m × 200m sports construction site in the sports activity area of the campus, which was a rare core area in such a mature campus environment. On the one hand, the gymnasium was likely to be the most prominent buildings in this area than teaching buildings and other school buildings around the site in terms of volume and function. Therefore, the physical volume and landscape processing methods

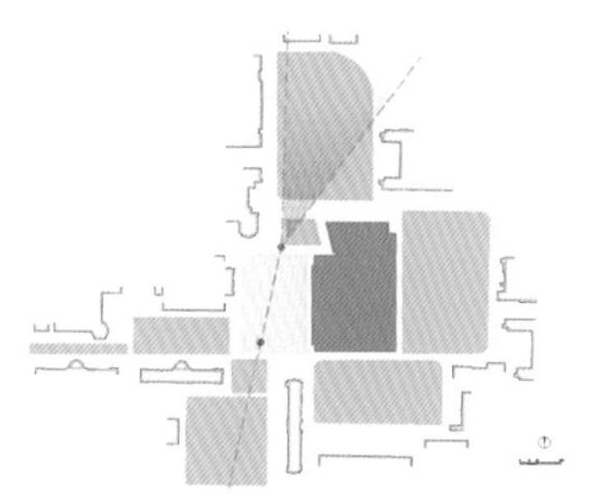

图 3-30　中国农业大学体育馆校园开放空间及周边建筑界面分析

Fig. 3-30　Analysis of the open space at the campus and of the interface of surrounding building of the China Agricultural University Gymnasium

中心则布置在用地北端，和主体建筑、附属用房一起，与周边建筑形成协调关系。室外空间设计上与建筑处理统一，一方面适当突出体育馆的主体地位，另一方面与体育馆一起营造校园中心区空间场所。场地西侧留出开敞空间作为体育馆的集散广场，并和主校道西侧原有的绿带一起形成校园中心区的主要空间节点（图 3-29 ~ 图 3-32）。

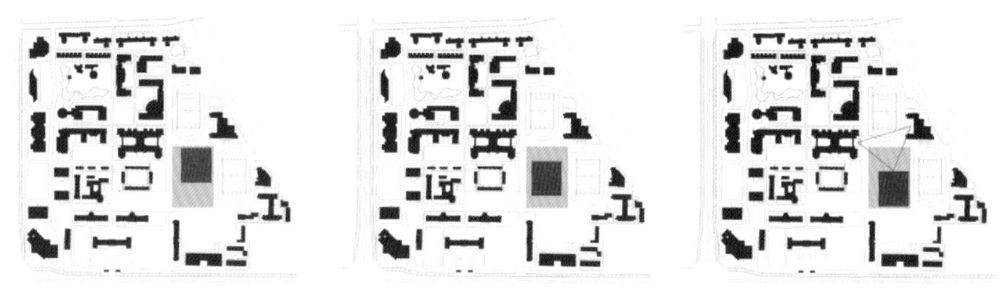

图 3-31　中国农业大学体育馆多方案比选

Fig.3-31　Comparison of schemes of the China Agricultural University Gymnasium

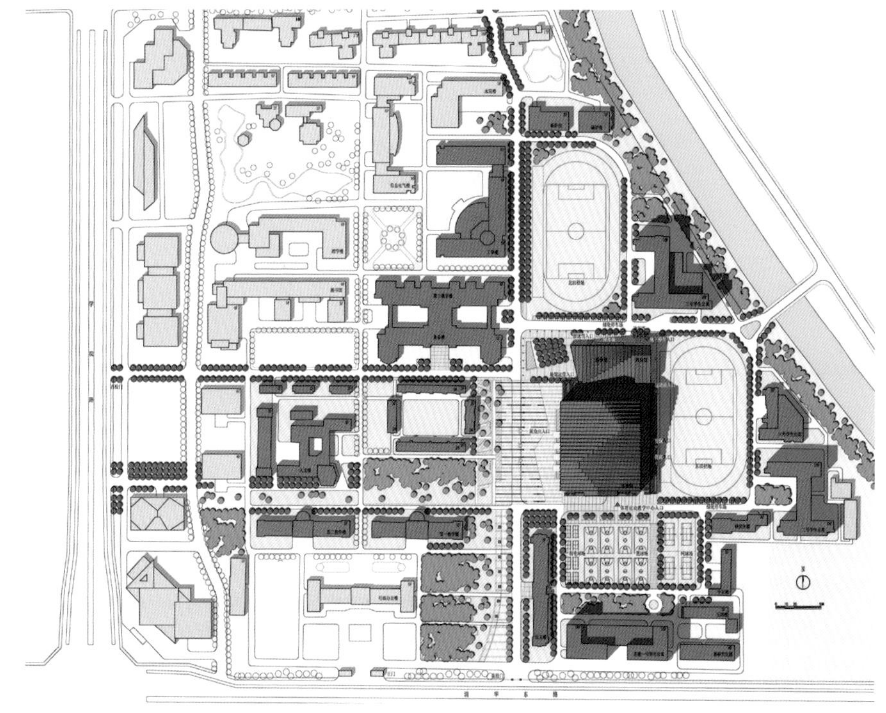

图 3-31　中国农业大学体育馆最终总体布局方案

Fig.3-32　Finalized overall layout scheme of the China Agricultural University Gymnasium

should be able to properly highlight the landmark of this main building; On the other hand, taking into account the strict arrangement and compact layout of the surrounding buildings, the gymnasium should ensure self-discipline in form and volume, to avoid publicity, and make the gymnasium to be coordinated with the surrounding relatively small-scale buildings. Therefore, balancing the landmark of the gymnasium as the main building and its coordination with the campus environment was the key element to the overall layout and physical design of the project.

Based on the above analysis, we carried out the layout of the gymnasium from the perspective of the overall planning of the campus. Taking into account the existence of the No. 3 student apartment and the food laboratory on the north side, placing the gymnasium on the central or north side of the site would make the overall space cramped. By contrast, relocating it to the south of the site could not only effectively avoid the pressure on the surrounding environment, but also enable several open spaces that were originally looser to be organized and defined, and thus highlights the role of the gymnasium as the area's overall focus. The swimming training center was located at the northern end of the site, which was in conjunction with the main building and ancillary room, forming a coordination relationship with the surrounding buildings. The design method of the outdoor space was unified with the gymnasium. On the one hand, the central position of the gymnasium was appropriately highlighted. On the other hand, a space in the center of the campus was created together with the gymnasium. An open space on the west side of the gymnasium was reserved as an evacuation square, and the main space node of the campus center zone was formed together with the original green belt on the west side of the main school road (Fig. 3-29 ～ Fig. 3-32).

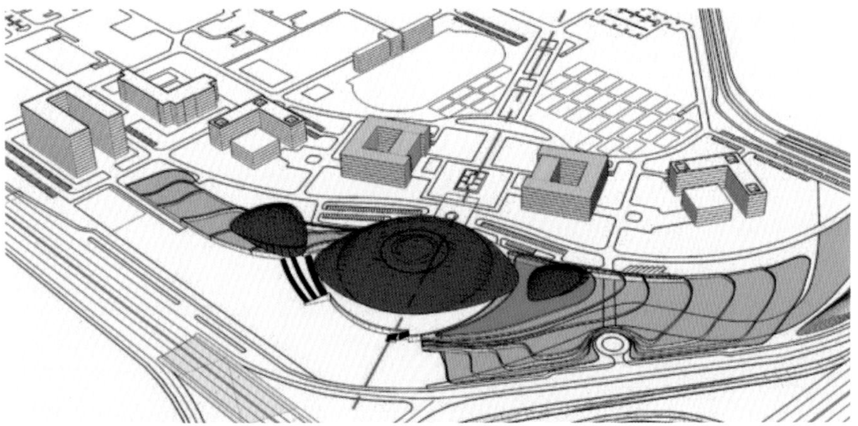

图 3-33　北京工业大学体育馆校园空间关系分析
Fig.3-33　Campus spatial relationship analysis of the Gymnasium of Beijing University of Technology

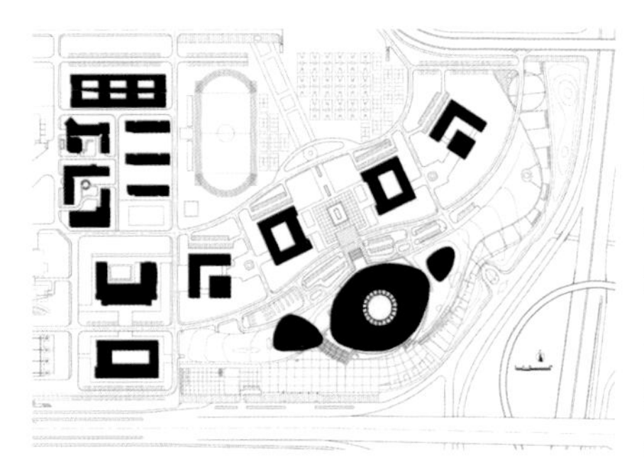

图 3-34　北京工业大学体育馆图底关系分析
Fig.3-34　Ground relationship analysis of the Gymnasium of Beijing University of Technology

北京奥运羽毛球馆，建设于北京工业大学校内，是唯一兴建于北京东南部的奥运场馆。羽毛球馆的招标要求主要包括 3 部分的内容：10,000 座位的主比赛馆，作为羽毛球训练基地的训练馆和赛时热身馆。我们在分析地形与研究功能使用后，确定了 3 部分分别形成体量，以一大带两小为体量组合的基本策略。在国际竞赛的初评阶段，我们的方案与集中布置的方案共同入围。随后而来的奥运场馆优化、瘦身过程中，体育馆规模首先进行了调整，规模从万人馆减少到 8000 席，训练中心也取消了，由于我们的设计中三个体量单独处理，每个体量的取消与缩减都比较容易。甚至在后期，规划变更，羽毛球馆的建设用地西移，用地规模减小的重大变更出现时，我们的设计都没有受到太大影响，原方案的创意一直保持，最终实施完成。而集中式的布局方案则在瘦身阶段就出现极不适应的问题，最终落选（图 3-33、图 3-34）。

Beijing Olympic Badminton Hall, built on the campus of Beijing University of Technology, is the only Beijing Olympic venues in the southeastern of the city. The tendering documents required three parts for the design project: a 10,000-seat main competition hall, the training base as a badminton training center, and a warm-up hall. After topography and program study, three independent masses, of which one large and two small, composed the general plan. During the preliminary evaluation of the competition, our project and a project with a centralized plan made it to the final. In the following modification of the Olympic venue construction requirements, the gymnsaium scale was reduced from 10,000-seat to 8,000-seat, and the training center was canceled. The independency of the three masses in our design allowed shrinkage and removal flexible, even when the site was compressed and moved westwards at later stage as the master-planning changed. Throughout the process, our design have not been severely affected, and the original concept survived; whereas our rival fell behind as their centralized plan was proved less capable of subtraction (Fig. 3-33, Fig. 3-34).

4. 结合自然，实现整体环境协调

体育场馆布置于城市公园是另一种常见的布局模式。在这一情况下，山体、水体等自然要素往往成为环境主角，一方面可能成为体育场馆布局和体量控制的制约因素，另一方面也可能成为建筑“依山傍水”、减小工程量并烘托建筑形体的环境基础。因此，能否处理好

4. Integrate nature to achieve overall environmental coordination

The arrangement of sports venues in city parks is another common mode. In this case, natural elements such as mountains and water always become the protagonists of the environment. On the one hand, it may become a constraint factor in the layout of sports venues and volume control. On the

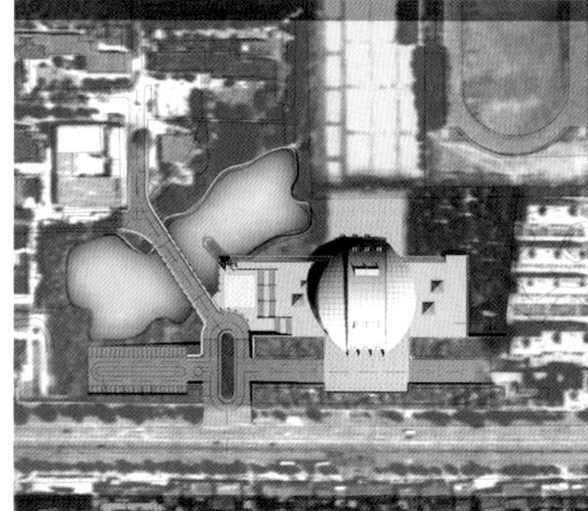

图 3-35　华中科技大学体育馆总体环境关系

Fig.3-35　The overall environmental relations of the gymnasium of Huazhong University of Science and Technology

体育场馆的体量和布局与自然生态要素之间的关系，成为可持续设计成功与否的关键。

2002 年华中科技大学 8000 座体育馆设计竞赛。业主提出的校园规划中，一条道路直通城市道路，校园设立新的东南门，体育馆位于门的东侧。规划完全忽略了地段范围内几经填埋、不断缩小的湖面。经过对地形的分析，我们提出了利用地形，将公众出口设在二层平台，直接对接校园外侧城市道路，使建筑尽可能靠近校园与城市的边界，以便对外服务时的人员疏散。另一方面，建筑东移后，大部分

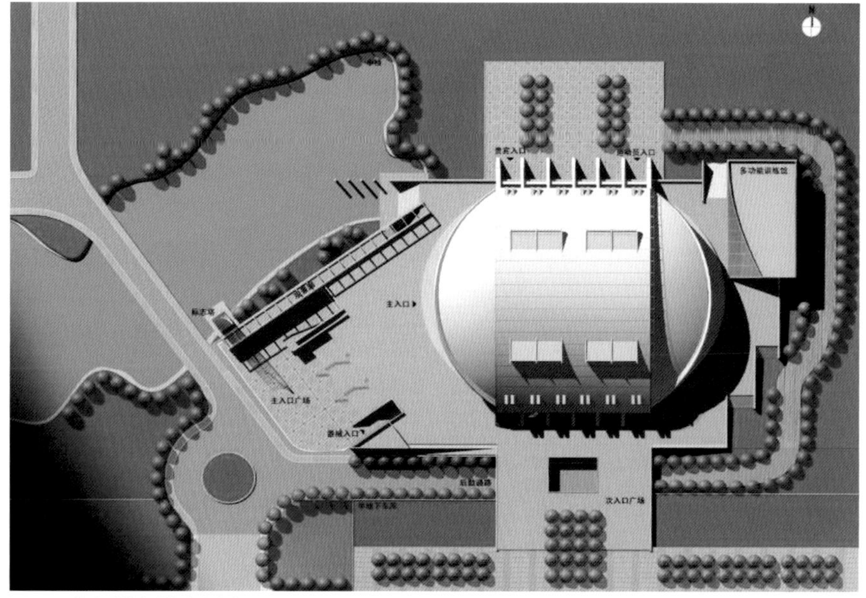

图 3-36　华中科技大学体育馆方案图

Fig.3-36　The design schemes of the gymnasium of Huazhong University of Science and Technology

other hand, it may also be the environmental basis for a construction 'with hill at the back and water in the front' and a reduction in the earthwork, and also serve as contrast to the building form. Therefore, the key to a successful sustainable design is whether it can handle the relationship between the volume and layout of the building and the natural ecological elements.

In the design competition for the 8000-seat gymnasium of Huazhong University of Science and Technology in 2002, the original campus plan proposed a path running directly into urban roads, a new gate in the southeast of the campus, and a gymnasium on the eastern side of the gate. The planning completely overlooked a lake that had suffered from repetitive reconstruction and subtraction. After topographic analysis, we utilized the terrain by setting the public exit on a second floor platform, pushing toe buildings towards the campus border to facilitate evacuation by directing the public to the off-campus urban traffic. Meanwhile, by nudging eastwards the gymnasium, campus paths would therefore twist sideways in order to integrate the entrance square and the evacuation square and to avoid encroaching lakeside spaces. The elevational droped from the urban environment to the campus provide ground floor accessibility for students and athletes on campus. The design of the gymnasium served both the campus and the city, made efficient use of the land and

湖面得以保存，校园道路略加扭转，大门广场与体育馆疏散广场结合，为保留湖面争取了用地。由于城市道路与校园存在高差，学生及运动员可以方便地从校园进入体育馆一层空间。这样形成了同时服务校园与城市、紧凑利用土地、对城市与校园景观均做出积极贡献的体育馆设计与布局（图 3-35、图 3-36）。

由我们主持设计，2012 年竣工的梅县体育中心处在自然城市环境中，梅县体育中心位于梅县梅花山下，人民广场旁，项目包括一个体育馆（梅县文体中心）和一个体育场（梅县曾宪梓体育场），与梅花山和人民广场一起成为梅县市民体育休闲的中心活动场所。设计巧妙地通过竖向设计和看台设置，很好地解决交通组织和人员疏散问题，成功处理好山体、

contributed to both the urban and campus landscape (Fig. 3-35, Fig. 3-36).

The Meixian Sports Center completed by us in 2012 was located in a natural urban environment. It was located next to the Meihua Mountain in Meixian, next to People's Square. The project included a gymnasium (Meixian Sports Center) and a stadium (Meixian Zengxianzi Stadium). Together with Meihuashan and People's Square, it would become a leisure center for the Meixian citizens. The design skillfully solved the problem of traffic organization and evacuation by vertical design and stands setting, and successfully handled the relationships among the mountains, stadiums, and cities to achieve harmony between the buildings and the overall environment.

Following the original terrain of the mountain, the stadium and gymnasium were arranged from west to east with the decline of Meihua Mountain. The stadium that could accommodate 20,000 spectators made use of the terrain conditions, set up the stand on the mountain, so that minimized the impact of the project on the mountain and reduced the amount of earthwork, made the building integrate into the natural mountain, reduced shelter effect of the stadium on the Meihua Mountain, which ensured the visual connection between the city and nature mountains. In terms of the details, the exterior of the stadium used a

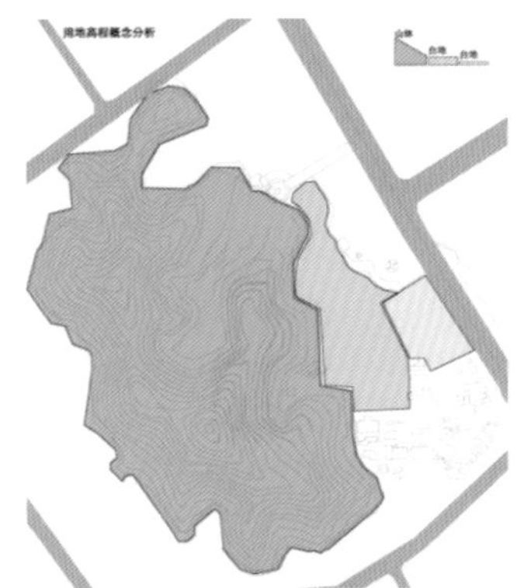

图 3-37　梅县文体中心：a. 基地高程分析
Fig.3-37　Meixian Cultural and Sports Center : a. Base elevation analysis

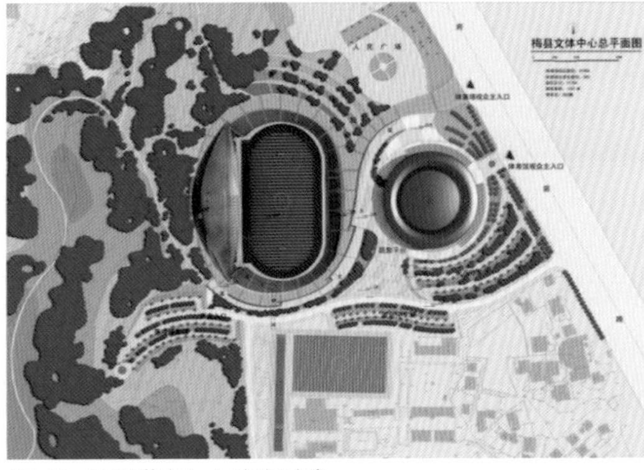

图 3-38　梅县文体中心：b. 总平面方案
Fig. 3-38　Meixian Cultural and Sports Center ; b. General plan

图 3-39　梅县文体中心：e. 建成效果
Fig.3-39　Meixian Cultural and Sports Center: e. as-built View

体育场馆和城市三者的关系，实现建筑和整体环境的协调。

设计遵从原有的山体地形，随着梅花山山势走低，由西向东布置体育场和体育馆。体育场可容纳观众2万人，利用地形条件，依山设置看台，最大限度减小项目对山体的影响并降低土方量，使得建筑融入自然山体，减小体育场对梅花山的遮挡，保证城市和自然山体之间的视线联系。在细节处理上，体育场外墙采用石笼墙，石笼墙内装当地石材，使得建筑体量更好地融入环境（图3-37 ~图3-39）。

gabion wall, which was filled with the local stone, allowing the building volume to be better integrated into the environment (Fig. 3-37 ~ Fig. 3-39).

The gymnasium could accommodate 7,000 audiences and host sports competitions, theatrical performances and exhibitions. Because of the stadium on the west side of the gymnasium, the conventional traffic organization and personnel flow arrangements were restricted by insufficient area of the evacuation square, and too much pressure on the city caused by the concentrated traffic flow of the venues. In our design, the entrances of all types of

高程设计多方案比选　　表3-1

方案	体育场地高程取值	土方工程量	节约用地效果	场馆体量与山体的协调性	最终采用方案
方案A	89.70米	较大	最好	不协调	×
方案B	85.91米	中等	较好	较协调	√
方案C	82.38米	较小	不佳	协调	×

Comparison of elevation design schemes　　Tab. 3-1

Scheme	Sports ground elevation	Earthwork	Land conservation	Coordination of the venues mass and the mountain	Final schemes
Scheme A	89.70 m	Large	Maximum	Uncoordinated	×
Scheme B	85.91 m	Moderate	Good	Quite coordinated	√
Scheme C	82.38 m	Minimum	Poor	Coordinated	×

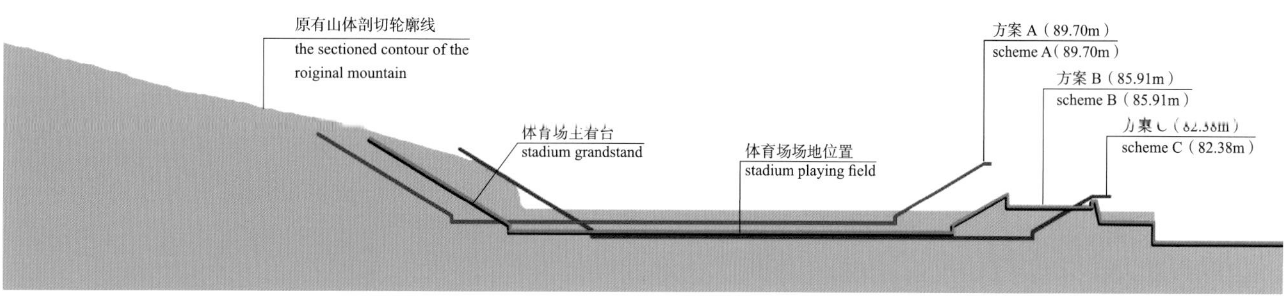

图3-40　梅县文体中心：c.高程设计多方案比选

Fig.3-40　Meixian Cultural and Sports Center : c. comparison of elevation design schemes

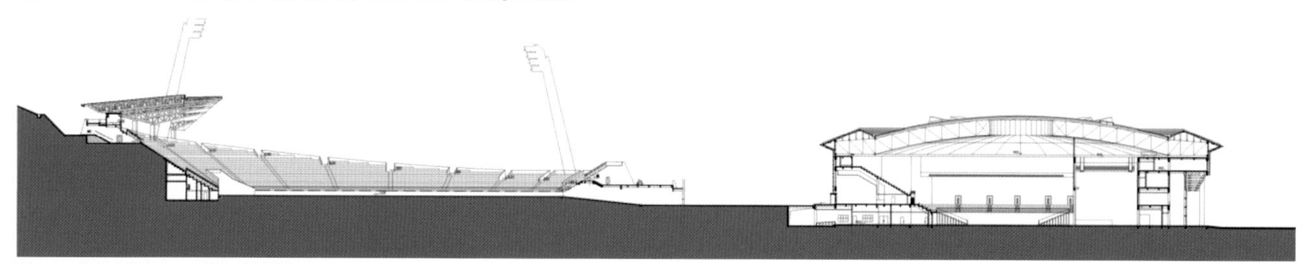

图3-41　梅县文体中心：d.最终确定的剖面

Fig.3-41　Meixian Cultural and Sports Center : d. finalized section

体育馆可容纳观众7000人，可举办体育比赛、文艺演出和展览活动。由于体育馆西侧尚有一个体育场，使得常规的交通组织和人员流线安排容易造成集散广场面积的不足，以及集中的人员流量对场地和城市的交通压力过大。设计中将各类人员出入口分布于一、二层，且将主要的观众出入口设置在体育馆西面靠近山体一侧，利用地形高差，从二层平台进入体育馆。这一处理方式将观众疏散的流线延长，在保证安全疏散的同时也有效地缓解了集中人流对城市交通的冲击（图3-40、图3-41、表3-1）。

personnel were distributed on the first and second floors, and the main entrances of the audience were arranged on the west side of the gymnasium that was closed to the mountain, which made use of the height difference of the natural terrain and the second floor platform to enter the gymnasium. This approach extended the evacuation flow line of the audience, which could effectively mitigate the impact of concentrated traffic on urban traffic under the premise of ensuring safe evacuation (Fig. 3-40, Fig. 3-41, Tab. 3-1).

体育场利用场地地形起伏建设看台的方式曾在我国早期的体育建筑实践中得到大量应用，如始建于1950年的广州越秀山体育场和建成于1953年的南京五台山体育场。尊重自然地形条件、因势利导的方法即便是在经济实力大幅提高到今天仍然具有参考并推广的价值，这对节约造价、保护环境，实现体育场馆和城市的可持续发展具有现实意义。

The use the natural terrain to construct the stands has been used in many of the early sports facilities construction practices in our country, such as the Yuexiu Mountain Stadium in Guangzhou in 1950 and Wutaishan Stadium in Nanjing in 1953. The method of respecting the natural terrain conditions and making use of the advantages of the situation is still worthy of reference and promotion even if the economic strength is greatly improved nowadays. This has practical significance for saving costs, protecting the environment and realizing the sustainable development of the sports venues and the cities.

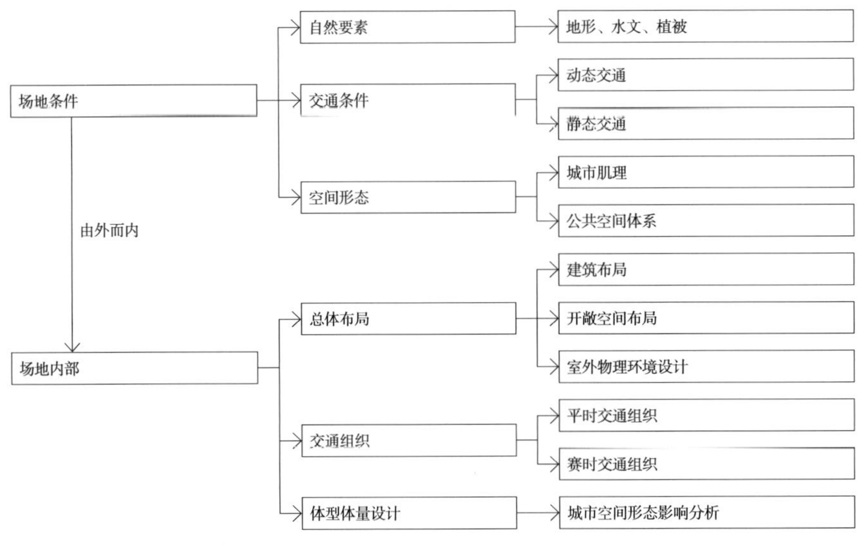

图3-42　基于城市环境的体育建筑设计方法

可持续的体育建筑设计应突破局限于用地红线、关注自我、由内而外的思维方式，适当地将视角扩大到城市范围，关注城市整体环境、结合进由外而内的思维方法，将体育建筑和城市环境进行功能和空间的整体考虑。体育场馆布局和建筑设计应从环境分析入手，梳理场地内外自然要素、交通条件、功能类型、空间形态等条件，在此基础上进行方案推敲，对应地进行总体布局、体型设计、功能设置、交通组织等工作。以城市设计的思路指导体育建筑的规划和设计，最终促进体育建筑和城市整体的可持续发展（图 3-42）。

Sustainable sports facilities design should break through the restriction of the red line, get rid of the self-consciousness, the single inside-outside thinking mode, appropriately expand the perspective to the urban scope, pay attention to the overall environment of the city, combine in the design path from the outside to inside, and take into account the overall functional and spatial considerations of the building and the urban environment. The layout and architectural design of the venues should start with environmental analysis, and scrutinize the conditions of natural elements, traffic conditions, function types, and space forms inside and outside the site. Based on this, the scrutiny should be accordingly carried out on the overall layout, shape design, function setting, and traffic organization. The planning and design of sports facilities should be guided with the idea of urban design, so as to ultimately promote the sustainable development of sports facilities and the overall city (Fig. 3-42).

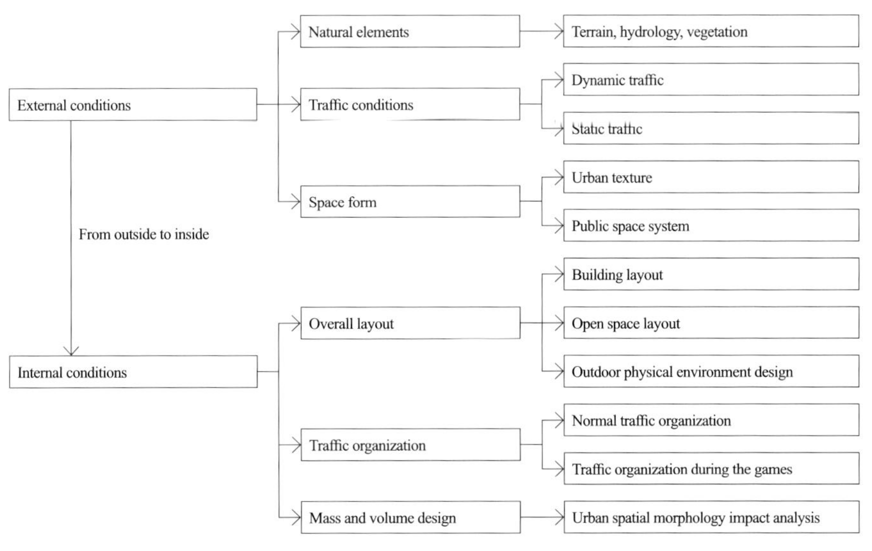

Fig. 3-42 The method of sports architecture design based on urban environment

四、功能理性

VI. FUNCTIONAL RATIONALITY

体育建筑功能究竟为何？体育的功能为什么需要建筑角度的研究？许多建筑，复杂如医院建筑，其性质、运行都是有特定明确定义和运行机构的。酒店也是有特定的专业公司运行，管理、标准一应俱全，明确了酒店管理公司，建筑师就不需要为功能头痛了。唯独体育建筑，在我国没有特定明确的机构。作为政府部门的体育机构更像是举国体制的代言人和管理者，无法对日常体育设施的管理运行提出系统的管理办法，这也是许多体育场馆运行问题百出的原因。体育协会，各类专项协会甚至包括奥委会这样的机构，大都是为自己特定赛事做出规划，赛事结束，对场馆连回头的力气都不想花。只有中国篮协将国际篮协的体育馆建设指引进行了中文翻译出版，以期作为俱乐部参考。其他寥寥无几的职业体育的机构也无暇参与设施的相关管理与探讨。这其实就是中国体育建筑的现实。因此，体育建筑设计不难，因为如果造型被决策者认为过得去了，没有人会细究功能，国内体育设施因为功能配置出现重大缺失导致设施无法使用的情况屡屡发生。而体育建筑又是最难做的，因为满足了体育工艺才是入了门，离真正满足使用还相距很远，不是许多著名的场馆完全依赖于开一次赛会，输一次血，然后又进入半休眠状态，等待下一次的赛会。特定场馆的良好运营状态又有着不可移植的偶然性。比如，公认状态不错的天河体育中心，也因为同城相似机构的竞争而如履薄冰。中国的体育建筑现状需要建筑师负责任解决许多问题。

What is the function of sports facilities? Why should sports functions be architecturally studied? Most buildings, even those as complex as hospital buildings, their nature is well-defined and their operations are handled by specific operational organizations. E.g., hotels are operated by specific professional operation companies where management and regulations are in place. Hotel designers need not to be responsible for functions once hotel management companies take over. However, sports facility is an exception, as there is no specific organization for its management in our country. Sports organizations, as a government section, is more like the spokesmen and manager of the nationwide system; they are not in a position to decide on the systematic management for the daily operation of sports facilities, which is why many sports venues are suffering from various operation problems. Sports associations and even the Olympic Committee, just do what they have to for their events. Once the events are over, they could not care less for the venues. The Chinese Basketball Association is the only one that has translated and published the guidelines for the design of sports venues by the International Basketball Association for domestic clubs' information. Other few professional sports organizations are too busy to engage in the management and discussion of facilities. This is the cold reality of sports facility in China. In summary, sports facility is easy, as long as the decision-makers approve, no one cares for the program and function. As a result, domestic sports facility recurrently fails from dysfunction. Sports facility is hard. Meeting the professional construction standard is merely the stepping stone. The major difficulty is to fully fulfill operational needs. More often than not, famous venues would rely on holding a game and quickly alternate to semi-hibernation. The successful few have peculiarity too incidental to be duplicated. For instance, the Tianhe Sports Center, which is generally recognized to be in a good operational state, is still at stake

近年来单就大型体育馆建设的策划就呈现非常尴尬的局面。从网上随便搜索一下可以发现，自从2008年五棵松体育馆被定义为NBA标准的场馆以来，先后声称以NBA为标准策划并建设完成的大型体育场馆有：1.8万座的广州国际演艺中心、1.8万座的上海原“世博演艺中心”、现在的梅赛德斯 - 奔驰文化中心、1.8万座的上海浦东东方体育中心、1.0万座的上海宝山体育中心体育馆、1.8万座的深圳龙岗体育中心体育馆、1.3万座的深圳湾体育中心体育馆、1.25万座的青岛国信体育馆、1.5万座的东莞篮球中心、2.0万座的南京青奥公园体育馆。总计万人以上体育馆超过10座。筹划中的还有重庆、佛山、合肥、厦门等城市。不夸张的说，如果把美国NBA赛事全部搬到中国来，以国内现有大型场馆的数量，完全可以胜任了（图4-1）。

所谓NBA标准，其实是个非常弹性变化的笼统概念，长期以来，NBA在市场运营方面的成果使场馆规模逐渐攀升，近年新建场馆达到了1.8万人的容量。一座NBA球馆，其实同时也是冰球队的场馆，因为NBA最初就是冰球队老板们为了填补冰球赛季空缺才联手打造的。有的NBA球队甚至合用一座球馆，如著名的斯台普斯体育馆就是湖人和快船共有的主场。在这种背景下，NBA球馆不担心年开放场次，更关心如何在有限的时间内完成不同功能的转换。所以，NBA球馆的篮球地板是活动的，因为要为冰球场的转换提供便利。NBA球馆结构是等所谓“平面结构”为主的，因为投资的俱乐部不需要展示体型变化，需要实实在在的便利功能，网架桁架便于各种设备吊装。NBA球馆最准确的原型就是个高举高打的黑盒子，因为NBA赛事日程紧张完全没有需要考虑节能降耗，他们需要的是以小时为单位计算场地转换的灵活性问题。这样下来看，区别来了：国内最好的体育场馆，比赛加演出的场次也大多50场左右，能够过

in competition with similar facilities of the city. The problematic status quo of sports facility in China needs to be salvaged by architects.

In recent years, the planning and building of large-scale venues has been caught in a very embarrassing situation. Electronic data via Internet prove that ever since the Wukesong Indoor Gymnasium was qualified as the NBA standard gymnasium in 2008, many large sports gymnasiums claimed to meet the NBA standards, e.g., the 18,000-seat Guangzhou International Sports Arena, the former 18,000-seat Shanghai EXPO Performing Arts Center or the current Mercedes - Benz Arena, the 18,000-seat Shanghai Pudong Oriental Sports Center, the 10,000-seat Gymnasium of Baoshan Sports Center, the 18,000-seat Gymnasium of Shenzhen Longgang Sports Center, the 13,000-seat Gymnasium of Shenzhen Bay Sports Center, the 12,500-seat of Qingdao Guoxin Gymnasium, the 15,000-seat Dongguan Basketball Center, and the 20,000-seat Gymnasium of Nanjing Youth Olympic Park. In total, there are more than a dozen gymnasium, each has a seating of 10,000 and above. More are planned to be built in Chongqing, Foshan, Hefei, Xiamen and other cities. It is no exaggeration to say that even if all the NBA games are moved to the exiting domestic large gymnasiums in China, there will be more venues to spare.

The so-called NBA standards, in fact, are very elastic. The achievcment of market operation encourages new arenas built in recent years to reach a capacity of 18,000. An NBA arena, in fact, is also an ice hockey gymnasium, for NBA was the brainchild of hockey games when hockey club owners decided to make use of the empty hockey gymnasium in the days without hockey games. Some NBA teams even share an arena, such as the famous Staples Center, which is the home arena to the Lakers and the Clippers. In such a context, NBA arenas have less concerns about the opening hours of the venue than the

百场的十分罕见。一年中换一个场平均有三五天时间，如果是个黑盒子，所有准备工作还要人工照明，那建筑设计的时候做不做天然采光，道理就非常清楚了。

我国体育建筑普遍存在功能单一、使用率低和场馆空置的问题，体制机制问题是首要问题，但除此之外，如何在决策设计阶段提高精明程度，实现可持续发展是中国建筑师无法推卸的责任。体育建筑建设的科学决策关系着项目的投资效益，对建筑全寿命周期的可持续运行有着关键影响，但目前我国体育建筑建设在决策层面尚存在不尽人意之处。体育建筑运营管理等经验，应该提前进入到建设决策阶段，避免由于建设的不合理而造成业主和经营者的长期负担。同样地，体育建筑设计不应该仅仅是在建设决策之后的执行过程，而也应当提前参与到建设决策过程中来。当下的体育建筑

switch between different functions within a limited time. Thus, the floor of an NBA basketball arena is removable to facilitate its conversion into a hockey gymnasium. NBA arenas are mainly of the so-called flat structure, since the investors do not require formal transformation. What they need is substantially useful features, hence truss structure and grid structure are commonly used to facilitate lifting. The most accurate prototype of NBA arena is a heavy-duty black box, since the intense schedule of NBA tournaments neglects energy saving. Instead, what requires attention is the flexibility that allows the arena floor to be transformed within hours. The difference between NBA arenas in China and the USA is that a top sports gymnasium in China would hold around 50 games or shows annually. It would be more than rare if one could hold more than 100. For such a gymnasium it has three to five days to change its floor and auxiliary facilities. If built within a black box, artificial lighting has to be counted into preparation tasks. Therefore, it is doubtless that whether natural light is part of domestic sports facility design.

The sports facilities in China generally have the problems of single function, low utilization rate, and vacancy of venues, on top of which is the institutional problem. It is the architects' inexorable responsibility to assist with decision-making in terms of sustainability. The scientific decision-making of sports facilities construction has a bearing on the investment benefits of the project, and it has the key influence on the sustainable operation of the entire life cycle. However, at this level, the sports facilities construction in our country is still unsatisfactory at present. Experiences in the operation and management of sports facilities should be advanced to the construction decision-making stage to avoid the long-term burden on the owners and operators caused by unreasonable construction. Similarly, sports facilities design should not only

图 4-1　可灵活进行功能转换的体育场馆

Fig.4-1　Sports venues of flexible functional conversion

设计存在片面迎合上级意志、重视形体形象等外在表现、忽视内在的功能性研究，容易导致体育场馆缺乏灵活性而无法适应功能需求的不断变化。在功能设置上，提高体育场馆使用的灵活适应性是体育建筑设计和运营所面临的长期课题，关系到体育场馆的可持续发展（图 4-1）。

be the implementation process after the construction decision, but also be involved in the decision-making process in advance. The current sports facilities design has problems such as catering to the superior will, excessively attaching importance to the physical image while ignoring the intrinsic function, which easily leads to the lack of flexibility in the venues and adaptability to the constant changes in functional requirements. In terms of the function setting, improving the flexibility and adaptability of the sports venues is a long-term task being faced with the design and operation of sports facilities, which relates to the their sustainable development (Fig. 4-1).

体育场馆的功能需求特点

体育场馆的功能需要主要有体育活动和非体育活动两大类，具体类型上前者包含群众日常体育锻炼、体育比赛等，这其中也包括了体育活动的观演部分，后者包含文艺演出、展览、集会等集体活动。为了在全寿命周期内提高场馆的利用率，可多功能利用的策略成为许多场馆实际运营的必然选择。然而在体育场馆全寿命周期内，其使用需求必然会随着国民经济和体育产业发展而不断动态变化，使得体育设施使用需求上具有动态、不确定性的特点。这使得在制定设计任务书时，由于体育场馆运营需

图 4-2　基于灵活适应性的设计策略研究内容

The Characteristics of the Functional Demand of Sport Facilities

The function demands for the sport facilities could be divided into sports and non-sport activities. The former includes the daily physical exercise and sports competitions of the masses, which also includes watching section for these activities, and the latter includes theatrical performances and exhibitions, and other group activities such as collective meetings. In order to increase the utilization rate of the venues throughout the life cycle, the strategy of multi-functional utilization has become an inevitable choice for the actual operation of many venues. However, during the entire life cycle of the sport venues, the function demand will inevitably change dynamically with the development of the national economy and the sports industry, giving dynamic and uncertain characteristics to the use of sports facilities. These factors make it difficult to form a clear project orientation due to the lack of market investigations and planning work for the sport facilities, and the difficulty to guarantee the reliability of future judgments even when they are carried out. The adaptability of sports facility to the dynamic and uncertain functional requirements in the present and the future is an important issue for the sustainable design of sport facilities.

求市场调研和策划工作的缺乏，或者即便进行调研和策划，也难以保证对未来判断的可靠性，因此造成项目定位模糊。让体育建筑具有面对现在和未来动态、不确定性功能需求的适应性，是体育建筑可持续性设计的重要议题。

与其他类型的建筑相比，体育场馆的大跨度结构和大空间体量使其自身的技术要求和建设成本高，并且建成之后的拆除和改造难度均较高。因此体育建筑建设应从决策阶段考虑场馆的多功能利用，并在设计阶段做出充分的灵活适应性设计，才有可能使场馆具备多功能利用的条件，在全寿命周期内提高场馆的使用率，实现体育场馆的可持续目标。

应对思路

通过空间和设施利用的灵活性设计策略，为解决体育场馆功能的可持续发展提供有效途径（图 4-2）。

空间设计策略

1. 功能构成关系

体育场馆内部功能构成主要包含体育活动和非体育活动，体育活动主要包含体育比赛及其观演、群众日常锻炼和学校师生体育活动等。作为大空间建筑，体育场馆还可以通过空间灵

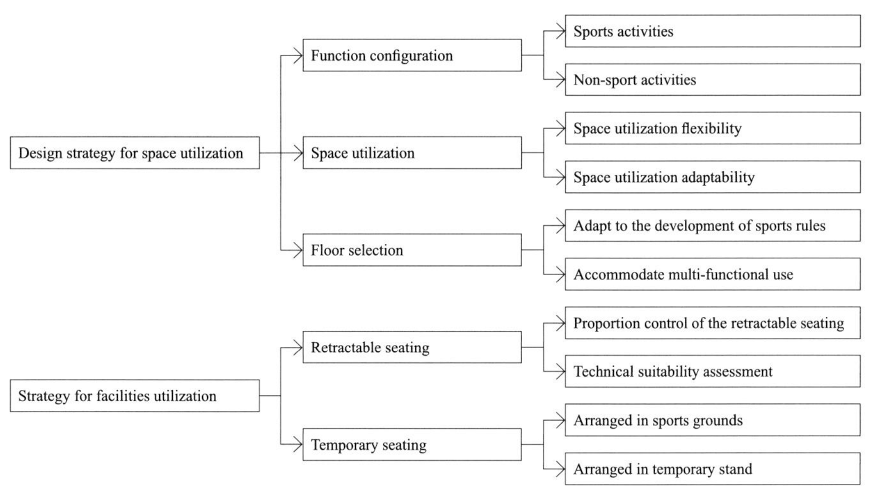

Fig.4-2 Research of design strategy based on flexibility and adaptability

Compared with other types of buildings, the large-span structure and large volume of sport venues pull up its own technical requirements and construction costs, and the difficulty of demolition and reconstruction. Therefore, the construction of sports facilities should take into account the multi-functional use of the venues in the decision-making stage, and adequate flexibility and adaptability in the design stage. Only in this way can the venues have the conditions for multi-functional utilization and increase the use rate of the sport facilities throughout the life cycle, and achieve the goals of sustainable development.

Solutions

Designing strategies through the flexibility of the use of the space and facilities provides an effective way to achieve sustainable development of the sports facilities' functions (Fig. 4-2).

Design Strategy for Space

1. Function configuration

The internal functional composition of the sport venues mainly includes sports activities and non-sport activities. The former mainly include sport s competitions and their performances, daily exercise of the masses, and school sports activities for teachers and students. As a large-space building, sports venues can also be flexibly arranged to carry out entertainment, celebration, conferences, collective meetings and exhibitions and other activities. There is a different space-time combination of functions in the sport venues, which can be broadly classified as:

The combination of space dimension refers to the simultaneous appearance of various functions in different spaces at the same time. Such as Nantong Sports Center, Changzhou Olympic Sports Center and Jiangmen Binjiang Sports Center, where the venues and the exhibition center are

活布置，开展文娱、庆典、会议、集会及展览等各种活动。不同功能在体育场馆时空上存在组合关系，这种关系大体上可以归为：

空间维度上的组合，指的是同一时间内，各种功能在不同空间并列出现。如南通体育中心、常州奥林匹克体育中心和江门滨江体育中心，将体育场馆和会展中心组合形成体育会展中心，从而实现体育功能和会展功能的并置；上海体育场和广东奥林匹克体育中心，结合体育场看台设置了宾馆功能，从而实现体育功能和酒店服务功能的并置。就体育场馆自身而言，即便是在同一块运动场地内，不同体育活动也可以通过场地临时划分和简单布置，实现不同活动项目的组合（图 4-3）。

时间维度上的组合，指的是同一空间内，各种功能在不同时间先后出现。这一过程不对体育场馆做大的改造，而主要对活动场地稍微调整以适应不同功能需求，典型的一个例子是美国职业球队的主场，通常由两支以

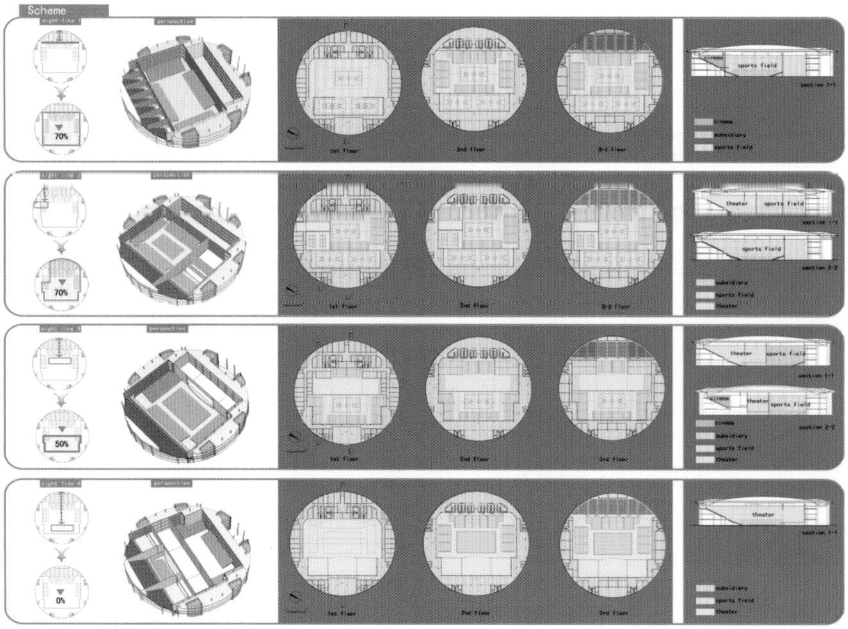

图 4-3　梅县体育馆主体空间利用研究

Fig.4-3　Study of the utilization of the main space of Meixian Gymnasium

combined to form a sports exhibition center, in order to achieve the juxtaposition of sports and the exhibition functions. The Shanghai Stadium and the Guangdong Olympic Sports Center, combine hotel functions in the stadium stands setting to achieve the juxtaposition of sports and the hotel service functions. As far as the sport venue itself is concerned, even in the same venue, different sports activities can be achieved through the temporary division and simple arrangement of the sport field to achieve a combination of different activities (Fig. 4-3).

The combination of time dimension refers to various functions occurring at different times in the same space, and this process does not carry out a major transformation of the venue, but mainly adjust it slightly to meet different functional demands. A typical example is that, for the American professional teams, the home venue is usually shared by more than two teams, of which a common pattern is that the basketball team and the ice hockey team share a home gymnasium. Since the two seasons are staggered and do not interfere with each other, it can increase the utilization rate of the venues throughout the year. The large-span space provided by the stadiums and gymnasiums can carry out various kinds of theatrical performances, exhibitions, and group activities during the period when there is no sports activity, which contributes a significant proportion of the post-game operation activities to the sports venues.

The combination of space-time dimensions refers to the appearance of various functions at different times in different spaces. In the whole life cycle of the sport venues, it is common for them to carry more than one type or even all types of activities. This situation is particularly prevalent with an overall or partial renovation of the venues to adapt to different functional demands. For example, in 1996, the main stadium of the Atlanta Olympic Games was originally

上的球队共用，一种常见的模式是篮球队和冰球队共用一个主场，两者赛季错开，互不干扰，可提高场馆全年的使用率。体育场馆所能提供的大跨度空间，可以在无体育活动的时间段开展各类文艺演出、展览和集会活动，这为体育场馆的赛后运营活动贡献了较大的比重。

时空维度上的组合，指的是在不同空间，各种功能在不同时间先后出现。在体育场馆全寿命周期内，往往需要承载以上多种甚至所有类型的活动，因而这种情况尤为普遍，伴随此过程的是对体育场馆为适应不同功能需求而作的整体或局部改造。如1996年亚特兰大奥运会主体育场，原本是可容纳8.5万观众的田径运动场，赛后改造为可容纳4.5万观众的职业棒球场。2000年落成的日本埼玉县体育馆，可以承办体育比赛、音乐会和大型展览活动，建筑由巨型可以移动的模块构成，通过“变动”可以实现大型体育场（37000座）、中型体育馆（22500座）和小型音乐厅（6000座）三种基本空间类型之间的切换，满足美式橄榄球比赛、篮球、排球、冰球等赛事和音乐活动，不同功能在时空上的组合为满足社区多样化使用，以及场馆自身可持续运营创造了有利条件。

当前我国体育场馆普遍面临严峻的运营压力，我国体育产业尚处于起步阶段，文艺演出市场尚不发达，竞技活动和文艺观演活动也缺乏群众基础，赋予场馆多功能性以保证赛后可以容纳更多活动类型的策略被逐渐接受。体育场馆的空间是其功能的载体，多功能的实现有赖于空间利用策略的合理性。

2. 空间利用策略

体育建筑内部各种功能关系可能同时存在，也可能随着使用需求的变化而有不同的表现，这给体育建筑的功能设计带来较大的不确定性。灵活适应性的空间利用策略有利于应对体育场馆多功能使用的发展趋势（表4-1）。

an athletic field that could accommodate 85,000 spectators. After the games, it was transformed into a professional baseball stadium that could accommodate 45,000 spectators. The Saitama Gymnasium in Japan, which was completed in 2000, can host sports competitions, concerts, and large-scale exhibitions. The gymnasium consists of giant mobile modules that can be transformed into and switched between the three basic space types including the large stadiums (37,000 seats), medium-sized gymnasiums (22,500 seats) and small concert halls (6000 seats), so as to meet the demands for different events such as the American football games, basketball, volleyball, ice hockey and musical activities, etc. The combination of different functions in time and space meets the diverse demands of the community and create favorable conditions for the sustainable operations of the venue itself.

At present, the sport venues in our country are generally facing severe operational pressures, since the conditions that the China's sports industry is still in its infancy, the market for theatrical performances is still underdeveloped, and the sports events and theatrical performances are also lacking in mass bases, therefore the strategies of giving venues multi-functionality to ensure that they can accommodate multi-activities after the game are gradually accepted. The space of the venues is the carrier of its function, and the realization of multi-function depends on the rationality of the space utilization strategy.

2. Strategy for space utilization

The various functional relationships within a sports building may exist at the same time, or may have different characteristics with the change of the use demand, which brings greater uncertainty to the functional design of sports facilities. Flexibility and

三大类型体育场馆功能空间组合策略　　表 4-1

场馆类型		体育活动			非体育活动			
		体育比赛和训练	体育训练	全民健身	文艺演出	会议集会	会展	商业经营
体育场	主体空间	√	√	√	√	√	√	
	辅助用房		√	√			√	√
体育馆	主体空间	√	√	√	√	√	√	
	辅助用房		√	√			√	√
游泳馆	主体空间	√	√	√				
	辅助用房		√	√				√

Combination strategies of space and function for three major types of sports venues　　Tab. 4-1

Type of the venues		Sports activities			Non-sport activities			
		Sports competition and training	Sports training	National fitness	Artistic performance	Conference and gathering	Exhibition	Business operation
Stadium	Subject space	√	√	√	√	√	√	
	Auxiliary rooms		√	√			√	√
Gymnasium	Subject space	√	√	√	√	√	√	
	Auxiliary rooms		√	√			√	√
Natatorium	Subject space	√	√	√				
	Auxiliary rooms		√	√				√

以我们设计的江门市滨江体育中心为例，项目位于广东省江门市滨江新城核心地段，作为新城开发的启动区，定位为集体育运动、商贸会展、休闲娱乐、文艺演出为一体的大型综合性活动中心，其中包含体育场、体育馆、游泳馆和会展中心等重要单体建筑。几大单体建筑采用了复合型空间利用策略，在用地北区将体育馆和会展中心进行组合，南区则将体育场和游泳馆组合一起并将两者扩展经营用房集中沿西侧城市道路布置。这种组合方案，一方面利用了体育馆和会展中心大空间的共性特点，为后期运营过程中的体育活动和会展活动相互借用空间的使用可能性提供了基础。另一方面各大单体建筑功能用房集约布置，有机地组织在一起，有利于节约空间和形成整体运营条件。本案中结合单体建筑空间和使用特点，进行分解和重组，为场馆的功能并置提供了可能性，为城市体育设施的功能设计提供一个有益参考（图 4-4）。

adaptability space utilization strategies are conducive to the development of multi-functional use of the sports venues (Tab.4-1).

Taking the Jiangmen City Riverside Sports Center we designed as an example, the project was located in the core area of Jiangmen Riverside New District in Guangdong Province. It was a startup area for the development of the new district and was positioned as a large-scale activity center integrating sports, trade fair and exhibition, leisure, entertainment, and theatrical performance, and included several large single buildings, namely the stadium, gymnasium, natatorium, and the convention center. The design adopted a composite space utilization strategy. The gymnasium and the exhibition center were combined in the north area of the site, while the stadium and the natatorium were combined in the south area, in which the extended operational rooms of the two were arranged along the west side of the city road. This combination scheme, on the one hand, made use of the common characteristics of the gymnasium and the large space of the exhibition center, which provided the basis for the use of the mutual borrowing space for sports activities and exhibition activities during the later operation, on the other hand, the function rooms of each major monomer building were intensively and organically arranged, which was conducive to saving space and forming overall operating conditions. In this case, considering the construction space and use characteristics of the single buildings, the decomposition and recombination were carried out to provide possibilities for the function juxtaposition of venues, which provided a useful reference for the functional design of urban sports facilities (Fig. 4-4).

The Beijing Olympic Wrestling Hau (Gymnasium of the China Agricultural University) designed by us was used as a venue for wrestling and sit-in volleyball during the 2008 Beijing Olympic Games and Paralympic Games, and after the

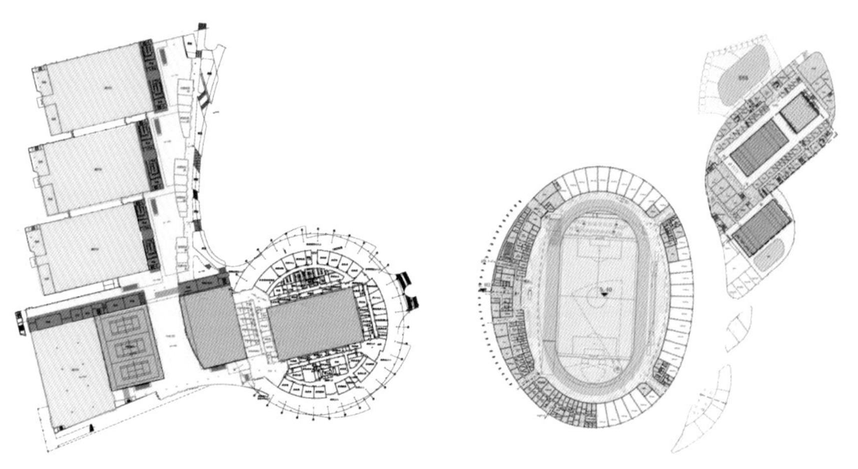

图 4-4　江门滨江体育中心：北区（左）和南区（右）
Fig.4-4　Jiangmen Binjiang Sports Center: north part (left) and south part (right)

我们设计的北京奥运摔跤馆（中国农业大学体育馆），在 2008 年北京奥运会和残奥会期间，分别作为摔跤和坐式排球比赛的场馆使用，赛后则移交给中国农业大学使用，因而场馆面临着不同类型的使用需求。对于此，我们采用了许多细节设计来满足功能转换。例如，对于主馆内运动员休息室的设计，考虑到对于摔跤比赛，运动员需要的休息室面积约为 20 平方米，而坐式排球比赛则需要更大的空间。我们对此采用了空间的可变设计，首先是把运动员休息室建成满足坐式排球比赛休息室要求的大房间，并设置了多个门，通过对大房间内的装修改建，增加隔墙，将大房间变为每间约 20 平方米的小房间，满足奥运会摔跤比赛期间的使用，尔后通过拆除隔墙，在短暂的时间内，恢复为残奥会坐式排球比赛期间的大休息室。赛后则再次进行少量改造，转换为给学校师生使用的活动用房。本案中，在设计阶段充分考虑到场馆功能转换的需求并对改造方法也预先设计好，使得实际运作过程中，在控制好改造的资金和时间成本的情况下，也能很好地满足赛时和赛后的不同使用需求（图 4-5）。

在我们设计的梅县体育馆（梅县文体中心）中，项目定位为集体育活动、大型文艺演

Games, it was handed over to the China Agricultural University for operation, therefore the project was faced with the different types of usage requirements. For this, we adopted many detailed design to meet the function conversion. For example, for the design of the athlete lounges in the main hall, taking into account the wrestling competition, the lounge area required by the athlete was about 20 square meters, while the sit-in volleyball game required larger space. For this purpose, we used a variable design method for the space. The first stage was to build an athlete's lounge into a large room that could satisfy the requirements of the sitting volleyball game lounge, then the large room was converted into a small room of about 20 square meters each, to meet the needs of the Olympic Games during the wrestling match, by modifications of setting up multiple doors and partitions. After the wrestling match, the partition walls were removed to return the space to the large lounge during the Paralympic sitting volleyball match in a short period of time. After the match, a small number of changes were made again and converted them into activity rooms for school teachers and students. In this case, the demand for conversion of venue functions was fully taken into account in the design phase, and the reconstruction method was also pre-designed so that in the actual operation process, under the premise of controlling the capital and time cost of the renovation, the different use requirements during the match and after the game could be well met (Fig. 4-5).

In the Meixian Gymnasium (Meixian Sports Center) we designed, the project was positioned as a multi-functional and comprehensive venue integrating sports activities, large-scale theatrical performances, exhibitions, and administrative services. Therefore, the venue design was required to satisfy the juxtaposition and compounding of various functions and the conversion between them. For this, we integrated multiple spatial strategies to meet different needs. For

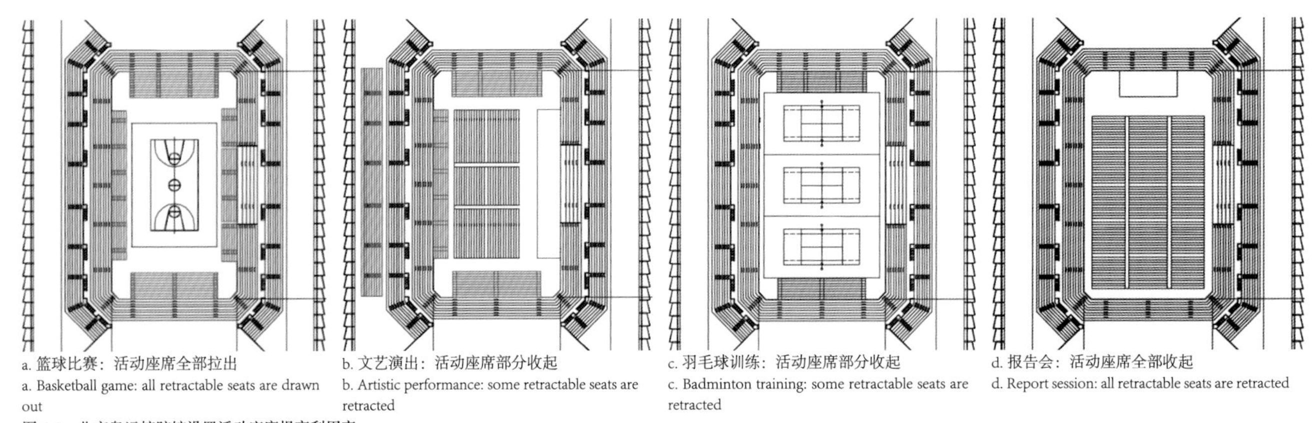

a. 篮球比赛：活动座席全部拉出
a. Basketball game: all retractable seats are drawn out

b. 文艺演出：活动座席部分收起
b. Artistic performance: some retractable seats are retracted

c. 羽毛球训练：活动座席部分收起
c. Badminton training: some retractable seats are retracted

d. 报告会：活动座席全部收起
d. Report session: all retractable seats are retracted

图 4-5　北京奥运摔跤馆设置活动座席提高利用率
Fig.4-5　The retractable seating of the Beijing Olympic Wrestling Hall improves its utilization

出、会展和行政服务于一体的多功能综合性场馆，因此场馆设计需要满足各种功能的并置、复合和转换。为此，我们综合多种空间策略以满足不同需求。例如，对于比赛厅和训练厅两个主要空间，在两者之间设置活动隔断，使得两空间可分可合，体育活动时可将两厅分隔开，展览活动时则可将隔断打开使得两厅空间合并，而文艺演出时则可将训练厅作为演出的舞台和后台；对于场馆座席布置，采用了不对称看台以保证未来大型演出和会议活动时更多的观众具有更好的观看质量；将辅助用房叠合为四层，集中布置在训练馆南侧（图 4-6、图 4-7）。

在项目推进过程中，在方案确定平面后，由于业主对使用要求的变化，场馆的辅助用房功能转移到靠山体一侧观众平台下的架空层，在举办高级别赛事时可通过搭建临时用房给予满足。而这部分集中布置的辅助用房的功能定位则经历了多次改变，从城市展览馆、青少年活动中心，直至建成后改造为行政服务大厅。这一过程中，其基本空间框架均能适应不同功能要求，也说明了体育场馆辅助用房的集中布置具有良好的灵活适应性，是对传统布置模式的一种突破。从建成后的使用效果看，梅县体育馆的功能和系统设置基本能满足各大型活动的需要，达到预期目标，也说明了本

example, for the two major spaces of the building, the competition hall and the training hall, a movable partition was provided between the two so that the two spaces could be separated and combined. They could be separated by closing the partitions during sports activities, while they could be combined during the exhibition activities by opening the partitions; the training hall could be used as a stage and backstage during theatrical performances; for the seating arrangements of the venues, we used asymmetrical stands to ensure that more audiences have better viewing quality during future large-scale performances and conference activities; the auxiliary rooms were overlaid into four floors and centrally arranged on the south side of the training hall (Fig. 4-6, Fig. 4-7).

After the plan was determined during the process of project, due to changes in the use requirements of the owners, the functions of the auxiliary facilities of the venues were transferred to the overhead layer under the audience platform on the side of the mountain, and temporary rooms could be set up to satisfy the functional demands during the holding of high-level events. The functional positioning of these centrally arranged auxiliary rooms had undergone many changes. It had been transformed into a city exhibition hall, youth activity center and finally the administrative service hall when it was completed. In this process,

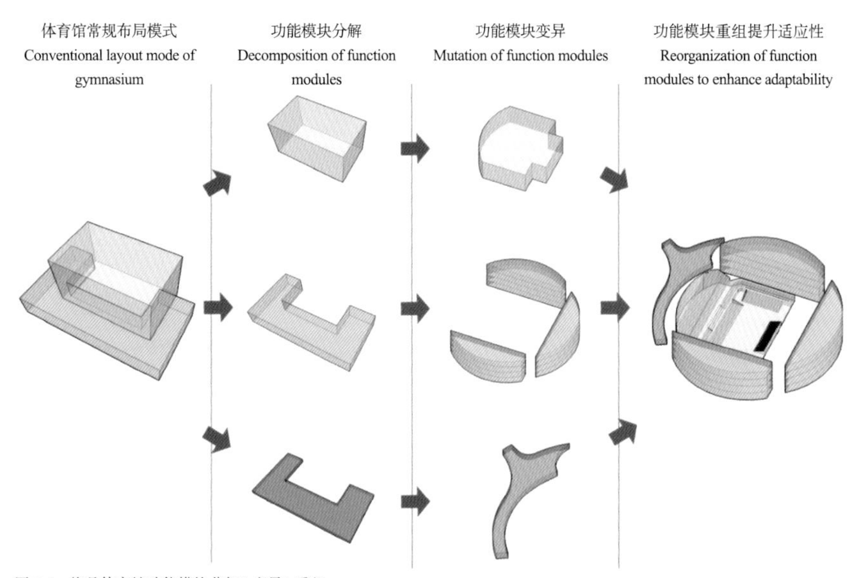

图 4-6 梅县体育馆功能模块分解 \ 变异 \ 重组
Fig.4-6 Functional module breakdown\variation\reorganization of the Meixian gymnasium

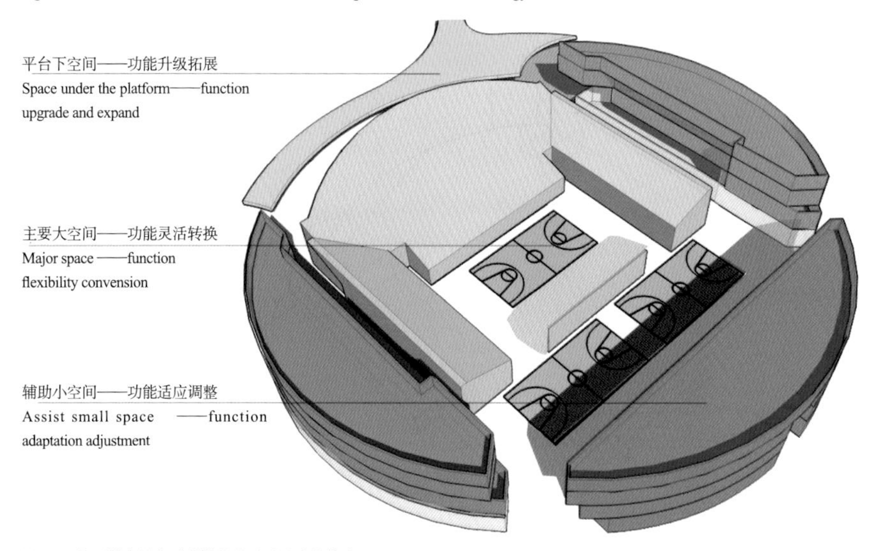

图 4-7 梅县体育馆各功能模块的灵活适应性策略
Fig.4-7 Flexible adaptability strategy for the functional modules of Meixian Gymnasium

案的空间利用策略具有良好的灵活适应性。

3. 场地尺寸选型

体育建筑的场地是其功能核心，是体育建筑特有的核心。体育建筑的每一种场地首先需要服从各自运动项目的规范要求，但每个体育馆在功能配置方面都有独特性，加上体育运动规范的不断发展变化，场地选型就成了体育建筑功能配置的第一要务了。

the basic space framework could adapt to different functional requirements, which also showed that the centralized arrangement of auxiliary facilities for gymnasiums had good flexibility and this was a breakthrough to the traditional layout mode. From the occupancy after the completion, the function and system settings of Meixian Gymnasium could basically meet the needs of large-scale activities and achieve the desired goals, which also showed that the spatial utilization strategy of this case had good flexibility and adaptability.

3. Selection for the sports field

The sports field of sports facilities is the core of its function unique in sports facilities. Sports venues shall first meet the specification requirements for their sporting events; yet due to the individual functional configuration of each venue and the evolving sports specifications, ground design has become the first priority in the functional program design of a sports facility.

关于体育馆的比赛场地选型，经过了一个发展变化的过程，其原因一方面是由于篮球等运动项目本身的发展。另一方面，也代表了对体育场馆灵活适应性的不断深入与发展的思考。从20世纪50年代开始，国内体育馆的场地尺寸设置主要以篮球场地为基准，然而70年代后期国际篮球运动规则发生变化，篮球场地尺寸从14m×26m扩大为15m×28m的过程中，许多国内体育馆都承受了痛苦的改造。在1980年代初，梅季魁教授带领团队在国内率先开展体育馆多功能场地的研究，提出（34～36）m×（44～46）m的多功能场地的建议尺寸，促进了场地选型的科学决策。1990亚运会主要场馆场地选型都没有出现遗憾。在2003年，我国第一部《体育建筑规范》由马国馨院士主持完成，较为全面地论述了相关的场地选型，明确了规范要求。

在高校体育馆方面，由于兼顾比赛和教学、课余活动，场地的适应性尤为重要。1988年，笔者在梅季魁先生的指导下，完成了关于高校体育馆的硕士论文，根据调研和分析提出34m×50m的高校体育馆场地选型方案（哈尔滨建筑工程学院硕士论文：高校多功能厅堂设计研究，学生：孙一民，导师：梅季魁，1988）。20世纪90年在哈工大体育馆首次运用34m×54m的场地选型。

进入新世纪，国际体操比赛场地要求发生变化，40m×70m的轮廓尺寸使大多数兴建于20世纪的体育馆无法满足国际体操比赛使用。2002年我们提出以40m×70m或48m×70m作为体育馆多功能场地选型的建议（华南理工大学硕士论文：体育馆多功能设计研究，学生：李玲，导师：孙一民，2002)。2002年，我们在华中科技大学体育馆的设计中采用40m×70m，使其成为武汉第一个满足国际体操比赛的场馆。2004年，北京奥运会摔跤馆、羽毛球馆作为奥运建筑，其场地的确定较为简

The field design of sports venues has undergone a developing changes, partly due to the development of sports themselves, or the insistent study of flexible and adaptable floors. Since the 1950s, the field size of venues for the gymnasium in our country was mainly based on the basketball courts. However, the rules of the international basketball movement changed in the late 1970s. Then the size of the basketball courts had expanded from 14m×26m to 15m×28m, which forced many domestic gymnasiums to have very difficult transformation. In the early 1980s, Prof. MEI Jikui and his team took the lead in carrying out multi-functional research on the field of sports venues, and proposed (34～36)m × (44～46)m as the recommended size of the multi-functional gymnasiums, which promoted the scientific decision-making on sport field size selection. There were no regrets in the sport field size selection for the main venues of the Asian Games in 1990. The Academician MA Guoxin presided over the China's first 'Sports Building Code' in 2003, which comprehensively analyzed the relevant field selection and defined the requirements of the specification.

As for the university gymnasiums, the adaptability of sports fields is particularly important, since they are expected to hold both games and campus activities. In 1988, the author of this publication completed a master's thesis under the instruction of Prof. MEI Jikui, in which a field of 34m×50m was proposed based on thorough investigation and analyses (Master Thesis of Harbin Institute of Architectural Engineering: Study of Designs for University Multifunction Halls, Student: SUN Yimin, Supervisor: MEI Jikui, 1988). In 1990s, 34m×54m field was first used in the design and construction of Harbin Institute of Technology Gymnasium.

In the new century, the requirements for international gymnastics competition changed so much that the 40m×70m dimension of most gymnasium built in the

单。但考虑到作为大学体育馆以及赛后多种利用的要求，我们以确保体育场馆最大的灵活性、适应性为原则，确定了 40m × 70m 的比赛场地，以便在赛后满足包括国际体操在内的各种比赛要求。随后我们在广州亚运体育馆工程的设计中也采用了相同的场地选型（图 4-8）。

通过比赛考察，场地的尺寸是能适应不同比赛场地以及各种仪式活动的，而对于教学训练使用，排球场地布置数量达到 5 个，羽毛球场地达到 18 个，乒乓球场地 25 个以上，可以适应高校的多种使用要求。而后我们针对奥运亚运赛事状况进行了实际观摩，并与其他尺寸场地的场馆进行了对比。某体操比赛馆，从设计渲染图可以看出其场地在 40m × 70m 的体操比赛场地之外空出了较为宽敞的空间。而实际使用看，过于宽敞的空间反而造成摄影记者

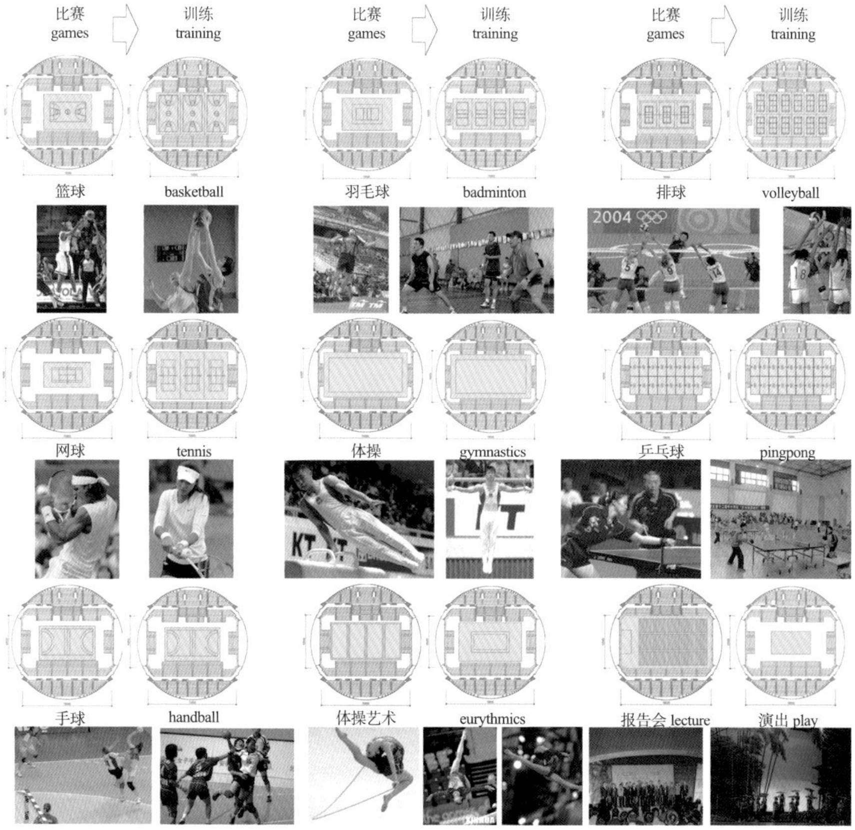

图 4-8　满足多功能使用的场地尺寸

Fig.4-8　Venue size meet multifunctional use

last century were no longer suitable for international gymnastics competitions. In 2002, we proposed 40m×70m or 48m×70m multi-purpose field for gymnasiums (Master Thesis of South China University of Technology: Study of Multi-purpose Design for Gymnasiums, Student: LI Ling, Supervisor: SUN Yimin, 2002). In the same year, we designed a 40m×70m field for the gymnasium of Huazhong University of Science and Technology, which was the first domestic venue for international gymnastics tournaments. In 2004, the fields for the Wrestling Arena and Badminton Hall for the Beijing Olympic Games were relatively easier to design, but considering their post-game use as university gymnasiums, we adhered to the principle of maximum adaptability and designed 40m×70m fields for the venues, so that they could meet the requirements for various tournaments, including the International Gymnastics Tournament after the Games. We also used the same type of field in the design for Guangzhou Asian Games Gymnasium (Fig. 4-8).

Competitions held on the venues proved that our field size is suitable for different kind of games and various ceremonial activities. To be specific, when it comes to teaching and training, the floor can accommodate up to 5 volleyball courts, 18 badminton courts, or 25 table tennis courts, let alone a variety of university multi-function requirements. We had compared the 40m×70m field with others during the Olympics and the Asian Games, among which a certain gymnastics competition hall, whose design rendering showed a spacious margin around the 40m×70m gymnastics court, brought visual interference between photojournalists and technicians at the tech-support platform. Moreover, the walkway behind the tech-support platform was not well utilized. The audience, distant from the players, might leave their seats for close-up photoshoots, which killed the atmosphere of the competitions. The 40m×70m field at Beijing University of Technology, is compact yet efficient, and more comfortable for the

图 4-9　华中科技大学体育馆的 40m × 70m 场地
Fig.4-9　The 40m×70m floor of the gymnasium at Huazhong University of Seience and Techaology

与技术台的视线交叉干扰。而技术台后部的人行通道利用率不高，观众又由于距离过远而离开座位递进拍照，使整个场馆空间气氛大打折扣。北京工业大学的 40m × 70m 场地，紧凑而有效，观众与场地更加容易互动，而华中科技大学体育馆的场地空间则说明，40m × 70m 的场地大小，为集会等多种活动提供便捷而高效率的空间布置（图 4-9）。

经过长期实际工程的应用，特别是亚运、奥运大型国际赛事的检验，我们认为 40m × 70m 的场地是多功能体育馆合理的场地选择。

audience to interact with the field. The field of the gymnasium at Huazhong University of Science and Technology shows that 40m×70m field is an efficient spatial arrangement for a variety of meetings and other activities (Fig. 4-9).

Practices over time, especially during the Asian Games and Olympic Games have convinced us that 40m×70m field is a reasonable choice for multi-functional sports gymnasiums.

设施利用策略

1. 活动座席

除体育比赛和文艺演出等有观演需求的活动类型外，体育场馆看台部分的日常使用率远远低于运动场地部分，过多的固定座席不但占据运动场地空间，而且容易造成长期空置现象。活动座席的设置可扩大体育场馆运动场地，在当今体育场馆中得到广泛应用，有些中小型场馆甚至可以全部采用活动座席(图 4-10、表 4-2)。

英国伯明翰的体育馆是一座可以用于室内田径比赛的大型场馆。这里每年举办的全英羽毛球公开赛是一项声名全球的重要赛事。羽毛球球场小、预赛阶段要求场地多，达到 6 块比赛场地，观众却比较少。到了决赛阶段，场地要求越来越少，最后只需要 1 个比赛场地和 1 处颁奖区域。而同时，观众越来越多。为了灵活调节空间，该馆大量使用活动座席，包括折叠拉伸和整体移动两种座席，同时用屋顶悬挂活动幕帘的方式根据赛程，围合出不同状态，不同规模的比赛观赏空间，这样的灵活变化状态贯穿全英赛的始终（图 4-11）。

Strategy for Facilities Utilization

1.Retractable seating

Except for the competitive sports events and theatrical performances that had a demand for watching area, the daily utilization rate of the venue stands is much lower than that of sports fields. Therefore, the excessive fixed seats would not only occupy sports space, but also cause long-term vacancy. The setting of movable seats can expand the sports fields and are widely used in today's sports venues. Some small and medium-sized venues can even use full movable seats (Fig. 4-10, Tab. 4-2).

The Barclaycard Arena in Birmingham, UK, is a large gymnasium that can host indoor track and field events. The All England Open Badminton Championships held in the arena annually is a major event of global fame. The difficulty in preparation for the event is that the preliminary stage games, watched by few audience, require up to six courts, whereas the final stage games, watch by a substantial amount of audience, require less field space and eventually down to one court and an award presentation area. To make flexible adjustment of the space, the arena makes extensive use of retractable seats, including slide-foldable and integral mobile seats. Movable curtains are hung from the roof to arrange different audience

图 4-10　活动座席
Fig.4-10　Retractable seating

各类活动坐席比较 表 4-2

类型	技术要点	土建和场地要求	应用
拆分组合式	将临时看台拆分为若干组，按单元移动至指定位置	需要预留一定的储存空间，地面设计要满足看台移动和站立的硬度需求	具有灵活性，适合兼顾文艺演出、集会会议功能的场馆使用
推拉折叠式	又称“壁纳式”活动看台，利用排间高差将看台折叠，一般是向后收回	收起后宽度约 1 米左右，可利用二层看台挑梁下空间给予隐藏	技术成熟，广泛应用于大中型体育场馆，可满足不同体育运动之间的转换
整体移动式	结合拆分组合式和推拉折叠式的特点，可将看台整体移动	对转换场地空间要求大，对土建要求也较高	实现转换的时间较长，适合大型体场馆，在国外应用较多
垂直升降式	通过升降、悬吊和翻转的方式实现转换	对土建要求高	技术相对不成熟，在国内外均较少应用

Comparison of various retractable seats Tab.4-2

Types	Technical points	Requirements for civil engineering	Application objects
split-combine	The grandstand can be split into a number of combination units that can be stored separately	It requires a certain storage space and the ground design to meet the hardness requirements for stand movement and standing	Flexible, and suitable for game venues with theatrical performance and conference functions
slide-foldable	Also known as the ‘Wanana’ event stand, the stands are collapsed using the height differences between the rows, and are usually taken back	After retracting, the width is about 1 meter, which can be used to hide the space under the two-story platform	Mature technology, widely used in large and medium-sized venues to meet the needs of different sports
integral mobile	Combined with the combination of split and push-pull folding, the entire stand can be moved	High demand for conversion space and civil engineering	It takes a long time for conversion and is suitable for large-scale venues. more applications abroad
vertical lift	Conversion by lifting, hanging, and flipping	High demand for civil engineering	Relatively immature technology, less applied at home and abroad

体育场馆总座席数中活动座席的比重成为评判场馆灵活性的重要指标之一，但目前活动座席应用还受到一定的技术限制。常用的活动座席包含拆分组合式和推拉折叠式两种类型，由于生产难度和安全隐患，其单元排数和座席数不能超过一定范围。此外，还有垂直升降式和整体移动式等类型，相对而言这些类型的活动座席在技术上尚不成熟，且对于土建和

图 4-11 英国伯明翰的英超赛场
Fig.4-11 The Barclaycard Arena in Birmingham

space for games of different importance according to the schedule. Such flexible adjustments are made through the entire Championships (Fig. 4-11).

The proportion of movable seats in the number of total seats in sports venues has become one of the important indicators to judging the flexibility of the venues, but the current application of movable seats for sports venues is subject to certain technical restrictions. The commonly used movable seats include types of ‘split-combine’ and ‘slide-foldable’. Due to production difficulty and security issue, the number of rows and seats cannot exceed a certain range. In addition, there are types such as ‘vertical lift’ and ‘integral mobile’, which are technically immature by comparison, and have special requirements for civil and site design. Therefore, their application range is far less than the former two types. With the trend of demand for expanding the sports fields, as well as advances in seats technology, movable seats have the prospect of promotion and application. This also requires the designers to make a reasonable assessment and selection on the levels of technology and application suitability based on the project conditions.

场地设计有特殊要求，因此应用范围远不及拆分组合式和推拉折叠两种类型。随着体育场馆运动场地扩大发展的需求趋势以及活动座席技术的进步，活动座席具有推广应用的前景，这也要求设计师根据项目情况对这一技术水平和应用适宜性作合理评估和选择。

2. 临时座席

除活动座席外，临时座席是提高体育场馆灵活适应性利用的另一个重要方法。对于这一方法的应用，通常存在两种情况，一种情况是在体育运动场地中布置临时座席，这种情况通常是在非体育类型的观演活动中出现。例如文艺演出和大型集会活动等，对于这类活动，表演或者演讲所需的舞台或者主席台空间远小于运动场地，而此时运动场地区域则具有观看舞台或者主席台的最佳视角。对于这类临时座席布置，采用轻型、便于搬动的桌椅根据需要进行摆设即可，但在方案平面设计时，需要预留用于储存这些桌椅的配套库房。另一种情况是搭建临时看台，这种情况通常是在举办体育比赛期间出现。由于赛时和赛后座席数量需求存在巨大差异，在赛时组装支撑结构和座席形成临时看台，赛后则将这些支撑结构和看台拆除和储存以备再次利用，这为解决体育比赛和日常使用之间的矛盾提供有效途径。

由于大型赛事在赛时和赛后在座席数量需求上存在巨大差异，上述的第二种策略越来越多地在因大型赛事而兴建的体育场馆中被采用，在 2000 年悉尼奥运会、2008 年北京奥运会和 2012 年伦敦奥运会的比赛场馆中，均采用了搭建临时座席满足赛事需求，并在赛后进行拆除的做法。

2000 年悉尼奥运会的国际水上运动中心设计，即以赛后功能为主，在单侧通过“加法”，设置了大规模的临时看台，这些看台在赛后拆除，并使得整个场馆的体量大幅缩小。

2. Temporary seating

In addition to the movable seats, temporary seats are another important method for improving the flexibility and adaptability of the sports venues. For the application of this method, there are usually two situations. One is the placement of temporary seats in the sports field, which usually occurs in non-sports activities, for example, theatrical performances and large-scale assembly activities. For such activities, the stage or podium space required for performances or speeches is much smaller than the sports fields, while the sports field area has the best view of the stage or the podium. For this type of temporary seating arrangement, those lightweight chairs and tables that are easy to move can be arranged according to the different demands, but in the case of plan design, there is a need to reserve a supporting warehouse for storing these tables and chairs. Another situation is to build a temporary stand, which usually happens during a sports event. Due to the huge differences in the number of seats during and after the game, the support structures and seats were assembled to form a temporary stand during the game, while after the game, these support structures and stands were demolished and stored for reuse. This provides an effective solution for the contradiction between the sports competitions and daily use.

Due to the huge difference in the number of seats in large-scale events during and after the games, the second strategy mentioned above is more and more used in the venues built for large-scale events. The venues for the 2000 Olympic Games in Sydney, 2008 Olympic Games in Beijing and the 2012

值的特别一提的是，本届奥运会采用了数个会展建筑改造为举重、摔跤和击剑，以及篮球、排球、羽毛球等比赛场馆，并为此配置大量临时座席，节约了建设成本（图 4-12、图 4-13）。

2008 年北京奥运会的国家游泳中心设计，在“水立方”中比赛池两侧布置 1.7 万个座席，赛后将高处的 1.1 万个临时座席拆除，改造为娱乐和健身等功能用房，保留低处的 6000 个固定座席，以满足日后运营需求，这一过程是在体育馆内部做“减法”，整个建筑外形不受影响（图 4-14、图 4-15）。

2012 年伦敦奥运会的水上运动中心设计，也是依据赛后功能为主，在比赛场地两侧通过“加法”，设置了 1.5 万个临时座席，并在赛后拆除，保留 2500 个固定座席满足日常运营需求，这一做法使得场馆体量大幅减小（图 4-16、图 4-17）。

Olympic Games in London have adopted the practice of setting up temporary stands to meet the demands of the events and to demolish them after the games.

The design of the 2000 International Aquatics Center at the Sydney Olympics took the post-game function as the main consideration object. Through the ‘addition‘ on one side, a large-scale temporary stand was set up, which were dismantled after the game and significantly reduced the entire volume of the venue One thing worth mentioning was that this Olympic Games has adopted several convention and exhibition buildings that were transformed into competition venues for weightlifting, wrestling and fencing, as well as basketball, volleyball, badminton and other activities, and allocated a large number of temporary seats for this purpose, which saved the construction costs (Fig. 4-12, Fig. 4-13).

The National Swimming Center of the 2008 Beijing Olympic Games was designed to arrange 17,000 seats on both sides of the

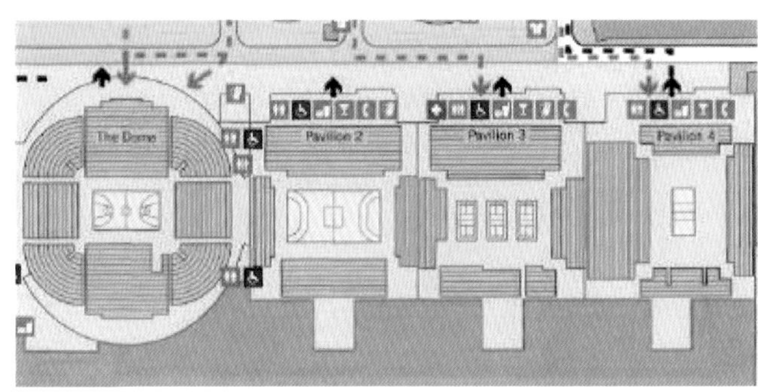

图 4-12　悉尼奥林匹克公园内皇家农业协会展馆赛时改造为篮球等比赛场地

Fig.4-12　The Royal Agricultural Society Hall at the Sydney Olympic Park is turned into venues for basketball and other games during the Olympic Games

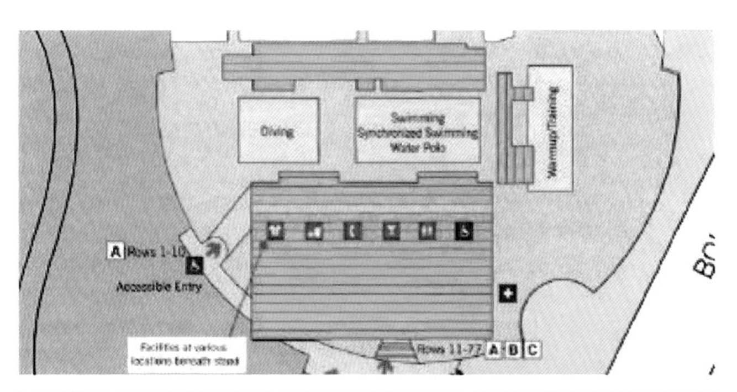

图 4-13　向公众开放的悉尼水上中心赛时为游泳跳水比赛场地，赛后向公众开放

Fig.4-13　The Sydney’s Aquatic Center open to the public provides venues for swimming and diving games during the Olympic Games and resumes to be open to the public match pool in the 'Water Cube'. After the match, 11,000 temporary seats of them at the higher position were demolished and transformed into function rooms for entertainment and fitness, etc. The 6,000 fixed seats in the lower position were kept to meet future operational demands. Compared with the former case, this process was doing 'subtraction' inside the gymnasium, while the entire building volume was not affected (Fig. 4-14, Fig. 4-15).

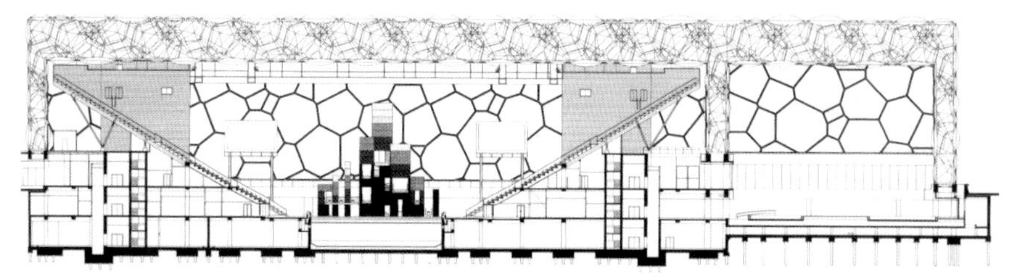

图 4-14　北京国家游泳中心奥运会赛时剖面
Fig.4-14　Sectional view of the Beijing National Aquatics Center during the Olympic Games

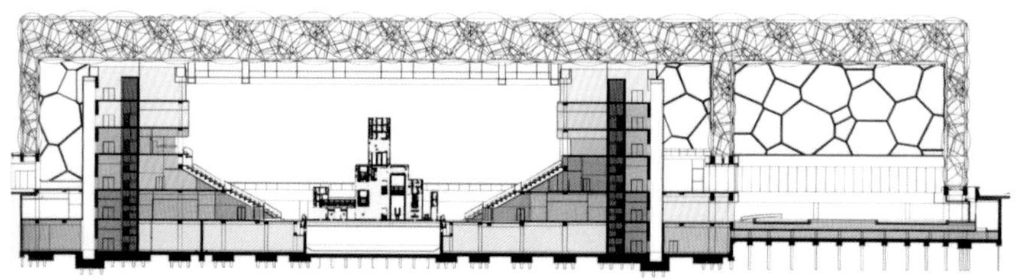

图 4-15　北京国家游泳中心奥运会赛后剖面
Fig.4-15　Sectional view of the Beijing National Aquatics Center after the Olympic Games

The design of the water sports center at the 2012 London Olympic Games was also based on the post-game function, where 'addition' was used to set up 15,000 temporary seats at the side of the venue, and they were demolished after the game, while 2,500 fixed seats were retained to meet the demand of daily operations. This approach had led to a substantial reduction in the volume of the gymnasium (Fig. 4-16, Fig. 4-17).

图 4-16　伦敦奥运水上中心主体空间与临时看台一体化设计
Fig.4-16　The Aquatic Centre for the London Olympic: the design of the integration of the main space and the temporary grandstand

图 4-17　伦敦奥运水上中心赛后将拆除临时看台
Fig.4-17　The Aquatic Centre for the London Olympic: remove the temporary grandstand after the Games

对比以上三个案例的具体策略，其中悉尼和伦敦奥运会的临时座席布置在体育场馆主体空间之外，在赛后将临时座椅及其附带的室内空间一起拆除，从而使得场馆的体量大幅缩小。北京奥运会的临时座席布置在体育场馆主体空间内部，赛后将临时座席拆除并替换为功能用房，但对于池厅仍然存在较大的体积浪费问题。相比之下，后者从外观上保持了场馆在赛时赛后的一致性，但这是否从本质上发挥了临时座席对场馆可持续运营的最大作用？仍有待斟酌。

Comparing the specific strategies of the above three cases, the temporary seats of the Sydney and London Olympic Games were arranged outside the main space of the gymnasiums, and these seats and the accompanying indoor space were demolished together after the game, resulting in a substantial reduction in the volume of the gymnasiums. The temporary seats of the Beijing Olympic Games were arranged in the main interior space of the gymnasium, and the temporary seats were demolished and replaced with functional housing after the match, and the large volume waste problem in the pool hall still existed. In contrast, the latter maintained the consistency of the pavilions after the match in appearance, but did it essentially play the greatest role of temporary seats in the sustainable operation of the venues? This question still waits to be answered.

五、技术理性

V. TECHNICAL RATIONALITY

体育场馆的技术需求特点

与其他类型的建筑相比，体育场馆的大跨度结构和大空间体量对技术设计和施工水平提出了更高的要求，往往成为同一时代最高建筑技术的代表之一，同时由于大量混凝土和钢材的消耗、大型结构和复杂设备系统的投入，大幅拉高其建设成本，也使得体育场馆往往成为一个城市最为昂贵的建设工程之一。体育建筑一旦启动并建设完成，拆除和改造难度和成本均较高，因而具有很高的不可逆性。

因此体育建筑建设特别地需要从“全过程”研究可持续性，从决策阶段充分考虑场馆建设和运营成本控制，在设计阶段进行结构、设备、体型、采光和通风的精明设计，才能在合理控制场馆建设成本的情况下，保证长期运营过程中的灵活适应性（图 5-1）。

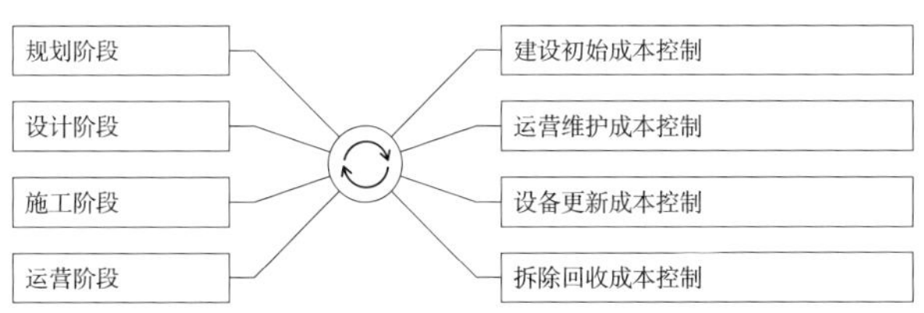

图 5-1　全寿命周期成本控制原则应贯彻各个环节

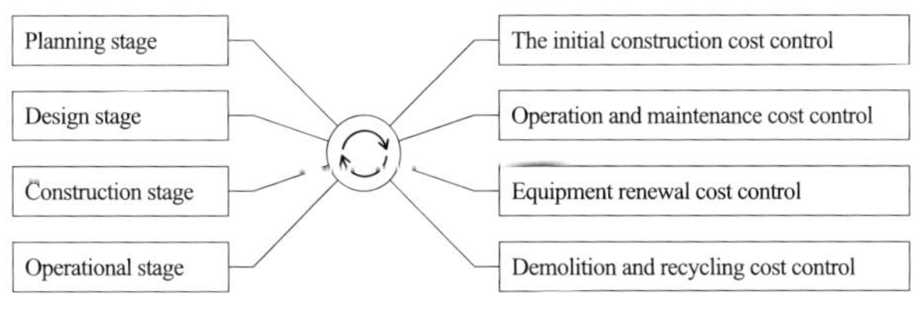

Fig. 5-1　Life cycle cost control principles should implement all aspects

Technical Requirements of Sport Venues

Compared with other types of buildings, the large-span structure and large space volume of sports venues place higher requirements on the technical design and construction standard, which make them usually become one of the representatives of the highest building technologies in the same era. Meanwhile, due to the consumption of a large amount of concrete and steel, and the investment in large-scale structures and complicated equipment systems, which have drastically increased the construction costs, and have also made the venues one of the most expensive construction projects of the city. Once the construction of sport venues is started and completed, the difficulty and cost of demolition and reconstruction are high, which make them highly irreversible. The demand for sports venues are divided into that for sports activities, including sports events and group activities, and non-sports activities, including cultural performances, meetings, exhibitions. The explicit and implicit demands makes it even harder to plan and design.

Therefore, the construction of sports facilities particularly needs to study the sustainability from the ‘whole process’, fully consider the control of construction and operation cost from the decision-making stage, and perform smart design of structure, equipment, volume, lighting, and ventilation at the design stage, so that the flexibility in long-term operations can be ensured under the premise of the reasonably controlled construction costs (Fig. 5-1).

应对思路

体育建筑在全寿命周期中，由于运营能耗、改造、设备系统维护和更新花费资金巨大，使得运营成本往往远高于初始建设成本。以 50 年的建筑使用年限作为计算周期，研究表明体育建筑建设初始投资占全寿命周期成本的 20% ~ 25% 左右。这也使得仅仅从建设初始投资进行成本控制，虽然可能在短期内节

Solutions

In the entire life cycle of sports facilities, the operating costs are often much higher than the initial construction costs due to the huge capital expenditures for operational energy consumption, renovations, maintenance and updating of the equipment systems. Taking 50 years of construction life as the calculation period, research shows that the initial investment in sports

省资金，但对于场馆长期运营而言，未必能保证总体成本的合理控制。因而需要在全寿命周期内系统考虑体育建筑建设成本，重视体育建筑“全过程”的精明控制。

我国体育产业发展尚处于起步阶段，运营市场尚未成熟，因而体育场馆的建设应集约节约，采用适宜性技术。一方面，在决策环节应立足于场馆建设作为城市公共服务设施的基本需求，尽量集约节俭地建设，进行科学合理的规模和建设标准定位，既要避免过度追求高标准和标志性，也不能为压低初始建设成本而造成功能的不足和运营成本的提高。另一方面，在技术手段上应结合我国国情，优先采用适宜性技术，实现体育建筑全寿命周期内节地、节能、节材、节水和环境保护，结合具体地域条件，采取科学实用的结构技术和体育工艺，综合被动式和主动式技术，既要避免盲目迷信昂贵的高技术和设备，又要避免技术配备不足而影响场馆的正常使用和可持续运营。

集约适宜的原则应贯穿到体育建筑立项、可行性研究分析与评估、决策、设计、施工到运营使用的整个过程中。进行精明决策和设计，

construction accounts for about 20%-25% of the entire life cycle. This also makes it not comprehensive to control the costs only from the initial investment in construction. Although it may save money in the short term, it may not guarantee reasonable control of the overall cost for the long-term operation of the sport facilities. Therefore, it is necessary to systematically consider the cost of sports facilities construction during the entire life cycle, and pay more attention to the smart control of the ‘whole process’ of the sports facilities construction.

The development of sports industry in our country is still in its infancy, and the operating market is not yet mature. Therefore, the construction of sports venues should be intensive and economical, and appropriate technologies should be adopted. On the one hand, the decision-making should be based on that the venues are constructed as the basic demands of the urban public service facilities, built intensively and thriftily as far as possible, positioned scientifically and reasonably in terms of the scale and construction standard, and avoid either the excessive pursuit of high standards and iconic, or a lack of functionality and increased operating costs caused by depressing the initial construction cost. On the other hand, the technological measures should be combined with China’s national conditions, and priority should be given to adopting appropriate technologies to achieve land-saving, energy-saving, material-saving, water-saving, and environmental protection throughout the entire life cycle of the sports facilities, adopting scientific and practical structures technologies and sports technologies in combination with specific local conditions and integrating both passive and active technologies, and avoid either blindly superstitious of expensive high-tech and equipment, or the lack of technical equipment that may affect the normal use and sustainable operation of venues.

The principle of intensivism and suitability

研究范围
设计策略
可持续目标
大跨度结构选型
普遍的结构形式
特别的建筑形式
设备与系统设置
集约适宜的设备系统选型
灵活适应的设备系统设计
容积与体积控制
单一功能体量的容积控制
多组功能体量的分解重组
自然采光和通风
主体空间自然采光
辅助空间自然采光
风压通风和热压通风
机械辅助式自然通风
全寿命周期内节地、节能、节材、节水和环境保护

图 5-2　集约适宜的技术设计策略

关注大跨度选型、设备与系统设置、容积与体积控制、自然采光和自然通风的可持续设计策略（图 5-2）。

should be applied throughout the entire process of the establishment of project, analysis and evaluation of feasibility studies, decision-making, design, construction, and operation. Smart decision-making and design should be conducted, focusing on the sustainable design strategies of large-span selection, equipment and system arrangement, capacity and volume control, natural lighting and ventilation, etc. (Fig. 5-2).

复杂与简单：大跨结构的精明选型

自从“现代建筑”产生，建筑与结构的关系就在纠结之中。“形式追随功能”、“形式反映结构”，被认为最基本的准则。有意思的是，形式似乎总是第一要务。体育建筑为代表的大空间建筑，受到结构技术的制约最多，产生的划时代杰作也最多。现代奥林匹克运动的发展基本上伴随了“现代建筑”运动的发展，一部体育建筑史，几乎是最富有特色的现代建筑结构发展历史。

二战后的体育建筑，大跨空间结构体系成为体育建筑形式创新的根本。许多建筑师的优秀作品中，结构师的作用显而易见。而其中最有趣的是意大利结构工程师奈尔维，结构工程

Complex and Simple: a Smart Selection of Large-Span Structures

Ever since the emergence of modern architecture, the relationship between architecture and structure has been in a tangle. ‘Form follows function’, ‘form reflects structure’ are considered the most basic criteria. Interestingly, form always seems to be the top priority. Large space buildings, represented by sports venues, though most crippled by the constraints of structural technologies, have produced the most of epochal masterpieces. As the development of modern Olympic movement has basically come along with the development of modern architecture movements, the history of sports facility is the most distinctive history of the development of modern architectural structure.

Among sports buildings erected after WWII, the structural system of large span space has underlain the innovation of forms for sports facilities. Structure engineers could take apparent credit for the outstanding works of many architects. And one of the most interesting ones is the Italian structural engineer Nervi, who, always outshined his architect partners. His well-known Palazzeto Dellospori of Rome is a perfect combination of structure, form and even construction process. The Florence stadium, though less known, is distinct for its totally architecturalized structural members, which is the marriage of structural engineers and architects (Fig.

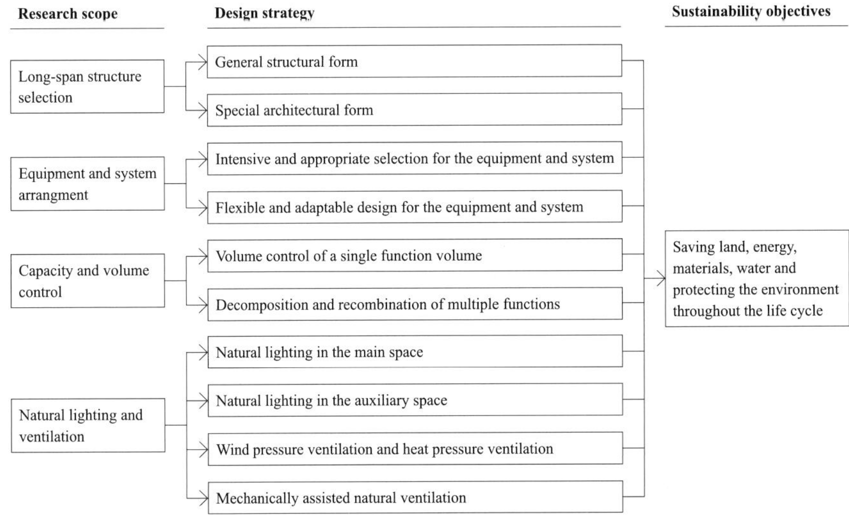

Fig. 5-2 Research framework for design strategy based on intensiveness and appropriateness

图 5-3　佛罗伦萨体育场
Fig. 5-3　The Florence stadium

图 5-4　东京代代木游泳馆
Fig. 5-4　The Tokyo Yoyogi natatorium

师出身的他，总是让他的建筑师合作者黯淡失色。在奈尔维的作品中，家喻户晓的罗马小体育馆将结构与造型乃至施工工艺都完美的结合。而鲜为人知的佛罗伦萨体育场则是将结构构件完全建筑化，结构工程师与建筑师完全融合（图 5-3）。东京代代木游泳馆用大跨度结构表现东方传统的意境，建筑上获得成功，但结构师却无法用悬索完成理想的曲面屋盖形式，最终采用劲性钢梁获得屋盖的曲面效果，作为结构工程师一直认为是遗憾之作（图 5-4）。同时也让丹下健三留下了虚假现代主义的尾巴。

今天，技术的进步已经到了让建筑师可以为所欲为的地步。以参数化为代表的数字化工具到来，建筑师更加沉浸在形式的提出和选择之中，而同时由于在工程中的合同地位，建筑师提出各种奇思妙想都会有结构师来实现。讲功能、讲造价几乎成了思想保守的同义词。以致于建筑师与结构师的合作关系濒临瓦解。

建筑设计关注形式是基本的。在结构技术受到各种限制的时候，产生过许多经典的杰作（图 5-5）。当结构技术发展到不受形态制约的时候，如果仍然侧重以形态作为建筑学基本价值取向，就会带来前所未有的问题。

在中国，“鸟巢”、国家大剧院与央视大楼的中标兴建，让建筑师追求形式的心态得到鼓舞。2003 年后的中国大型公共建筑设计投标

5-3). As an opposite example, the Tokyo Yoyogi natatorium swimming, which has a traditional oriental exterior through a large span structure, is an architectural success, but always a pity in the eyes of structural engineers, who eventually compromised the curved roof with stiff steel beams structure, as they failed with catenary structure (Fig. 5-4), for which the architect, Kenzo Tange, is sometimes criticized as a pretended modernist.

Today, the progress of technology have gone as far as architects can do whatever they want. The rapid development of computer technology and the advance of digital tools led by parameterization have indulged architects in novel forms so much that they overlook the reasonability in structural technology Architects are in an advantageous position in contracts to have their whimsy ideas realized by structural engineers. As a result, respect for function and cost have almost become conservative thinking, which pushes the cooperation between architects and structural engineers to the verge of collapse.

It is normal for architectural designers to pay attention to form. When structural technologies were under various constraints, many classic masterpieces were produced (Fig. 5-5). But when structure technologies have grown out of the formal constraints, if we still see form as the fundamental value of architecture, it will bring up unprecedented problems.

In China, the bidding and construction of the ‘Bird Nest’, the National Theatre and the CCTV building have encouraged architects to pursue their success in designing forms. The design bids submitted for large public buildings in China since 2003 have shown a very obvious trend of formalization. Design rely on forms rather than functions, leading to extravagant shells and irrational structures.

Internationally, the star effect within the

图 5-5　日本岩手县体育馆（左）、麻省理工学院小礼堂（右）
Fig.5-5　Japan Iwate Gymnasium (left), MIT chapel (right)

的形式化倾向十分明显，忽略结构理性，关注表皮的设计层出不穷。

国际范围内，建筑师的明星效应让建筑师追求的目标日益时尚，早年现代主义建筑师那种理性精神让位于当代建筑师的恣意与主观。2012 年伦敦奥运游泳馆在临赛前发现总数近 5000 名观众在跳水比赛的时候，将无法看到跳台，导致不得不退票的结果。分析图纸可以发现，部分屋盖的厚度缺乏结构依据，而体育馆建筑设计为了保证曲线的流畅感觉，采用了较为封闭的吊顶方案，使大跨空间结构的逻辑关系完全隐藏，更加加剧了视线遮挡（图 5-6，图 5-7）。

无独有偶，扎哈事务所的 2020 年日本奥运体育场建筑方案，是该公司在伦敦奥运游泳馆后又一个重要的体育建筑项目。日本东京 2020 奥运主体育场基地现场局促的城市空间并没有限制建筑师的畅想，结构技术也不足以影响建筑大师的强悍表现欲望，完全超出尺度的设计方案得以顺利提出，并奇迹般的获得评委认可。这一即刻引起轩然大波的设计，还没有深化就争议一片。抛开形式本身的争议，即便是作为奥运主比赛场的基本体育功能、结构选型的合理性都存在令人质疑之处。而后续的发展，也依然没有根本改变的迹象。

毕竟日本建筑界的工程技术思想是根深

architecture circle fed architects' desire for stylish and fashionable designs. The rational spirit cherished by early modernistic architects has given way to the arbitrary and subjective creation of contemporary architects. It is not accidental that in 2012, the Aquatic Natatorium was found shortly before the London Olympic Games that nearly 5,000 spectators would not be able to see the platform in the diving competition, to which the final solution was an embarrassing refund of tickets. An analysis of the drawings shows us that the thickness of a certain part of roof lacks structural analysis. In order to maintain the smoothness of the curved shell, the designer of the Natatorium has adopted a relatively enclosed ceiling plan to completely hide the structurally logical relationship within the span, which was proved to be hideous for visual connection inside the venue (Fig. 5-6, Fig. 5-7).

Coincidentally, the project developed by Zaha Hadid Architects for the 2020 Japan Olympic stadium, another important sports construction project of the firm after its London one, had provoked massive controversies even before the project developed. The cramped urban space of the site for the main stadium of the Japan 2020 Tokyo Olympic Game did not circumscribed the imagination of the architect; nor did structural technology stand in the way of her burning desire to express, which stimulated a design completely beyond the dimension, yet miraculously approvable to the judges. Aside from the controversies associated with form, the availability of the project's basic functions as an Olympic venue, as well as the reasonability of structure design, is already questionable. The subsequent development showed no sign of fundamental change.

After all, engineering and technical spirits were ingrained in the Japanese architecture circles. In the face of the tough and persevere star architect, the Japanese architect elites raised a collective objection

图 5-6　伦敦奥运游泳馆剖面

Fig.5-6　A sectional view of the London Olympic Aquatic Natatorium

图 5-7　伦敦奥运游泳馆屋顶

Fig.5-7　Roof of the London Olympic Aquatic Natatorium

图 5-8　日本东京 2020 奥运主体育场基地现场

Fig.5-8　The site of the main stadium of the Japan 2020 Tokyo Olympic Games

图 5-9　扎哈·哈迪德事务所方案

Fig.5-9　The scheme proposed by Zaha Hadid Architects

蒂固的，面对强悍固执的明星建筑师，日本明星建筑师们集体反对，终于推翻了扎哈事务所脱离城市实际的设计成果，让大型公共工程的建设重新获得符合建筑工程基本逻辑的机会（图 5-8、图 5-9）。

大跨结构选型设计是体育建筑创作的重要环节。合理的结构选型设计是体育建筑实现可持续发展的基础。但同时应认识到，结构技术只是建筑表达的手段而不是建筑表达的目的。作为手段的选择，其灵活适应性原则十分重要。好的结构选型，往往更注重立足于场馆本身整体的灵活适应性。脱离使用需求、不顾施工水平、片面追求所谓“结构的先进性”、脱离国情不顾施工难度的设计与可持续发展的理念相悖。

当结构技术的发展没有出现革命性的内容，而建筑形式却日益“表皮化”时，建筑师滥用材料，将自己没有逻辑的建筑形式奢侈的伪装包裹，也给结构和设备工程师提出昂贵复杂的难题，社会需要更加沉重的代价来建造实施。许多重大工程，由于建筑师的缘故人为加剧复杂程度，为我们提出了一个严肃的问题：

that overthrew the Zaha design which disobeys with the reality of the city. As a result, the Japanese peers regained the chance for large public works to follow the basic logic of building and construction (Fig. 5-8, Fig. 5-9).

The design for large-span structure is an important part of architectural creation. A reasonable design of structure is the basis of the sustainable development of sports facilities. But simultaneously, it is noteworthy that the structure technology is only an expressive means, rather than a purpose, of architecture. Therefore, to select a construction method, the principle of flexibility and adaptability is the primary guideline. A good structure tends to be more focused on the flexibility and adaptability of the venues as a whole. Designs divorced from user needs, regardless of the level of construction, in pursuit of the so-called structural advancement, neglect of construction difficulties in spite of the reality in China, are inconsistent with the concept of sustainable development.

While the development of structural technology has not shown anything revolutionary, and architectural forms are becoming increasingly surficial, architects tend to abuse materials and wrap their illogical architectural forms with extravagant camouflage, which leave expensive and complex challenges to structure and equipment engineers and a heavy price to be paid by the society for construction and operation. Many major projects over-complicated by architects bring up a serious problem: the complexity of construction should be something designs resolve, rather than something designs produce. Due to the advanced technology available nowadays that liberates architects, the complexity of works are exacerbated, engineering difficulties mounting, consumption of materials and resource growing, calling on us to adjust and substantiate the traditional architecture concepts.

建筑工程的复杂性，应该是精明设计去化解的目标，而绝非设计的结果。当今天先进的技术手段让建筑师无所约束，作品的复杂性加剧，工程难度日益提高，对材料和资源的消耗也持续增长，传统建筑学的理念到了必须调整充实的时候了。

一方面，结构设计要充分考虑功能使用的灵活性。以体育馆的多功能设计为例，为提高使用率，大中型体育建筑的赛后使用要求存在着相当大的不确定性，体育馆比赛厅除了进行体育比赛外，多数还兼顾文艺演出功能。香港红磡体育馆的文艺娱乐演出使用率占全年使用率 80% 以上，远大于体育比赛的使用场次。从国内大量体育场馆运营来看，文艺演出的功能需求是体育场馆重要的功能构成之一。随着文艺演出形式的多样化，演出舞台的设置方式也日趋丰富，这就给设计条件带来不确定性因素。正因为此，屋盖在预留荷载与灵活悬吊等方面对结构形式的选择提出新的要求。比赛厅大跨度屋盖结构所承担的荷载除了应考虑自重及相应照明音响等设备的重量外，还应具备一定的弹性余量，以满足运营过程中进行文艺演出搭设舞台吊挂相关设备的可能性。某著名体育馆建筑与结构设计整洁一体，但未能充分预计屋面荷载，特别是未来文艺演出的功能需求，在荷载计算时未留余量，建成使用后无法吊挂演出活动的额外灯光、电器设备，大大制约了场馆未来使用的灵活性（图 5-10）。

另一方面，结构构思应权衡技术条件、施工水平、工期要求、造价投资等方面的因素，结构选型构思上要充分考虑适应性和可选择性。主要有两类方法应予关注：

对于普遍使用的结构形式，应运用建筑手段予以特别调整，改变习惯认识，形成新的形象。

On one hand, structural design should fully consider functional flexibility. Take the multifunctional design of venues for example, for the purpose of better utilization, there is considerable uncertainty in the post-game use requirements of medium- and large-sized sports architectures, since the game halls of venues are expected to hold sports games and, a majority of them are also designed stages for theatrical performances. For instance, the Hong Kong Coliseum, of which theatrical performances account for more than 80% of the usage, heavily outweighing the percentage of sporting events. Judged from the status of many domestic sports venues, the functional need of theatrical performance is one of the most important functions of the venues. While the forms of theatrical performances vary, stage settings have also become increasingly diversified, which brings uncertainty to design conditions. Thus, requirements for reserved load and flexible configuration of ceiling suspension and other aspects have presented new challenges for the formal structure. In designing the load on the large span roof structure of a game, designers should consider the dead load and the load of lighting and audio devices, and leave a certain amount of flexible margin for possible hanging members over the stage during theatrical performance. In a certain well-known gymnasium, the architectural and structural designs are neatly integrated, but the failure to leave margin in the calculation of load in anticipation of the functional requirements of future theatrical performances, including additional lighting and electrical devices for performances from hanging over the stage, greatly restricts the flexible future use of the venues (Fig. 5-10).

On the other hand, technical conditions, the level of construction, schedule requirements, cost and investment and other aspects should be evaluated in structure designs, as structure designs require thorough consideration of suitability and availability. Primarily, there are two types of methods.

图 5-10　未能充分预计屋面荷载的某体育馆

Fig. 5-10　A gymnasium case that failure to leave margin in the calculation of load and to fully anticipate the roof load

图 5-11　中山市体育馆
Fig.5-11　Zhongshan Gymnasium

1992 年完成初步设计的中山市体育馆，在综合考虑地方施工水平与造价标准后，提出了桁架组合的建筑形象，利用桁架自身高度形成 4 处垂直布置的天窗。在主体结构外侧，用小桁架形成坡度较大侧面转折，丰富了屋面形体。总体效果突破了惯常的网架、桁架结构的平面性特征，而创新的造价成本并不高，并且便于建造（图 5-11）。

奥运摔跤馆同样考虑到紧凑的体型、节省的造价、可持续的低造价运行需求以及简便、易施工建造等多方面因素，采用了门式钢架的基本结构体系，利用桁架的高差形成采光和外部造型的元素。最终摔跤馆成为奥运造价最低的场馆（图 5-12）。

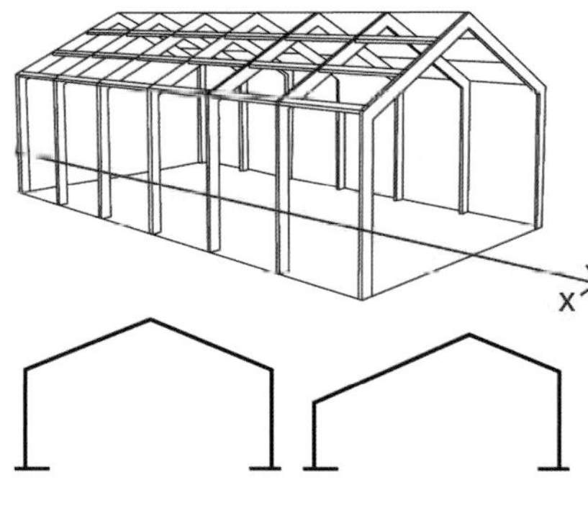

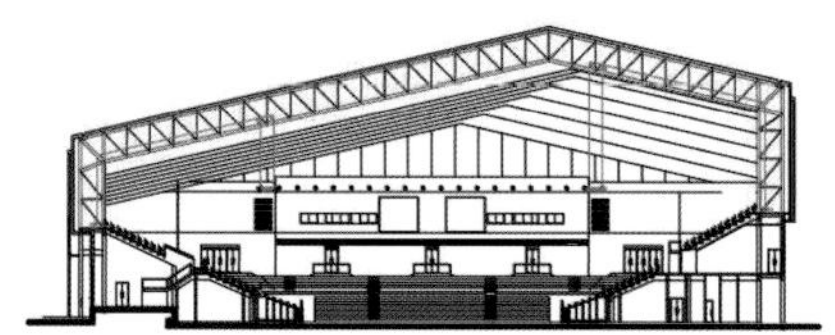

图 5-12　奥运摔跤馆结构图
Fig. 5-12　The stucture of the Olympic Wrestling Hall

亚运游泳跳水馆则是采用直线的桁架单元，经过参数化的调整，形成曲线和变化的层次感，辅以结合结构构件高差形成天窗系统，结构与材料的分解却极为简单，同样是极低的造价、简陋的施工，不仅在赛时使用顺利，而且是亚运工程中赛后利用率最高的场馆（图 5-13）。

For commonly used structure forms, constructions should employ special adjustments to escape from empirical thinking and to develop creative style.

Zhongshan Gymnasium, whose preliminary design was completed in 1992, a truss system, whose height could create 5 vertical skylights, was proposed according to local construction level and investment constrains. Outside of the main structure, a smaller truss system created a steep-sloped corner to enrich the roof design. While the overall effect broke away from the flatness of common grid and truss structures, the cost remained modest and the construction was simple (Fig. 5-11).

The designer of the Olympic Wrestling Hall, after careful evaluation of factors such as compact size, sustainable low-cost operation, convenient and simple construction, adopted a basic structural system of moment frame with additional trusses for lighting and exterior styling elements. The careful design makes the Wrestling Hall eventually the least costful of all Olympic venues (Fig. 5-12).

The Swimming and Diving Hall of the Asian Games used linearly arrayed truss units which, after parametric adjustments, formed a curve of layered variation. The trusses, combined with the elevation difference of structural members, formed the skylight system. Yet the adjustments did not make the decomposition of the structure and the materials any less simple, nor did the extremely low cost and simple construction trouble the smooth functioning of the Hall during the Games. On the contrary, the hall turned out to be the most frequently used venue after the Asian Games (Fig. 5-13).

对于特别的建筑形式，应该考虑运用成熟简单的结构单元与明确的几何关系作为母本，复杂形体尽可能简单处理。这样的好处是，结构单元清晰，便于计算。复杂形体的达成又可以适应我国的施工水平。

以奥运羽毛球馆为例。根据规模和奥运设计大纲的要求，设计确定了圆形的建筑基本体量，并努力形成飘逸轻盈的外观。作为由高校投资建设的奥运场馆，造价的控制极其严格，因此羽毛球馆方案在投标阶段确定了采用球网壳的结构形式。但基本结构单元是采用球型网壳与桁架结合，核心部分是球型网壳，两侧采用桁架以便形成边缘的轻盈形态。在方案深化阶段，业主为追求结构技术的先进性，提出了采用预应力张悬穹顶结构方案。由于核心结构单元的完整几何形态，使这一改动非常简便。最终，羽毛球馆屋盖主结构选择新型预应力弦支穹顶结构体系，成为2008北京奥运场馆中最为先进的结构形式，建成时是世界上当时跨度最大的弦支穹顶结构，工程用钢量62kg/m^2，造价并没有攀升，据北京市公布的资料，羽毛球馆的造价是第二节省的（图5-14）。

For special building forms, constructions should rely on proven and simple structural units and clear geometric relationship as reference to handle sophisticated shapes as simply as possible. The advantage is that the neatness of structural units will make calculation an easy job, while complex shapes may be adapted to the level of construction in China.

Take the Olympic Badminton Hall for example. Subject to requirements of scale and the design brief of the Olympic Games, the designers defined a circular massing in attempt to create an elegant and light appearance. As an Olympic venue invested by a University, the cost control of the gymnasium was extremely strict, hence a reticulated shell structure was chosen at the tendering stage. The basic structure was a combination of spherical reticulated shell and trusses, where trusses were arranged on both sides of the central shell to emphasize the lightness at the edges. As the project developed, the employer, in pursuit of advance structural technology, proposed to adopt a pre-stressed suspend-dome. The geometry of the core structure made this change of structure simple. Eventually, the new structural system of the pre-stressed suspend-dome was selected as the main structure of the roof of the Badminton Hall. It became the most advanced structure among the 2008 Beijing Olympic venues. When the gymnasium was completed, the structure was the world's largest span suspend-dome structure at the time of a relatively low cost, with a consumption of 62kg/m^2 structural steel. It was the second most economic efficient according to statistics released by the Beijing municipality (Fig.5-14).

Another example of rational structure design is the Nansha Gymnasium of the Guangzhou Asian Games. In light of the complex loading conditions of the Nansha Gymnasium (such as large wind load and other uneven live loads, etc.), the design proceeded from the formal and functional

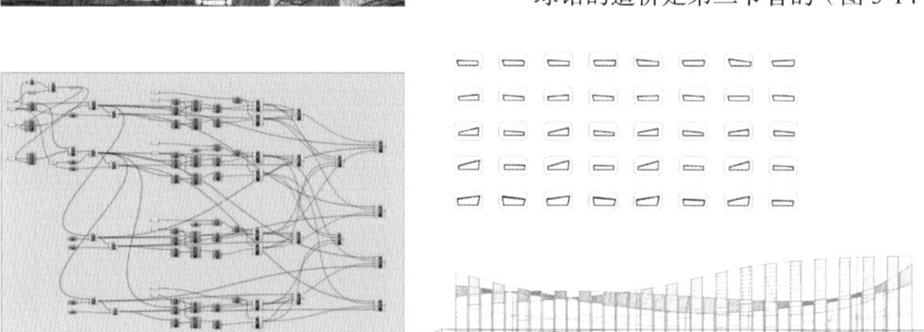

图5-13 亚运游泳跳水馆结构图
Fig.5-13 The structure of the Swimming and Diving Hall of the Asian Games

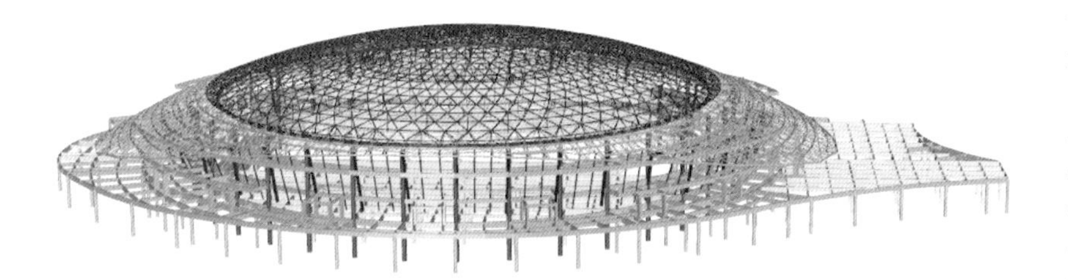

图5-14 奥运羽毛球馆结构图
Fig.5-14 The structure of the Olympic Badminton Gymnasium

广州亚运南沙体育馆的设计，建筑造型和体育功能的需求出发，结合南沙体育馆荷载条件复杂（比如较大的风荷载和其他不均匀活荷载等）的特点，取意于滨海地段的海螺造型，但结构选型非常理性。核心单元的球面与外围屋面的桁架共同形成曲折生动的造型，屋顶的采光带蜿蜒浮动。在计算机辅助下，外观复杂多变的屋面，结构简洁合理，采用了天然具有的优良整体结构稳定性的新型双重肋环 - 辐射形张弦梁结构，让屋盖空间显得轻盈通透，从而取得了较好的经济效益和社会效益（图 5-15）。

requirements and ended with conch shape symbolizing the coastal nature of Nansha. The spherical surface of the core and the truss of the peripheral roof join to create a convoluted and vivid shape and a winding lighting band on the roof. With computer technologies, the complex and variable shape of the roof is resolved with a simple and reasonable structure. The new double-rib radial beam-string structure, with intuitive structural stability, gives the roof a light and transparent spatial quality, thereby producing great economic and social benefits (Fig.5-15).

灵活与适应：设备与系统的精明设置

采暖空调、给排水、电气、照明、通风和智能化等专业为体育建筑运行提供了设备系统支持。建筑设备与系统的精明设置对场馆内部声、光、热等物理环境和建筑能耗等有关键性影响，对场馆的灵活适应性运营和可持续发展至关重要。

Flexible and Adaptable: Smart Configuration of Equipment and System

The professional cooperation with the heating and air-conditioning, water supply and drainage, electrical, lighting and ventilation and intelligence provide equipment and system support for the operations of sports facilities. The smart arrangement of construction equipment and systems has a crucial influence on the physical conditions such as the sound, light and thermal environment, and the building energy consumption of the venues, and is crucial to the flexible and adaptable operations and sustainable development of the venues.

1. 注重集约适宜的设备系统选型

竞技体育，尤其是高等级的体育比赛项目，对场馆室内温湿度、采光和通风条件具有各自的特殊要求，且通常远比日常的体育活动严格。按照这等规格兴建的体育场馆，一方面由于对于模式过于专业化的设备和系统，将不适合用于其他类型的项目中；另一方面赛后

1. Intensive and suitable selection of equipment and system

Competitive sports, especially those high-level sports competitions, have their own special requirements for indoor temperature, humidity, lighting and ventilation, and are usually far more stringent than the daily sports activities. Sports venues constructed in accordance with these specifications, on the one hand, will not be suitable for other types of activities as the equipment and systems are too specialized. On the other hand, the equipment and system configuration standards have exceeded the standard of daily operations after the match, which also increases the cost of

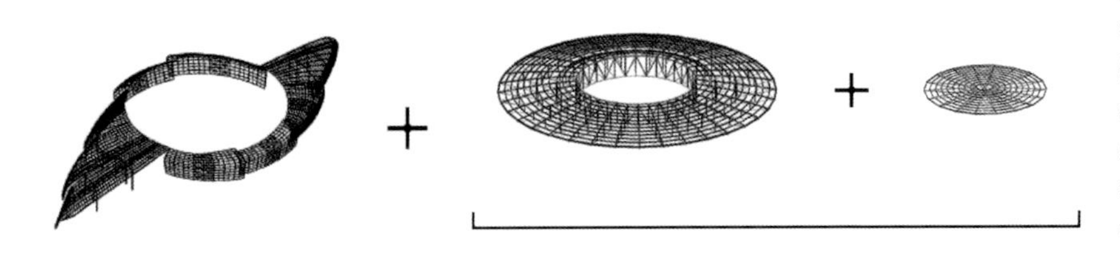

图 5-15　广州亚运南沙体育馆结构图

Fig.5-15　The structure of the Nansha Gymnasium of the Guangzhou Asian Games

日常运营而言，设备和系统的配置标准过高，导致维修和替换的成本也随着提高。注重体育场馆分级化，采取集约适宜的设备和系统，是应对以上问题的有效途径。

以北京奥运会羽毛球馆为例，该馆为承办 2008 年奥运会羽毛球比赛而兴建，赛后作为北京工业大学和专业训练队的活动基地。奥运会羽毛球比赛对比赛场地风速和照度等有着严格要求，如风速需要控制在 0.2m/s 以下，为满足电视转播，要求水平和垂直照度都达到 1200lx，色温大于 5000K，这等规格的需求与赛后的训练活动和师生日常教学与锻炼活动需求之间存在较大差异。为此，工程师设置了上下分层分区的空调系统，以及可以在比赛和训练两种模式之间切换的照明系统，成功地化解了以上矛盾，满足了赛时赛后对设备系统的不同需求（图 5-16）。

设备系统选型的集约适宜是实现建筑可持续性的基本策略之一，这对于体育建筑而言尤为重要，作为大型公共建筑和城市投资项目，

图 5-16　北京工业大学体育馆施工

Fig. 5-16　The Construction of the gymnasium of Beijing University of Technology

maintenance and replacement. Focusing on the grading of sports venues and adopting intensive and appropriate equipment and systems are effective ways to deal with the above problems.

Taking the Beijing Olympics Badminton Hall as an example, it was built to host the 2008 Olympics badminton competitions, and it serves as the base for the activities of Beijing Industrial University and the professional training teams after the match. The Olympics badminton competition had strict requirements on the wind speed and illuminance of the competition fields. For example, the wind speed must be controlled below 0.2m/s, and to meet the television broadcast, both horizontal and vertical illuminance were required to reach 1200lx and the color temperature was greater than 5000K. There was a big difference between the post-game training activities and the daily teaching and training demands of the teachers and students. To this end, engineers set up an air-conditioning system with separated sub-zones of the upper and lower layers, and a lighting system that could switch between the competition and training modes, which successfully resolved the contradictions, and met the different requirements for equipment systems after the game (Fig. 5-16).

The intensive and suitable selection of equipment and systems is one of the basic strategies for the sustainability, which is particularly important for sports facilities. As a large-scale public building and urban investment project, the capital investment is large, and the government attaches great importance and often places great hopes, so that care should be taken to avoid the problem of over-positioning due to the subjective decision-making. The intensive and suitable selection of equipment selection and system design is an important part for the sustainable operation of the venues, so it needs to be fully discussed in the decision-making stage. This process can also provide reference and guarantee for

资金投入大，受政府重视并往往寄予厚望，尤其要注意避免由于主观决策带来定位过高的问题。设备选型和系统设计的集约适宜是场馆运营可持续性的重要组成部分，因而需要在决策阶段给予充分的探讨，而这一过程也能为项目的合理定位提供参考和保证。在2009年的梅县体育馆（梅县文体中心）项目中，建设方一开始提出“国际标准”的定位。作为设计方，我们考虑到当地近期举办高级别国际赛事的概率不大，因此结合使用方实际需求，确立“按乙级场馆定位，未来通过较小代价改造实现甲级场馆功能”的原则。并基于此进行了合理的整体方案设计，避免由于盲目拔高建设标准而造成场馆建设和可持续运营的不必要负担。

the reasonable positioning of the project. In the Meixian Gymnasium (Meixian Sports Center) project in 2009, the constructors put forward the position of ‘international standards’ at the beginning. As a designer, we considered that the probability for this project to hold a high-level international competition in the near future was small. Therefore, we combined the actual will of users, and proposed the principle of ‘position according to the Grade B venues, but meet the demands of functions of Grade A through a lower-cost transformation in the future’. Based on this, a reasonable overall plan was designed to avoid the unnecessary burden of construction and sustainable operation of the venues due to blindly raising the construction standards.

2. 注重灵活适应的设备系统设计

随着体育运动的不断发展，体育赛事对体育场馆物理环境要求也不断地精细化，这也对场馆设备系统提出越来越高的要求。尤其对于为举办高级别比赛项目而兴建的体育场馆，往往需要配置昂贵的设备系统，而实际情况是绝大多数体育场馆在其寿命周期内并没有多少举办体育比赛的机会，这就可能给体育场馆赛后日常运营造成设备闲置和浪费问题，甚至是维护和更换的长期负担。另外一种可能是，体育场馆为了适应多功能利用的趋势而做出一定的空间架构调整，而此时的设备系统已无法跟上步伐。

随着城市环境发展、社会经济进步、体育运动自身的规则演进以及多样化的非体育活动对大空间的使用需求，体育建筑必须具备灵活适应性以满足需求的不断变化。灵活适应性的设备系统设计是重要策略之一，对这一策略的重视和提倡实际上在《体育建筑设计规范》等设计要求和指导中已有一定的强调，如在《规范》的10.2.5中“比赛大厅有多功能活动要求时，空调系统的负荷应以最大负荷的情况计算，并满足其他工作情况时调节的可能性”；

2. Flexible and adaptable design of equipment and system

With the continuous development of sports, the physical conditions of sports facilities are constantly being refined, which also puts forward higher and higher requirements for the equipment and systems of the venues. For those venues built to hold high-level competition projects, expensive equipment systems are especially required. However, the actual situation is that most sports venues do not have much chance of holding sports competitions during their life cycle, which may cause problems of idle and waste of equipment to the daily operation of the venues after the games, and even the long-term burden of maintenance and replacement. Another possibility is that the even venues realize that they have to make some adjustments to the space in order to adapt to the trend of multi-purpose utilization, the equipment and system are unable to keep pace at this time.

With the development of urban environment, socio-economic progress, the evolution of the rules of sports and the use of diverse non-sports activities in large spaces, sports facilities must be flexible and adaptable to meet the changing

10.2.6中“大型体育馆比赛厅可按观众区与比赛区、观众区与观众区分区设置空调系统”和“乙级以上游泳馆池区和观众区也应分别设置空调系统”；10.2.7中“体育馆比赛大厅当采用侧送喷口时，宜采用可调节角度及可变风速的喷口。特级、甲级体育馆大厅的气流组织，应满足不同比赛时进行调节的可能性”；10.3.4中“体育建筑和设施的照明设计，应满足不同运动项目和观众观看的要求以及多功能照明要求；在有电视转播时，应满足电视转播的照明技术要求；同时应做到减少阴影和眩光，节约能源、技术先进、经济合理、使用安全、维护方便”，分别指出了空调系统和照明设计等方面设备系统设置的灵活性设计需求。

在实践中，为场馆配置面向日常运营的设备系统，而在体育比赛、大型文艺演出等需求高峰时段，租赁临时设备可以有效地化解赛时赛后设备系统需求的矛盾。这些特殊活动往往开展时间短促，对设备系统需求陡增，且配置还要求有一定的针对性，因此针对特定活动租赁专用灯具、音响、活动计时记分牌和活动电源等临时设备可以有效满足活动需求。例如2005年在北京首都体育馆举办苏迪曼杯羽毛球比赛时，就租赁了2002年在广州举办汤尤杯羽毛球比赛时所用的专用灯具。临时租赁设备的方法有利于满足多样化的功能需求，提高设备利用率，降低设备建设、维修和改造的成本，从而减轻场馆运营负担。

除设备系统的灵活设置外，体育场馆设计和建设中，预留好临时设备系统的安装条件是发挥其功能的必要前提。这也要求场馆在设计和建设时做好充足的准备，保留设备系统的开放性，预留临时设备的接口，适当加大设备容量参数，保证设备系统的可更新性。例如，为了满足比赛大厅不同类型活动的不同照明和音响等需求，常见的做法是在屋盖结构上设置可以移动的设备器材吊挂系统。设备系统可以

demands. Flexible and adaptable design of the equipment and system is one of the important strategies, which has been actually been emphasized and promoted in the design requirements and guidance of the ‘Sports Building Design Code’, such as the article 10.2.5 ‘When the competition hall has a multi-functional activity requirement, the load of the air-conditioning system should be calculated with the maximum load and meet the possibility of adjustment in other work conditions’; in article 10.2.6, ‘Air-conditioning system of the large gymnasium competition hall shall be set up for the division, based on the division between the competition area and audience area, and among the different audience areas’ and ‘Air-conditioning system of the pool area and audience area for the swimming pool above Class B shall also be set up for division’; and article 10.2.7 ‘When using the side vent air supply in the competition hall of gymnasium, the air-conditioning system should be set with adjustable angle and variable wind speed vents. The air-flow organization of Class superfine and Class A gymnasium halls should meet the possibility of adjustment during different competitions’; article 10.3.4 ‘Lighting design of sports facilities should meet different sports programs, audience requirements for viewing and multi functional lighting requirements; in the case of television broadcasts, the lighting technology requirements for television broadcasting should be met; meanwhile less shadows and glare, energy saving, advanced technology, economical rationality, safe use, and easy maintenance should be achieved’, which points out the need for flexible design of equipment of system for the air conditioning and lighting system.

In practice, for venues that are equipped with equipment and system for daily operations, the lease of temporary equipment, in the peak hours of sports events, large-scale theatrical performances, etc., can effectively resolve the contradiction in equipment requirements after the game.

随着吊挂系统的灵活控制，实现水平移动或者垂直升降，使得声光系统可以根据使用功能灵活变换，这为场馆的多功能利用，特别是对于文艺演出活动的开展提供了必要条件。日本横滨室内比赛场在屋顶顶棚设置了600个吊点，每个吊点可承载负荷为1-9t，可以根据需要布置不同的音响和照明设备，满足不同使用需求的灵活切换。我们设计的中国农业大学体育馆，是为2008年北京奥运会摔跤比赛而兴建的，赛后移交学校运营。在设计之初即考虑到部分办公附属用房赛后改造为游泳馆的可能性，因此在地下预留好泳池和水处理机房，并敷设好相应的设备管线，保证赛后改造的便利性，节约了时间和资金成本（图5-17、图5-18）。

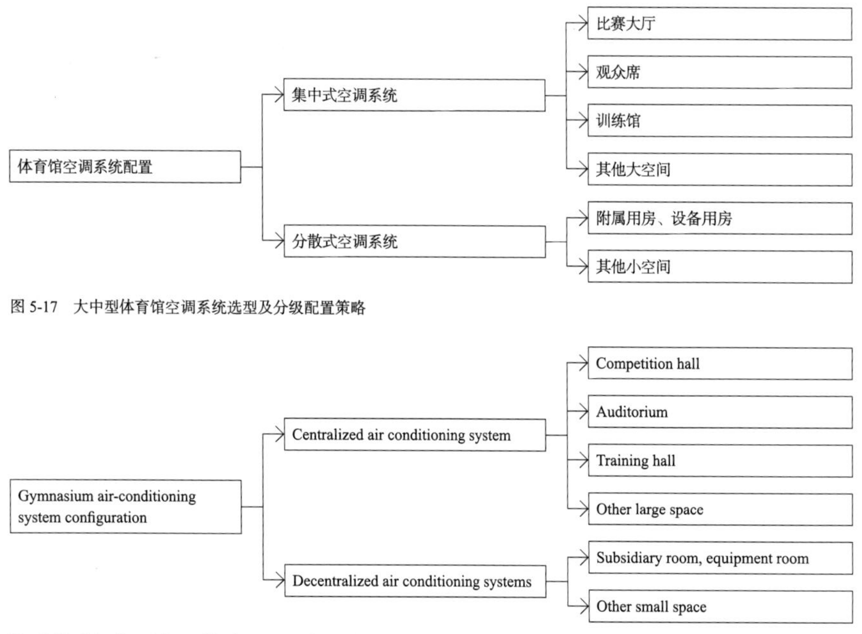

图5-17　大中型体育馆空调系统选型及分级配置策略

Fig. 5-17　Selection of air conditioning systems for medium- and large-sized gymnasiums and strategy of graded configuration

These special activities are often carried out for a short period of time, their demand for equipment and system increase sharply, and the configuration also requires certain pertinence. Therefore, temporary equipment such as special lighting fixtures, audio equipment, activity timing scoreboards and active power supplies for specific activities can effectively meet the demand for such activities. For example, when the Sudirman Cup badminton game was held in the Beijing Capital Gymnasium in 2005, the special lamps used in the 2002 Tongyu Cup badminton game held in Guangzhou had been leased. The method of temporarily leasing equipment is conducive to meeting diversified functional requirements, increasing the utilization rate of equipment, reducing the cost of equipment construction, maintenance and renovation, thereby reducing the burden on the operation of the venues.

In addition to the flexible setup of the equipment and system, in the design and construction of the sport venues, the installation conditions of the provision for a temporary equipment and system are necessary prerequisites for its function. This also requires that venues be adequately prepared during the design and construction process, preserve the openness of equipment systems, reserve interfaces for temporary equipment, appropriately increase equipment capacity parameters, and ensure the renewability of equipment and system. For example, in order to meet the different lighting and audio demands of different types of events in the competition hall, a common practice is to install a movable equipment hanging system on the roof structure. The equipment system can be moved horizontally or vertically according to the flexible control of the hanging system, so that the acoustic and optic system can be flexibly changed according to the different demands, which provides the necessary conditions for the multifunctional use of the venue, especially for the development of theatrical performances.

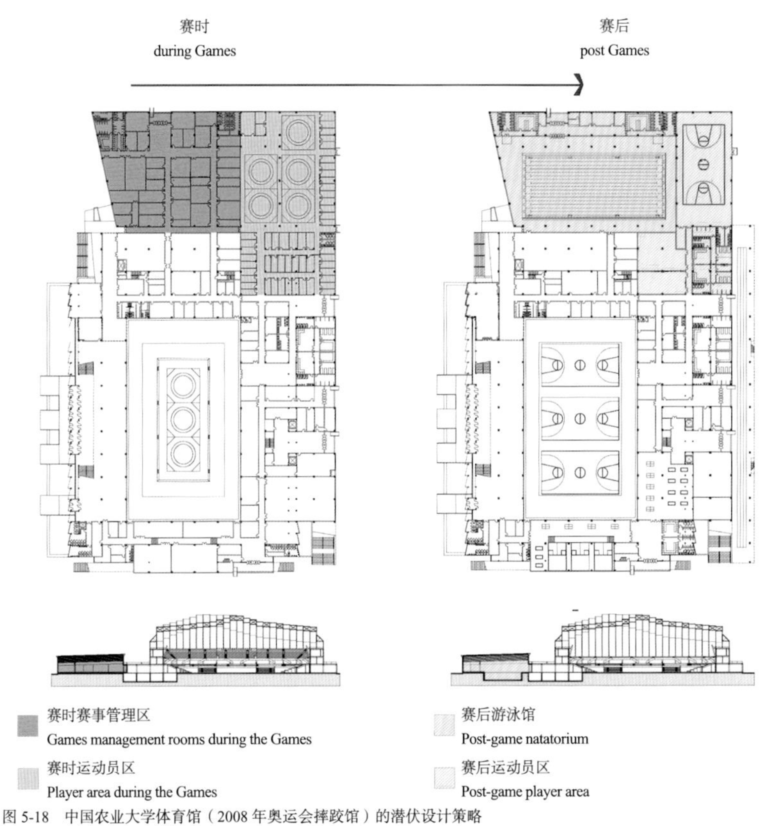

图 5-18　中国农业大学体育馆（2008 年奥运会摔跤馆）的潜伏设计策略

Fig.5-18　The latent design strategy for the gymnasium at China Agricultural University (2008 Olympic Games Wrestling Hall)

The Yokohama indoor competition venue in Japan has 600 hanging points on the roof of the building and each point can carry a load of 1-9t, so that different sound and lighting devices can be arranged as required to meet the flexible switching requirements for different applications. The China Agricultural University Gymnasium we designed was built for the 2008 Beijing Olympic Games wrestling competition and handed over to the university after the match. At the beginning of design, the possibility of converting some office-subsidiary houses into swimming pools after the match was considered. Therefore, pools and water treatment rooms were reserved in the underground, and appropriate equipment pipelines were laid to ensure the convenience of the post-game reconstruction, which saved considerable time and money costs (FIg. 5-17, Fig. 5-18).

Appropriate and Slim: Smart Control of Capacity and Volume

The capacity is closely related to the indoor physical environment control and energy consumption of the venues. Some sport activity projects have particularly strict control requirements for hygrothermal environments, so that too larger indoor space will increase the difficulty of HVAC systems and lead to the raise of energy consumption. For an active project with special requirements for the acoustic environment, an excessively large volume will also cause an increase in room acoustics control difficulty and equipment costs. Compared with the investment in technical equipment, the strategy of the capacity control not only saves initial construction investments, but also helps to reduce the daily operating costs (Fig. 5-19).

The main space of a sport venue is often a single large-span space, which is different from ordinary civil buildings. Even if the sport venue exceeds 24m in height, it is not considered as high-level either. With the increasing construction investment and the

得体与修身：容积与体积的精明控制

体育场馆的容积与室内物理环境控制和能耗密切相关。有的体育活动项目对热湿环境有特别严格的控制要求，因而室内空间过于高大将会增加暖通空调系统难度，并且导致能源消耗的提高。对于声环境有特殊要求的活动项目，过大的体积同样会造成室内声学控制难度和设备成本的增加。相比于技术设备投入，场馆容积控制的策略不仅能节约初始建设成本，更有助于降低日常运营成本（图 5-19）。

体育建筑的主体空间往往为单层的大跨度空间，区别于一般民用建筑，体育建筑即使高度超过 24m，也不算作高层。在建设投入不断增加，标志性被过于强调的今天，出于造型目的，设计师容易先入为主地用表皮和外壳“罩住”主体空间，而这一手法容易导致体育

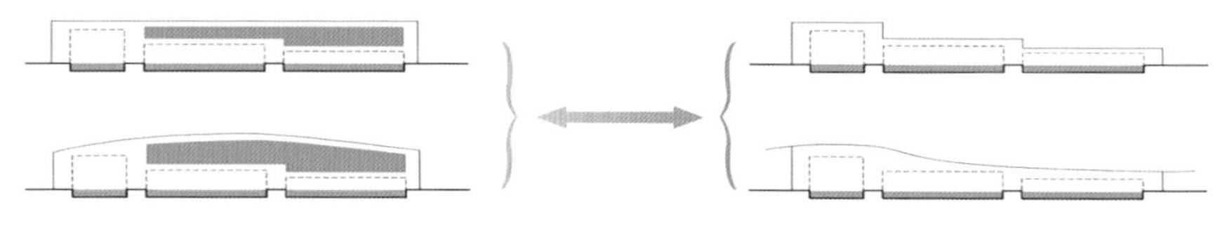

图 5-19 容积控制策略分析
Fig.5-19 Analysis of Volume Control Strategy

场馆的体量远远大于实际体育运动的空间需求。实际上，在满足体育运动项目基本要求的前提下，对主体结构进行跨度控制能显著减小建设成本，对主体空间进行容量压缩则能有效降低日常运行中的暖通空调能耗。这要求在建筑设计和深化阶段，在早期方案概念和造型意向确定后，对方案造型设计、平面布置、剖面设计和结构选型等方面做精细推敲，量体裁衣，推导出精明集约的设计方案。

在我们设计的北京奥运会两个场馆中，摔跤馆主体空间采用巨型门式刚架结构，通过渐变手法构成，这一过程中经过多轮精细推敲，最终完成在满足各种规范条件下的容积与体积优化，并产生丰富的具有韵律的场馆造型。由于容积控制得当，内部大厅在无特殊吸音处理的情况下满足了室内声环境要求，节省了声学处理的材料和施工成本。羽毛球馆的招标要求中主要包含一个主比赛馆、训练馆和赛时热身馆。方案设计过程中通过功能分析，将三部分功能的主体空间及其附属用房进行重组，使得每个馆的体量都能够在满足使用要求的情况下得到有效控制（图 5-20）。

值得一提的是，特别是对于游泳馆建筑而言，体型控制还是重要的建筑节能手段。在我们设计的广州亚运会游泳跳水馆中，基本设计构思就是让建筑高度符合功能需要。跳水功能处为最高，游泳池部分其次。训练池与比赛大厅隔开，便于赛后的训练池单独对外使用。同时，训练池水池提高。缩小容积，有利于对外开放使用时的节能降耗。

over-emphasis on the iconic nature, for the purpose of styling, it is easy for designers to use the skin and shell to 'cover' the main space, which can cause the volume of the venues to be much larger than the actual space requirements for sports activities. In fact, under the premise of meeting the basic requirements of sports events, the span control of the main structure can significantly reduce the construction cost, and the capacity and volume compression of the main space can effectively reduce the energy consumption of HVAC in daily operation. This requires that in the phase of architectural design and development, when the shape intentions of the early concept are determined, the design of the form, layout, section, and structural selection should be refined, the form of the venues should be cut according to the actual capacity and volume of the necessary space, so as to derive a design scheme for smart intensive development.

In the two venues of the Beijing Olympic Games that we designed, the main space of the wrestling hall used a giant portal frame structure, which was formed by a gradual approach. After many rounds of detailed scrutiny in this process, the volume of the wrestling hall was finally optimized to fulfill various regulatory conditions, and produced a rich rhythm of the venue's shape. Due to the proper volume control, the internal hall could satisfy the requirements of the indoor acoustic environment without special sound-absorbing processing, saving the material and construction costs for acoustic treatment. The tender requirements for the badminton hall mainly included a main competition hall, a training hall, and a warm-up hall during the match. In the process of designing the project, through the functional analysis, the main space of these three function and their auxiliary rooms were reorganized, so that the volume of each hall could be effectively controlled under the premise that the use requirements are met (Fig. 5-20).

It is worth mentioning that, especially for the natatoriums, volume control is still an important means of building energy conservation. In the Guangzhou Asian Games swimming and diving hall design by us, the basic design concept was to make the building height accord with the functional demands. The basic concept for the Asian Games Swimming and Diving Hall was to bring the building height up to functional needs. The diving function required the highest of elevation, followed the swimming pool. The training pool was isolated from the competition hall so that it could be elevated and put into separated public use after the Games. A reduced volume is conducive to energy saving when the pool is open to the public.

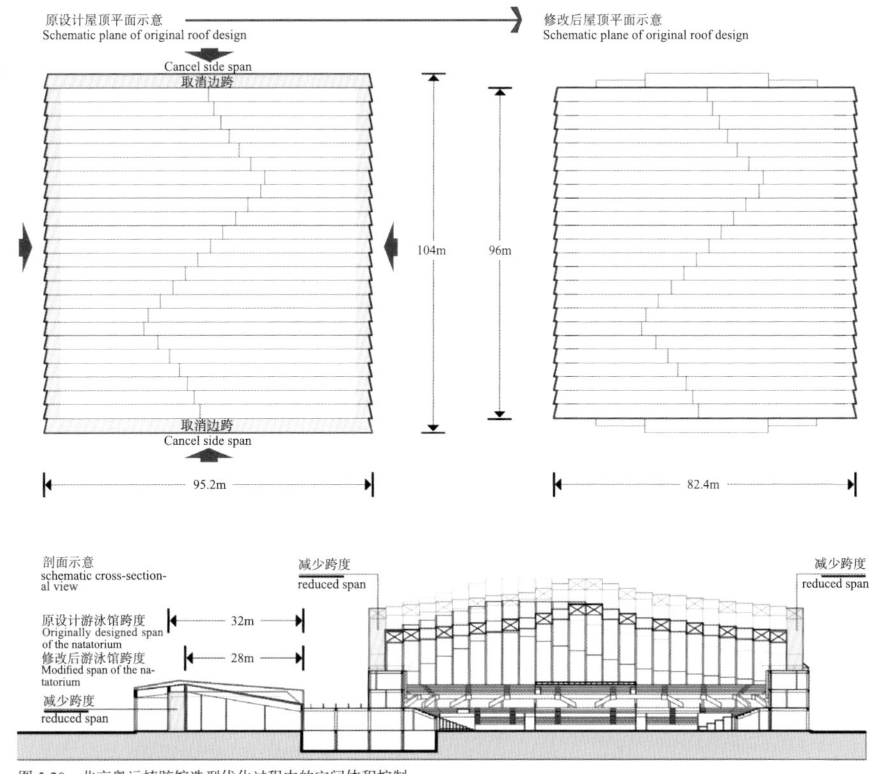

图 5-20　北京奥运摔跤馆造型优化过程中的空间体积控制

Fig.5-20　Spatial volume control in the modeling optimization process for the Beijing Olympic Wrestling Hall

通透与遮蔽：自然采光的精明转换

室内体育活动项目对光环境的要求，使得人工照明电量消耗成为众多体育场馆日常运营成本的主要部分之一，自然采光是实现体育场馆节能降耗的重要途径之一。除特定的体育活动项目外，体育场馆日常运营在大多数情况下允许引入自然光。自然光线所具备的提高室内光环境质量，改善室内人员心理感受的作用，也是人工照明无法取代的。同时，在赛时，由于体育比赛的苛刻要求，自然采光需要合理控制，对于像羽毛球这样的运动种类，天窗需要予以遮蔽。这就要求自然采光进行精明的设置，既要充足又要便于转换。

Permeation and Blockage: Smart Switch for Natural Lighting

The requirements of the indoor sports activities on the optic environment make the artificial lighting power consumption become one of the main constituents of the daily operation cost of many venues. Natural lighting is one of the important ways to achieve energy saving and consumption reduction of the venues. Except some specific sports events, the daily operation of the venues allows the introduction of natural light in most cases. Natural light has the effect of improving the quality of the interior optic environment and improving the psychological feelings of indoor people, which cannot be replaced by the artificial light. Meanwhile, subject to demanding requirements of sports events, natural lighting needs reasonable control. For sports like badminton, skylights will even be covered. This requires a configuration for natural lighting smart enough to provide sufficient and switchable natural lighting.

1. 主体空间自然采光

在20世纪50年代，我国兴建的第一批体育馆中大多考虑了比赛厅的自然采光，但实际上并没有取得理想效果，这其中有设计、构造和材料技术等问题，导致实际使用中多数天窗被封盖。此后兴建的许多大中型体育馆均无自然采光，如首都体育馆（1968年建成）和上海体育馆（1975年建成）等。20世纪80年代开始，天窗采光重新得到重视，如吉林冰球馆，在屋面上均匀布置了77个采光口，金字塔形的采光罩采用2mm厚的玻璃钢制成，这种材料透明度低但透光率高，使得室内采光系数达到5%，是当时采光面积较大的体育馆之一。然而，在当下许多体育场馆的建设实践中仍然无视自然采光的运用，造成很多“黑盒子”场馆。

与之相反的是另外一种极端，自然采光运用不当同样造成实际运营上的问题，例如过量光线造成眩光，或者伴随光线进来的热量造成室内制冷负荷的提高。如2001年建成的广州体育馆，为“九运会”体操、篮球等项目的比赛地，项目采用全透光的顶棚设计，屋面材料为双层半透明乳白色聚碳酸酯板材，透光率为10%，为室内创造了大空间漫射光效果，这在建成之初的一段时间内得到广泛好评。然而，在赛后的运营过程中逐渐暴露出来各种问题，例如在室内进行羽毛球运动时，由于乳白色屋面天光和羽毛球的颜色相近，造成光线干扰，白天进行比赛和训练时必须给羽毛球涂上荧光色才能让运动员判断球的位置。另外，大面积的无遮阳屋面采光，在广州这一热湿气候条件下，造成室内屋顶热辐射过大，尤其是在高处靠近屋盖的观众席温度过高（这其中也有由于空调系统送风管道隔热措施不足，造成冷空气运输过程中的冷量损失的原因），导致了体育馆空调能耗的增加。以上缺陷一定程度上也影响了该馆作为2010年广州亚运会比赛馆的使用（图5-21）。

1. Natural lighting for main space

In the 1950s, most of the first batch of the gymnasiums built in China considered the natural lighting of the competition hall, but in reality they did not play an ideal role. The reasons among these included that there were problems in design, construction, and material technology, leading to that the majority of the skylight in actual use is covered. Many large and medium-sized gymnasiums built since then had no natural light, such as the Capital Gymnasium (completed in 1968) and the Shanghai Gymnasium (completed in 1975). Since the 1980s, the skylight lighting had been re-emphasized, such as the Jilin ice hockey hall, where 77 light portals had been evenly arranged on the roof, and the pyramid-shaped lighting hood was made of 2mm thick fiberglass. This kind of material had low transparency but high light transmittance, enabling the indoor lighting factor reach 5%, which was one of the gymnasiums with large lighting area at that time. However, in the construction practice of many sport venues today, the use of natural lighting is still ignored, resulting in many ‘black box’ venues.

On the contrary, the other extreme is that the improper use of natural light also causes practical operational problems, such as glare caused by excessive light, or the increase of indoor cooling load of the air conditioning due to the heat coming in together with the light. For example, the Guangzhou Gymnasium built in 2001 was a venue for holding the gymnastics, basketball and other projects of the 9th National Games. The project used a fully translucent roof design, which was a double translucent milky white polycarbonate sheet with a light transmittance of 10%, and created a great amount of space diffuse light for the interior, which was widely praised for a period of time at the beginning of construction. However, various problems have gradually emerged during the post-game operation. For example, when badminton exercises are performed indoors, due to the light

图 5-21　全透光顶棚设计的广州新体育馆
Fig. 5-21　The all translucent roof designed for the Guangzhou New Gymnasium

interference caused by milky-white skylights and badminton colors, the shuttlecock must be coated with fluorescent color during daytime competitions and training, so as to allow the players to determine the position of the ball. In addition, under the hot and humid climate conditions in Guangzhou the large-area non-shading roof lighting, causes excessive heat radiation in indoor roofs, and especially high temperatures at the auditorium near the roof (this is also related to the reason of the insufficient thermal insulation measures of the wind pipe of the air-conditioning system, which causes the loss of cooling capacity in the transportation process of cold air), resulting in an increase in the air-conditioning energy consumption of the gymnasium. The above defects have also affected the use of the pavilion as a venue for the 2010 Guangzhou Asian Games (Fig. 5-21).

由此可见，忽略自然采光或者自然采光利用不当均不能有效解决问题，甚至会造成更大的尴尬。在提倡为体育场馆主体空间引入自然光线的同时，应该注意结合体育场馆的多功能性，设置能与之相匹配、灵活适应的采光策略，选择合适的采光构造和材料，必要时设置光线遮蔽和过滤的装置，并可以有效开启以适应不同天窗模式之间的切换，这样才能保证自然采光切实起到改善室内光环境，降低人工照明能耗的作用（表 5-1, 表 5-2）。

以我们设计的北京奥运会摔跤馆为例，考虑到馆赛后将移交中国农业大学供学校师生使用，节约日常运营和维护成本是其可持续性发展的关键。设计中将建筑层层错开的屋面与天窗设置相结合，通过北向天窗为室内引入自然光。同时，摔跤馆利用简易的电动窗帘实现了天窗的快速遮蔽，使室内光环境提高了适应性。从实际运行效果看，由于学校师生和附近社区居民日常体育教学和锻炼活动对光环境并无苛刻要求，自然采光提供了良好的室内光环境，在大多数情况下均能满足要求，这有效地降低了体育馆的照明耗电量。类似的采光策

It can be seen that both ignoring natural lighting and improper use of natural light are not able to effectively solve the problem and even create greater imperfections. While advocating the introduction of natural light for the main space of the sport venues, attention should be paid to combining the versatility of the venue, setting lighting strategies that can be matched and flexibly adapted to, selecting appropriate lighting constructions and materials, and setting up light shielding and filtering device as necessary, which can be effectively opened for switching between different skylight modes, so as to ensure that natural lighting can effectively improve the indoor optic environment and reduce the energy consumption of artificial lighting (Tab. 5-1, Tab. 5-2).

Taking the Beijing Olympic Games Wrestling Hall we designed as an example, considering that the gymnasium would be handed over to the China Agricultural University for use by teachers and students, saving the daily operation and maintenance costs was the key to its sustainable development. In the design, the staggered

各类运动对自然光线的要求　　表 5-1

运动项目	训练和娱乐活动	业余比赛、专业训练	专业比赛	TV 转播国家、国际比赛	TV 转播重大国际比赛	HDTV 转播重大国际比赛	空间利用特征	对光环境要求
篮球	300	500	750	1000	1400	2000	利用空间为主，视线经常向上，需考虑避免眩光	照度、均匀度提出了较高要求的同时，也要求顶棚反射率不小于 60%，最好达到 80%
排球	300	500	750	1000	1400	2000	利用空间为主，视线经常向上，需考虑避免眩光	在场地上方的一段空间内也应有一定的亮度和较高的均匀度；在场地球网的上空不应有高亮度的光源
羽毛球	300	750/500	1000/750	1000/750	1400/1000	2000/1400	利用空间为主，视线经常向上，需考虑避免眩光	球网上空至少 7 米高度内不应产生明暗光斑。羽毛球运动场所的墙面和顶棚的反射率应满足下列要求：后墙为 20%，侧墙为 40% - 60%，顶棚为 60% - 70%，墙面不应有花纹和图案
乒乓球	300	500	1000	1000	1400	2000	在 3 米以内的空间进行，视线以向下为主，应注意避免反射眩光	要求光线柔和，四周要有强烈的对比度，场地不得有明显的反光，要求在背景的衬托下能够看清球的整个飞行途径，要有较高的垂直照度和水平照度

资料来源：根据体育建筑设计资料集和体育建筑照明设计资料整理

Natural light requirements imposed by different kinds of sports　　Tab. 5-1

Sports events	Training and entertain-ment	Amateur and professional training	Professional competition	TV broadcast national and internation-al competition	TV broadcast major inter-national competition	HDTV broadcast of major inter-national competition	Space utilization feature	Light environment requirements
Basketball	300	500	750	1000	1400	2000	Space utilization-based. People look upwards a lot, so avoidance of glare should be considered	high standards are required for illuminance and uniformity, the ceiling reflectance is also required to be not less than 60%, preferably 80%
Volleyball	300	500	750	1000	1400	2000	Space utilization-based. People look upwards a lot, so avoidance of glare should be considered	A certain degree of brightness and a high degree of uniformity are also required for the space over the field and no highly bright light sources over the net
Badminton	300	750/500	1000/750	1000/750	1400/1000	2000/1400	Space utilization-based. People look upwards a lot, so avoidance of glare should be considered	No light and dark spot should be produced in the space 7 meters above the net. The wall finish and ceiling of a badminton venue shall have a reflectance meeting the following requirements: 20%, for the back wall, 40% -60% for the side walls, 60%-70% for the ceiling, and no pattern design should be on the walls
Billiards	300	500	1000	1000	1400	2000	Within the radius of 3 meters, and people look downwards a lot, so reflected glare should be avoided	The light is required to be soft, yet in a strong contrast with the surrounding. There should be no obvious reflected light in the court but the light against the background should be bright enough for people to see the entire flying path of the ball. A high degree of vertical illuminance and horizontal illuminance is required

Source: according to the sports architectural design data collection and sports architecture lighting design data

略还在广州亚运会游泳跳水馆和江苏淮安市体育中心的体育馆和游泳馆综合体中采用（图 5-22）。

以 2010 年广州亚运会游泳跳水馆为例，在方案设计过程中借助计算机参数化和物理环境模拟工具，进行了自然采光设计。方案采用穿插的体块形成层级渐变、缓和起伏的总体形态。对于高起的体块，在朝北方向设置可以开合和遮蔽的窗口，借助太阳散射光的引入改善室内光环境并有效降低照明能耗。此外，通过这些窗户可及时将室内过高的湿分排出，

roof of the building was combined with the natural lighting, and the natural light was introduced into the hall through the north skylight. At the same time, the project adopted a simple electric curtain to achieve a quick shuttering of the skylight, which improved the adaptability of the indoor optic environment. Judging from the actual operation effect, since the daily physical education and exercise activities of the teachers and students of the school and the residents in the nearby communities do not have strict requirements on the optic environment, the natural lighting provides a good indoor optic environment and can

自然采光方式和采光策略的适用性分析　　表 5-2

自然采光方式			应用特点	案例
顶面自然采光	顶向天窗	直接在屋面上开启采光窗或采光带	采光效率高，照度均匀，但应注意天窗遮蔽和节点防水处理	江苏盐城市体育中心 北京工业大学体育馆 广州亚运会武术馆
	北向天窗	结合屋面结构和造型，使光线经过折射后到达室内，	结合屋面结构和造型，采光效率高，但应注意光线均匀性	巴塞罗那篮球馆 中国农业大学体育馆 北京朝阳体育馆 江苏吴江体育馆 淮安体育中心体育馆 广东奥林匹克游泳跳水馆
	透射顶棚	采用透光性好的膜结构、阳光板，形成整体采光	光效亮度均匀、避免强烈的亮度反差造成炫光，但应结合遮阳隔热措施，避免能耗过大	佛山世纪莲游泳馆 广州新体育馆
侧面自然采光	侧窗	在侧向界面上开启高侧窗、落地窗等采光口	侧窗采光的照度沿进深方向下降很快，分布很不均匀，因而适用于小进深空间	广州大学体育馆 广东药学院体育馆
	开放侧向界面	在侧向采用大面积落地玻璃，将室外光线直接引入	将室外景观引入室内，对观众区域光环境改善效果大，但应避免过分明亮造成炫光和逆自然光区域座席的光污染	德国 Inzell 速滑馆 慕尼黑奥林匹克中心游泳馆

Applicability analysis of natural lighting modes and lighting strategies　　Tab. 5-2

Natural lighting type			Application features	Case
Top interface lighting	Top skylight	Directly provide daylighting windows on the containment system on the roof	High lighting efficiency, uniform illumination, but attention should be paid to sunroof obstruction and node waterproofing	Yancheng Municipal Sports Center in Jiangsu Gymnasium of Beijing University of Technology Guangzhou Asian Games Martial Arts Gymnasium
	North skylight	The variation of height of the roof structure create lighting openings	Combining the roof structure and shape, high lighting efficiency, but attention should be paid to light uniformity	Barcelona Basketball Hall Gymnasium of China Agricultural University Beijing Chaoyang Gymnasium Wujiang Gymnasium Jiangsu Gymnasium of the Huai’an Sports Center Guangdong Olympic Swimming and Diving Hall
	Transmissive ceiling	Polycarbonate panels and membranous materials are used to create the lighting effect of the entire roof	Brightness of light effect is uniform, and glare is not caused by strong brightness contrast, but shading and heat insulation measures should be combined to avoid excessive energy consumption	Foshan Shijilian Aquatics Centre Guangzhou New Gymnasium
Side interface lighting	Side window	Clerestories and foot windows are provided for lighting	Illumination of the side windows decreases rapidly along the depth, and the distribution is very uneven, making it suitable for small depth spaces	Guangzhou University Gymnasium Gymnasium of Guangdong College of Pharmacy
	Open side interface	Open the interface on one side and provide a glass curtain	Introducing the outdoor landscape into the interior will have a great effect on the improvement of the light environment of the audience, but it should avoid excessive light that causes glare and light pollution in the counter-natural light region seats	German Inzell speed skating hall Swimming Hall of Munich Olympic Center

Source: according to the sports architectural design data collection and sports architecture lighting design data

从而改善室内湿环境。方案优化过程中，在计算机参数化工具中加入了对高侧窗可开启面积的检测模块，使得在场馆整体体型优化的过程中，保证了高侧窗的可开启面积。从建成效果看，以上措施实现了关闭设备条件下室内的光、热物理环境，有利于场馆日常的可持续运营。

meet the requirements in most cases, which as effectively reduced the lighting power consumption of the gymnasium. Similar lighting strategies have also been used in the Guangzhou Asian Games Swimming and Diving Hall and the complex of gymnasium and swimming pool of the Huai'an Sports Center in Jiangsu (Fig. 5-22).

Taking the 2010 Guangzhou Asian Games Swimming and Diving Hall as an example, natural lighting strategy was carried out during the design of the project by means of computer parameterization and physical environment simulation tools. The program

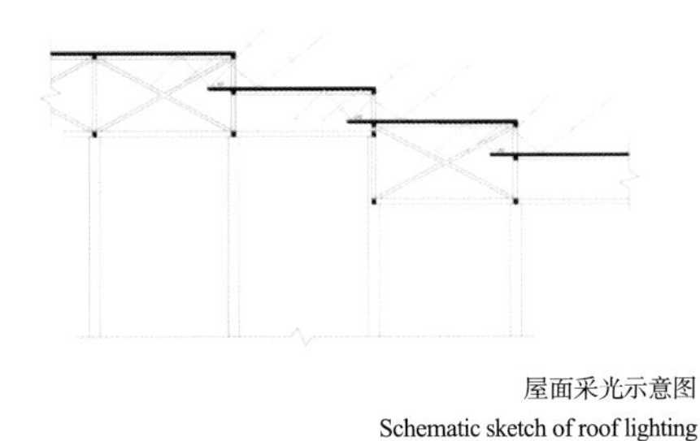

屋面采光示意图
Schematic sketch of roof lighting

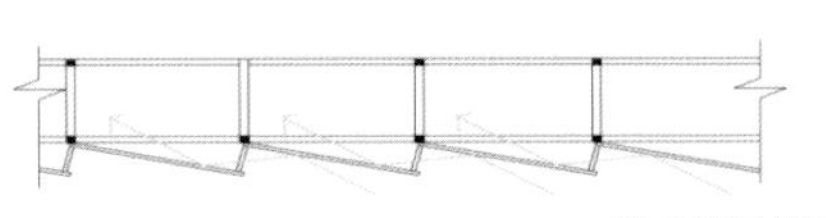

墙面采光示意图
Schematic sketch of wall lighting

图 5-22　北京奥运摔跤馆自然采光天窗构造
Fig. 5-22　The structure of the natural daylighting skylight of Beijing Olympic Wrestling Hall

used interspersed body masses to form a gradual and undulating overall form. For the raised block, a window that could be opened and closed and set in the north direction was provided, and the introduction of the scattered skylight improved the indoor optic environment and effectively reduced the lighting energy consumption. In addition, high humidity in the room could be expelled through these windows in time to improve the indoor hygric environment. During the optimization of the scheme, a detection module for the openable area of the high-side window was added to the computer parameterization tool, so that the openable area of the high-side window was ensured during the optimization of the overall form of the gymnasium. Judging from the results of the construction, the above measures have realized the indoor optic and thermal environment when the equipment is shut down and are conducive to the daily sustainable operation of the venue.

2. 辅助空间自然采光

体育场馆的赛时辅助用房通常包括赛事管理用房、运动员和贵宾的休息室等，日常运营时则为运营管理、办公甚至是经营出租用房。这些用房经常被布置在平台底下，而由于体育场馆的平台通常进深大，这给平台底部辅助用房的采光造成一定困难。实践中许多场馆未考虑这些辅助用房的采光问题，直接影响了日常使用，并造成人工照明耗电量的上涨。如佛山世纪莲体育场，设计方案中为了实现体育场与整个体育公园统一的景观效果，采用草坡覆盖了底部平台，这一方案导致底部用房无法自然采光，影响其日常运营（图 5-23）。

在我们的实践中尝试了许多种解决平台下辅助空间的自然采光问题。优先选项是优化平面设置，保证辅助用房至少有一面墙体具备开窗条件，如江门市滨江体育中心方案，在设计过程中通过平面优化，使得大部分辅助用房实现了自然采光。对于进深小的平台，可以利

2. Natural lighting of auxiliary spaces

The auxiliary rooms during the game in the sport venues usually include the event management room, the lounge for the athletes and the VIP, etc., while they include the operation management, office, or even the rented room in the daily operation. These rooms are generally arranged under the platform, and the platforms of the sport venues are usually deep, which makes it difficult for the lighting of the auxiliary rooms under the platform. In practice, many venues do not consider the lighting problems of these auxiliary rooms, which directly hinder the daily operation, and increase the artificial lighting power consumption. For example, in the design of the Foshan Century Lotus Stadium, in order to achieve the unified landscape effect of the stadium and the entire sports park, a grass slope is used to cover the bottom platform, which stop the bottom rooms from naturally lighting and affect the daily operation (Fig. 5-23).

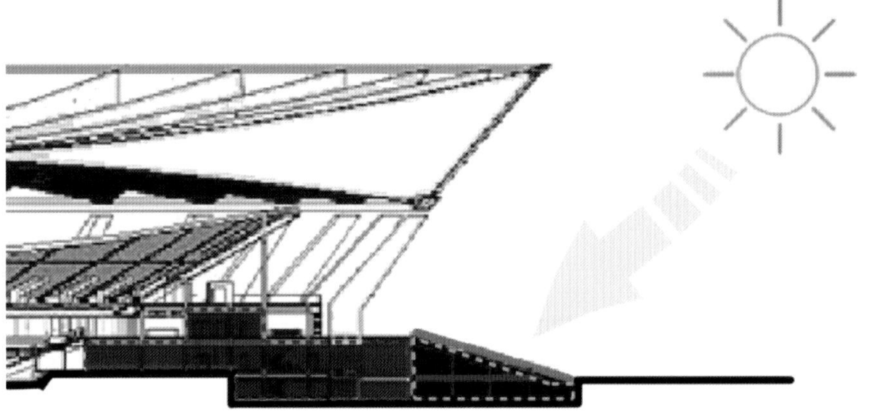

图 5-23　佛山世纪莲体育场平台下空间自然采光不足

Fig. 5-23　Insufficient natural lighting of the space under the platform of Foshan Century Lotus Stadium

用外立面开窗或者高侧窗的方式直接采光，如南沙体育馆，采用了与佛山世纪莲类似的草坡环绕的方法，但在草坡顶部预留高侧窗的采光带，使得场馆平台底部的辅助用房均可以通过高侧窗实现自然采光。对于进深较大的平台，可以利用天窗的设置为底部空间引入光线，也可以通过置入采光天井或者庭院的办法解决，如中山市体育馆，通过四个采光庭院的设置，保证了平台下大多数辅助空间的采光条件（图 5-24 ~ 图 5-26）。

随着自然采光新构造和新材料的不断研发，以及智能化控制系统的不断进步，自然采光设计和建造技术将会为体育场馆中利用好自然光提供更多的可能性。作为一项重要的可

In our practice, we have tried many solutions to the problem of natural lighting in the auxiliary space under the platform. The priority option is optimization of the plane setting to ensure that at least one wall of the auxiliary rooms has conditions to open a window, such as the Jiangmen Binjiang Sports Center, in which most of the auxiliary rooms achieve natural lighting through the optimization of the plane during the design process. For platforms with a small depth, direct lighting can be achieved through the use of façade windows or high-side windows. For example, the Nansha Gymnasium uses a grass slope that is similar to that of the Foshan Century Lotus, but a high side window is reserved at the top of the slope, so that the lighting belts allow the auxiliary rooms at the bottom of the gymnasium platform to achieve natural lighting through the high-side windows. For those deeper platforms, the settings of the skylight can be used to introduce light into the bottom space, the lighting patio or courtyard can be installed in to solve the problems, such as the Zhongshan Gymnasium, which ensures the lighting conditions for the majority of the auxiliary space under the platform through the setting of four daylighting courtyards (Fig. 5-24 ~ Fig. 5-26).

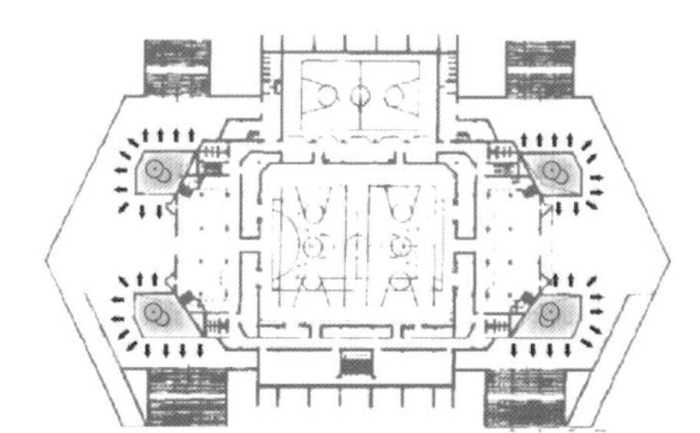

图 5-24　中山市体育馆利用天井实现自然采光

Fig.5-24　A patio is provided for Zhongshan Gymnasium to realize natural daylighting

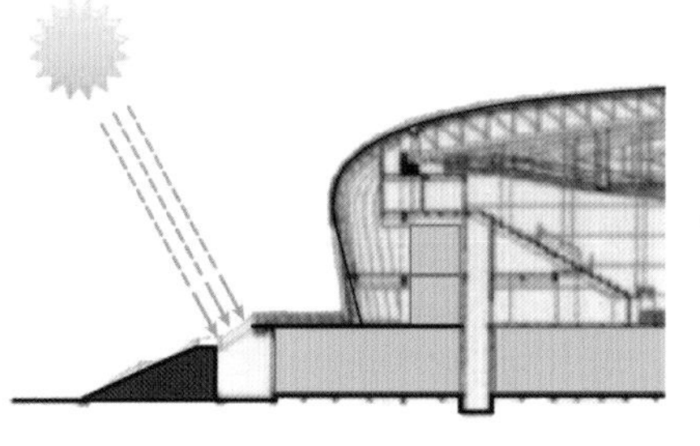

图 5-25　南沙体育馆利用高侧窗采光

Fig. 5-25　Clerestory windows are arranged for Nansha Gymnsaium to provide lighting

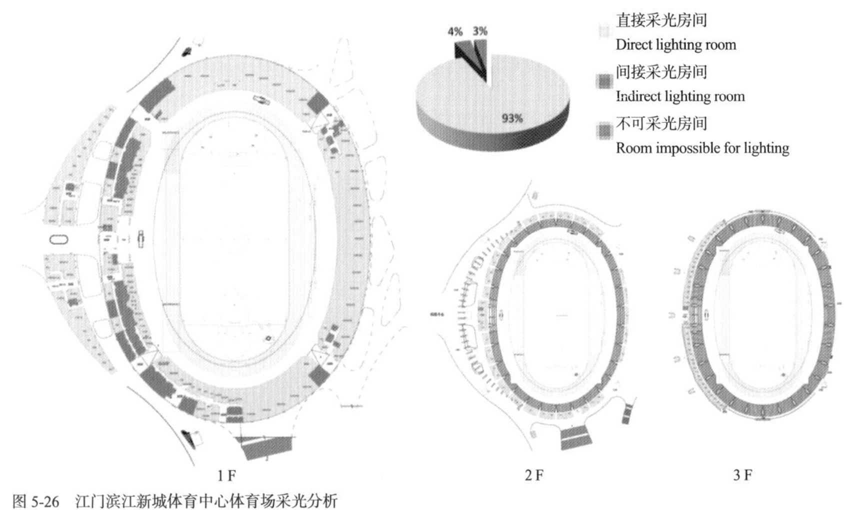

图 5-26 江门滨江新城体育中心体育场采光分析

Fig. 5-26 Lighting analysis of the stadium of Jiangmen Binjiang Sports Center

持续性设计策略，自然采光与体育建筑设计的结合值得我们不断探索。

With the continuous development of new constructions and new materials of natural lighting, and the continuous advancement of intelligent control systems, natural lighting design and construction technology will provide more possibilities for the use of natural light in sport venues. As an important sustainable design strategy, the combination of natural lighting and architectural design is worth exploring.

起承与转合：自然通风的精明疏导

除了羽毛球、乒乓球和滑冰等特定的体育比赛对室内风环境有特别要求之外，体育场馆的日常运营多数情况下允许自然通风。对于富于大众健身义务的体育建筑而言，自然通风，意味着低能耗开放时间的增加，意味着体育建筑的可持续发展。同时，自然通风能为室内提供新鲜空气，带走二氧化碳等气体，改善舒适性。与自然采光类似的，自然通风所具备的改善室内人员心理感受的作用，是机械通风无法取代的。

自然通风是指不借助机械动力，依靠自然驱动促使空气流动从而实现建筑室内外空气交换的一种通风方式。从技术原理上看，自然通风包括风压通风和热压通风两种基本形式，当两者共同作用时则形成风压与热压相结合的自然通风，当借助一定的机械方式实现或加强风压和热压通风时则形成机械辅助式自然通风。各种通风方式的适宜性受到多种因素影响，往往不限于仅采用其中的一种。自然通风

Generation and Transition: Smart Design of Natural Ventilation

Except the special requirements for indoor wind environment of some specific sports events such as the badminton, table tennis and skating, the daily operation of sport venues allows natural ventilation in most cases. For sports facilities that are full of public obligations, natural ventilation means that the increase opening hours with lower energy consumption, and means the sustainable development of sports facilities. At the same time, similar to natural lighting, natural ventilation can provide fresh air for the interior and remove carbon dioxide and other gases to improve the indoor comfort. This function of improving the psychological feelings of people inside is what the mechanical ventilation cannot replace.

Natural ventilation refers to a kind of ventilation method that makes use of the natural driving for the air flow without using mechanical power to achieve the air exchange between indoor and outdoor of a building. From the technical perspective,

还通常和其他生态手法相结合，如将送风道埋入地下，借助土壤进行新风的预冷或者预热，或者借助深井进行热量交换等。在选择自然通风方式时，应结合气候、地形、外部风环境和建筑自身条件，因势利导地组织自然通风的最佳方案（图 5-27）。

风压通风的原理是借助水平方向空气流的风压差对通风进行引导。当空气流受到建筑阻挡时，会在建筑迎风面产生正压，反之在背面形成负压。风压通风即是利用两者的压力差，使得空气由正压区向负压区流动。这一过程中，室外风速和风向、建筑物几何形状、建筑物与风向之间的夹角均会对通风效果产生影响。一般经验是，理想的外部风速一般不小于 3 ～ 4m/s，热湿气候区具有较大的风压通风潜力。对于建筑平面布局，应使建筑浅进深方向与主导风向平行。对于建筑几何形状，

natural ventilation includes two basic forms, namely the wind pressure ventilation and thermal pressure ventilation. When the two work together, they become the combined natural ventilation by wind and thermal pressure, and when a certain mechanical method is used to achieve or strengthen the ventilation by wind pressure or thermal pressure, they become the mechanically assisted natural ventilation. The suitability of various ventilation methods is affected by many factors and is often not limited to only one of them. Generally natural ventilation is also combined with other ecological practices such as burying the air delivery conduits underground, using the soil to precool or preheat the fresh air, or using deep wells for heat exchange. In the selection of the methods of natural ventilation, the best solution should be organized in accordance with the climate, topography, external wind environment and the conditions of the building itself (Fig. 5-27).

The principle of wind pressure ventilation is to guide the ventilation by the wind pressure difference of the horizontal air flow. When the air flow is blocked by the building, positive pressure will be generated on the windward side of the building, and vice versa. Wind pressure ventilation is the use of the pressure difference between the two, so that air flows from the positive pressure zone to the negative pressure zone. In this process, the wind speed and direction, the geometry of the building, and the angle between the building and the wind flow all affect the ventilation efficiency. The general experience is that the ideal external wind speed is generally no less than 3 ~ 4m/s, so that there is great wind pressure ventilation potential in the hot and humid climate zone. For the plane layout of the building, the depth direction of the building should be parallel to the dominant wind flow. For the building geometry, the depth should generally not be greater than 14m, but if it is unilateral ventilation, it should not be more than twice the storey height of the building.

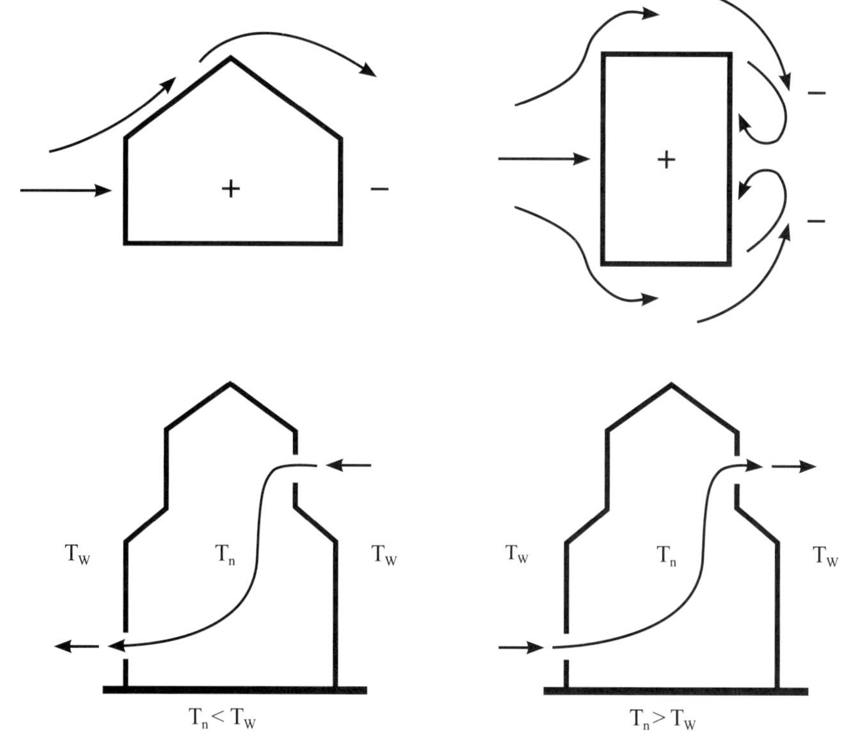

图 5-27　风力通风和浮力通风原理示意图

Fig. 5-27　Schematic sketch of wind-driven ventilation and buoyancy-driven ventilation

进深一般不宜大于 14m，而如果是单侧通风，则不宜大于建筑层高的两倍。此外，为了形成穿堂风效果，排风口应不小于进风口，也不能大于进风口的 3 倍。

In addition, in order to form a draught, the air outlet should not be less than the air inlet, nor should it be more than 3 times larger than the air inlet.

热压通风的原理是借助室内空气在垂直方向上，由于温度差的存在引起的空气流动，这一作用通常被称为“烟囱效应”。室内温度较高的空气由于膨胀而密度降低，向上浮起，在顶部形成负压区，反之温度较高的空气在底部形成正压区。当建筑顶部和底部存在开口时，温度较高的空气从顶部出风口排出，带动室内空气向上流动，并不断地将室外新风通过建筑底部开口吸入室内，从而形成室内自然通风。一般情况下，“烟囱效应”的强度与进出风口的高差、室内外空气温度差和室外空气流动速度有关，三者越大则造成室内热压通风效应也越明显。但热压通风不一定是由下而上的流动，也可能是其他方向，这由冷热压差的分布场决定。相比于风压通风，热压通风受到外部风条件的影响较小，对于室外风速不大的地区，热压通风可起到良好通风效果。此外，空气含湿量增大会造成对通风动力需求的增加，空气越干燥则越有利于提高通风风速。因此，干热气候区和温和气候区，热压通风具有较高的自然通风潜力。

The principle of thermal pressure ventilation is the use of temperature differences in the vertical direction of indoor air to drive the air flow, which is often referred to as the ‘chimney effect’. The air with a higher temperature in the room will decrease in density due to the expansion, float upwards and form a negative pressure zone at the top, whereas the air with a lower temperature forms a positive pressure zone at the bottom. When there is an opening at both the top and the bottom of the building, the air with higher temperature is exhausted from the top air outlet, which drives the indoor air to flow upward, and continuously draws outdoor fresh air into the room through the bottom air inlet of the building, thereby forming the natural indoor ventilation. In normal circumstances, the strength of the ‘chimney effect’ is decided by the height difference between the air inlet and outlet, the temperature difference between indoor and outdoor air, and the outdoor air flow speed. The larger the three factors, the more obvious the effect of the indoor thermal pressure ventilation. However, the thermal pressure ventilation does not necessarily have to flow from bottom to top, but may also be in other directions, which is determined by the direction of the thermal pressure difference. Compared to the wind pressure ventilation, thermal pressure ventilation is less affected by the external wind conditions. For areas with low outdoor wind speeds, thermal pressure ventilation can also provide good ventilation. In addition, the increase in the air moisture content will increase the demand for ventilation power, therefore the drier the air, the more favorable it is to increase the ventilation speed, so that there is a high potential for natural ventilation in the hot and dry climate zones and the temperate climate zones.

风压通风和热压通风也可以相结合使用，但当两者同时运用时要注意避免两者作用的相互削弱，这特别需要对外部气候和建筑自身的通风条件做充分评估。通常情况下，对建筑进深浅、层高低的部分，如办公辅助用房，采用风压通风为主，而进深大、空间高的部分，如体育馆比赛厅，采用热压通风为主。

对于体育场馆，或者如会展中心、商业综合体、地下车库等大型室内空间，由于通风路径过长以及空气流动阻力增大，单纯依靠风压和热压无法形成足够的通风动力，或者由于避免室外空气污染和噪声干扰等原因，不宜通过开窗进行直接的自然通风，可采用一定的机械设备进行辅助，以加强和改善通风效果。例如，对于大空间的室内空间，借助大空间由于高差引起的热压通风动力，同时进行进风口机械送风的加强和出风口机械排风的辅助，优化自然通风效果。

以广州亚运会摔跤馆为例，项目承担2010年广州亚运会期间的摔跤比赛项目，赛后作为华南理工大学大学城校区的综合体育馆，供学校师生日常教学和锻炼使用为主，因而借助自然通风和采光等策略，有助于赛后场馆运营的节能降耗。体育馆平面长轴和短轴尺寸分别达到97m和67m，考虑到场馆自身的大空间特点，具备进行热压通风的条件，方案设计中对此给予强化利用，在屋面设计中，采用四片混凝土扭壳组合而成的反曲结构，形成四周向中部升起的大空间，使得中间屋盖顶部和比赛场地之间的高差达32m，为热压通风创造了良好的高差条件。结合基地的地形高差条件，将功能用房置于地下，使得进风口引入的空气源获得约2℃的自然冷却，这也进一步拉大与体育馆顶部拔风口的气温温差。

作为亚热带地区，大空间的自然通风尚无法满足舒适性的要求，适当的机械辅助通风，可以增加空气的循环与流动，有助于减少空调开放的时间，对于节能降耗的意义更加重大。在本项目中，采用了机械系统的辅助，创造性地在顶部拔风口设置排风机以增强场馆的自然通风能力，一共使用了8台风机，加速顶部热空气的排出和底部空气向上流动，从而提高自然通风效率。采用的机械辅助热压通风系统包括进风口、风道以及用于送风和排风的风机，构成完整的通风体系。在这一体系中，室外新风从进风口吸入，经过地下室的自然冷却，通过风道引导到主空间看台活动座席后方的辅助用房，借助风扇将新风送到赛场。设计过程中

Wind pressure ventilation and thermal pressure ventilation can also be adopted in combination, but when the two are used simultaneously, care should be taken to avoid the mutual weakening of the two effects, which requires in particular a thorough assessment of the external climate and the ventilation conditions of the building itself. In normal circumstances, the wind pressure ventilation is mainly used for the space with small depth and low storey height, such as office auxiliary rooms, and to the contrary, the thermal pressure ventilation is the main method for the deep and high space, such as the competition hall of a gymnasium.

For sport venues, or other large indoor spaces such as the exhibition centers, commercial complexes, and underground garages, due to the long ventilation paths and the increased resistance to air flow, simply relying on wind pressure and heat pressure cannot provide sufficient ventilation power, or in other cases that need to avoid opening to outdoor for reasons such as air pollution and noise interference, direct natural ventilation through windows should not be used, and certain mechanical equipment can be used to assist in strengthening and improving the ventilation effects. For example, for a large indoor space, the natural ventilation effect is optimized by the thermal pressure ventilation power caused by the height difference due to the large space, the mechanical air supply at the air inlet, and the mechanical ventilation of the air outlet at the same time.

Taking the Guangzhou Asian Games Wrestling Hall as an example, the project was to assume the wrestling competition during the 2010 Guangzhou Asian Games, and after the game, it was transformed to serve as a comprehensive gymnasium for the University Town Campus of the South China University of Technology. Strategies such as natural ventilation and lighting could contribute to the energy saving and consumption reduction of operations after the game. The lengths of the major and minor axes of the plane were 97m and 67m

借助计算机模拟工具进行辅助，利用 CFD（计算机流体力学）的模拟结果，明确进风口的位置和大小、排风口风机总功率和台数（与噪声控制有关）的优化方案（图 5-28、图 5-29）。

respectively. Considering the large space of the gymnasium itself, the conditions for thermal pressure ventilation were available, so that in the design stage this was given greater utilization. In the roof design, four concrete twists were combined for the recurved structure to form a large space that rose from the center to the middle. The height difference between the top of the middle roof and the playing field was up to 32m, which created a good height difference condition for thermal pressure ventilation. Considering the conditions of the original terrain elevation of the site, the function rooms were placed underground, so that the air source introduced by the air inlet was naturally cooled at about 2°C, which further enlarged the air temperature difference with the air outlet at the top of the gymnasium.

In a sub-tropical region, natural ventilation for a large space is yet insufficient for the requirements of comfort. Appropriate mechanically assisted ventilation will improve air circulation, shorten the opening hours of air conditioning, which is significant for energy saving. In this project, a mechanical system was used to assist in the creative installation of exhaust fans at the top air outlet to enhance the natural ventilation capacity of the gymnasium. A total of eight fans were used to accelerate the exhausting of the hot air at the top and the upward flow of the air at the bottom, which improved the natural ventilation efficiency. The mechanically assisted

图 5-28　广州大学城华南理工大学体育馆排风口和新风口

Fig.5-28　The exhaust enginery and fresh air aperture of the Gymnasium of South China University of Technology, Guangzhou Higher Education Mega Center

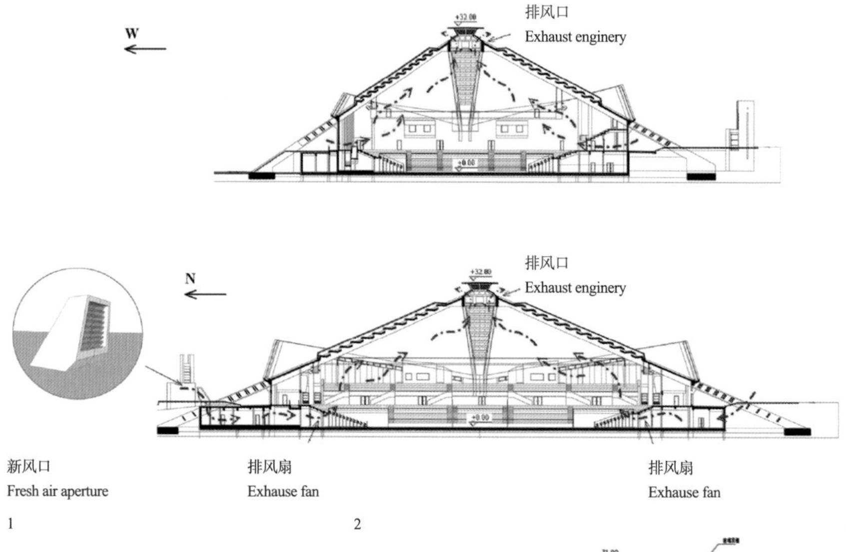

1

2

3

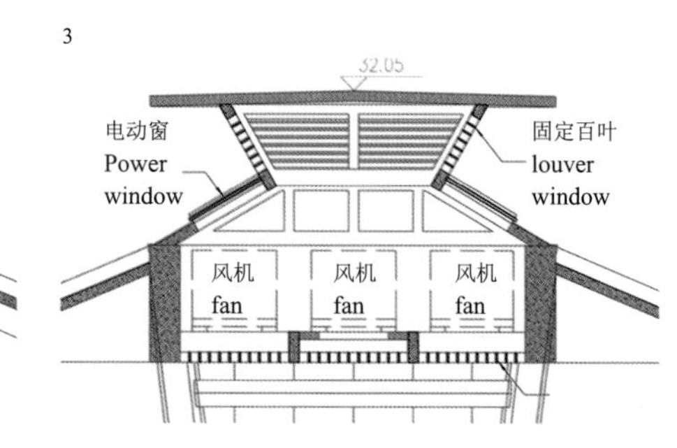

屋顶排风构件形式演变　Evolution of the form of roof exhaust member

图 5-29　广州大学城华南理工大学体育馆自然通风设计

Fig. 5-29　Natural ventilation design of the Gymnasium of South China University of Technology, Guangzhou Higher Education Mega Center

thermal pressure ventilation system included air inlets, air delivery conduit, and fans for air supply and exhaust, which constituted a complete ventilation system. In this system, outdoor fresh air was sucked in from the air inlet, cooled naturally through the basement, guided through the air delivery conduit to the auxiliary rooms behind the seat of the stand, and then the fresh air was sent to the competition hall by means of a fan. In the design process, computer simulation tools were used to assist, and the simulation results of CFD were used to define the position and size of the air inlet, the total power of the air outlet fan and the number of units, which were also related to the noise control (Fig. 5-28, Fig. 5-29).

设计策略应用

在我们的多个体育场馆设计中，重视大跨结构精明选型、设备和系统的集约灵活设置、场馆容积控制以及自然采光和通风策略的应用，借助相关计算机模拟工具和参数化工具的辅助，优化场馆设计方案，为场馆的可持续运营打下良好基础。

以我们设计的两个奥运场馆为例，2008年北京奥运会摔跤比赛馆（中国农业大学体育馆）和羽毛球比赛馆（北京工业大学体育馆）均采用了集约适宜的技术方案，从建设和运营的实际效果看，两者都收到了显著成效。这两个奥运场馆具有一定共性，并且具有国内许多大中型场馆建设的共同特点。一方面，两个场馆均是因为承办奥运会赛事需要而兴建，并且预先策划好赛后移交高校使用，由于赛时功能面向奥运会特定比赛项目，赛后功能则面向高校师生日常教学和体育锻炼活动，因而赛时赛后存在巨大的使用需求落差。另一方面，两个场馆的投资主体均为高校，投资额度严格受限，因而面临着通过低资金投入获取高效益目标的挑战。

Application of the Design Strategies

In many of the sports venues designed by us, attention is paid to the smart selection of long-span structures, the intensive and flexible arrangement of equipment and systems, the control of volume, and the application of natural lighting and ventilation, which are aided by relevant computer simulation and parametric tools to optimize the design program of the venues and lay a good foundation for the sustainable operation of the sport facilities.

Taking the two Olympic venues that we designed as an example, the 2008 Beijing Olympic Games Wrestling Hall (China Agricultural University Gymnasium) and the Badminton Hall (Beijing University of Technology Gymnasium) have adopted intensive and appropriate technical solutions. Judging from the actual results from construction and operation, both venues have achieved remarkable effects. The two Olympic venues have certain commonalities of many large and medium-sized venues in China. On the one hand, both venues were built for hosting Olympic events and were handed over to the colleges for operation after the game in the pre-

对此，我们采取了多方面的设计策略，具体而言，方案设计以结构选型优先，避免片面地在结构方面追求先进性；采用主动和被动相结合的技术，充分引入自然通风和采光，降低运营能耗成本；集约进行功能和体型控制，节能节材，降低初始建设和赛后运营成本；方案设计充分考虑到赛时和赛后的功能转换，确保改造过程只需要最小的时间和资金成本；进行土建装修一体化的设计和施工，有效减少二次装修成本；合理控制设备标准，在满足使用需求下节约初始投入和后期维修更新成本；根据不同规模的活动要求，采用分区分级的设备控制策略，有效降低日常的运营成本。

设计方案严格把握建设标准，在建设初始投入和赛后运营成本控制方面收到显著效果。例如摔跤馆每平方米造价为 6600 元，成为北京奥运会中“最省钱”的场馆。得益于自然通风和采光等技术应用，该馆赛后运营表现良好，尤其是在减小运营能耗方面表现突出。据使用方的反馈，每年可节省约 100 万元的运营电费。但与此同时，该馆在有些方面仍有改进的空间。如根据业主反馈，场馆的设备控制灵活性应加强，强弱电分区应该更加灵活，副馆的办公用房分区控制管理计量上也存在不足。

经专家评审，两馆相关研究成果《北京奥运摔跤馆和羽毛球馆可持续设计策略与技术应用》获得 2012 年华夏建设科学技术奖。两个场馆采用适宜技术，实现低成本和高效率的结构选型和节能降耗；综合运用了灵活适应性的设计方案和技术措施，很好地满足了赛时赛后功能的高效切换；两个场馆均较好地运用城市设计的理念，实现体育建筑赛后融入高校，回归城市的可持续发展目标。这为国内同类体育场馆的可持续性设计和建设提供有益参考（图 5-30、表 5-3）。

planning stage. Since the functions during the match were oriented to the specific Olympic Games competitions and the post-game functions were aimed at daily teaching and sports activities for college teachers and students, there was a huge demand gap between that during and after the game. On the other hand, the investors of both venues were colleges, whose investment quotas were strictly limited and they faced the challenge of obtaining high-efficiency targets through low capital investment.

In this regard, we have adopted a variety of design strategies. In particular, we prioritized the structural design to avoid the unilaterally pursuing advancement in the structure; adopted active and passive technologies to fully introduce natural ventilation and lighting and reduce the operational costs of energy consumption; intensively performed functional and volume control, energy and materials saving to reduce the costs of initial construction and post-game operating; fully took into account the functional transitions during and after the game in the plan design to ensure that the renovation process required only minimal costs of time and funds; integrated the design and construction of the civil engineering and the decoration works to effectively reduce the cost of the second renovation; controlled reasonably the equipment standards to save the initial investment and the cost of later maintenance and renewal under the premise of satisfaction to the use demand; adopted district-grading equipment control strategies according to the requirements of different sizes of the activities to effectively reduce the daily operating costs.

The design plan strictly controlled the construction standards, which has achieved significant results in terms of initial investment in construction and the operation cost after the match. For example, the wrestling hall costed 6,600 Yuan per square meter, becoming the ‘most cost-effective’ venue for the Beijing Olympic

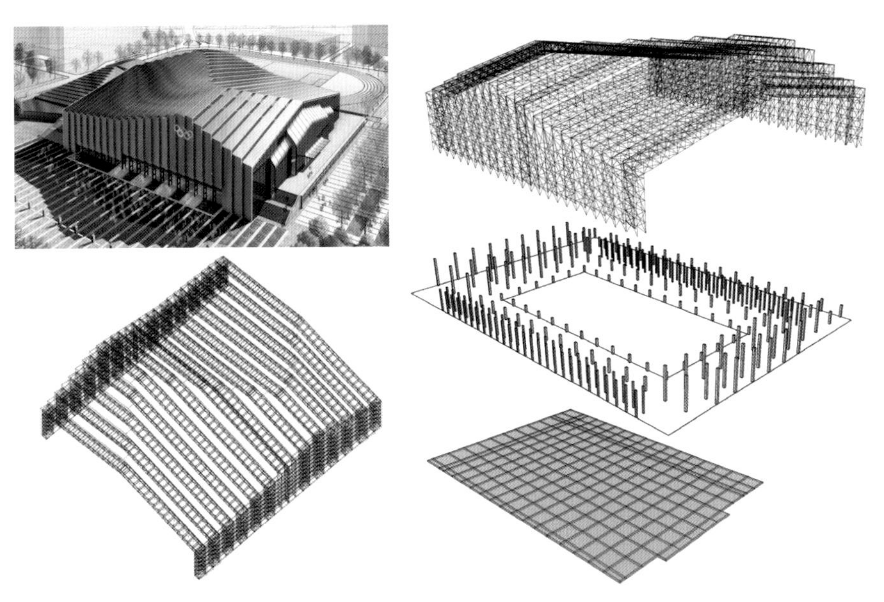

图 5-30　北京奥运摔跤馆采用成熟门式钢架结构技术实现设计创新

Fig. 5-30　The design innovation realized with the proven technology of portal frame structure for the Wrestling Hall of Beijing Olympic Games

四所高校奥运场馆工程造价比较　表 5-3

体育馆	建筑面积（m^2）	座席数（座）	工程造价（亿元）	单位面积造价（元/m^2）
北京大学体育馆（北京奥运乒乓球馆）	26525	8000（其中固定6000，活动 2000）	2.60	9802.07
中国农业大学体育馆（北京奥运摔跤馆）	23950	8500（其中固定6000，活动 2500）	1.58	6597.08
北京科技大学体育馆（北京奥运柔道馆）	24662	8000（其中固定4000，活动 4000）	2.20	8920.61
北京工业大学体育馆（北京奥运羽毛球馆）	24383	7500（其中固定5800，活动 1700）	2.15	8817.62

Comparison of construction cost of the four university Olympic venues　Tab. 5-3

Gymnasium	Total construction area (m^2)	Number of seats (seats)	Construction cost (hundred million Yuan)	Cost per unit area (Yuan/m^2)
Peking University Gymnasium (Beijing Olympic Badminton Hall)	26525	8000 (fixed 6000, retractable 2000)	2.60	9802.07
China Agricultural University Gymnasium (Beijing Olympic Wrestling Hall)	23950	8500 (fixed 6000, retractable 2500)	1.58	6597.08
Beijing University of Science and Technology Gymnasium (Beijing Olympic Ping-pong Gymnasium)	24662	8000 (fixed 4000, retractable 4000)	2.20	8920.61
Beijing University of Technology Gymnasium (Judo & Taekwondo Gymnasium for Beijing Olympic Games)	24383	7500 (fixed 5800, retractable 1700)	2.15	8817.62

Games. Thanks to the application of natural ventilation and lighting, the gymnasium has performed well after the game, especially in reducing the operational energy consumption. According to the feedback of the owner, the operating electricity fee of about 1 million Yuan can be saved every year. However, at the same time, the gymnasium still has room for improvement in some areas. For example, according to feedback from the owner, the equipment control flexibility of the venues should be strengthened, the strong and weak electric power districts should be more flexible, and there are also deficiencies in the zoning controlled measurement and management of the office rooms of the auxiliary venue.

After review by experts, the relevant research results of the two venues, "The Sustainable Design Strategy and Technical Application of Beijing Olympic Wrestling Hall and Badminton Hall" won the 2012 China Construction Science and Technology Award. The two venues adopt appropriate technologies to achieve low-cost and high-efficiency structural selection and energy saving; comprehensively use flexible and adaptable design solutions and technical measures to well meet the efficient switching of the functions after the match; both venues use the concept of urban design to realize the integration of sports facilities into the universities and return to the city's sustainable development. This provides useful reference for the sustainable design and construction of the similar sport venues in China (Fig. 5-30, Tab. 5-3).

作品
Works

2008 年奥运会摔跤比赛馆（中国农业大学体育馆）中国，北京，2004

项目背景

中国农业大学体育馆位于中国农业大学校区内。北京 2008 年奥运会作为摔跤比赛用馆，奥运会后成为中国农业大学室内综合体育活动中心，经改造后除原比赛大厅外，包括一个标准篮球训练馆以及一个拥有标准游泳池的游泳馆。在保证继续承接各类体育赛事的同时，满足教学、文艺、集会以及学生社团使用。

本项目于 2004 年 3 月举行国际招标，6 月评选出华南理工大学建筑设计研究院，川口卫（日本）、澳大利亚与北京建筑设计研究院联合体 3 个优秀入围方案。2004 年 7 月至 12 月，经过多轮优化比较，其中包括北京奥组委、中国摔协和国际摔跤联合会等组织在内的专家评审，对规模、功能及技术细节进行了分析与必选，于 2005 年元月确认华南理工大学建筑设计研究院为中标实施单位。历经 3 年，2007 年 6 月中国农业大学体育馆落成。

2008 OLYMPIC WRESTLING HALL (CHINA AGRICULTURAL UNIVERSITY GYMNASIUM) BEIJING, CHINA, 2004

Background

China Agricultural University Gymnasium, located in the east campus of CAU, was used as the wrestling competition venue at the 2008 Beijing Summer Olympics. After post-Olympics reconstruction, it accommodates the original competition hall, a standard basketball training hall and a natatorium with a standard pool. The gymnasium opens to various sports events and put into use for P.E. teachings, theatrical performances, assemblies and student association activities.

The international tender for the Gymnasium was initiated in March, 2004. From June, 2004 to January, 2005, projects submitted by Architectural Design & Research Institute of SCUT, Kawaguchi & Engineers (Japan), and Architectural Design & Research Union of Australia & Beijing went through rounds of conscientious evaluations on scale, function and technical details by professional judges from associations including the Beijing Olympics Committee, Association of Wrestling of China and United World Wrestling; eventually, the design of Architectural Design & Research Institute of SCUT was approved for execution. Three years later, the gymnasium was completed on campus in June, 2007.

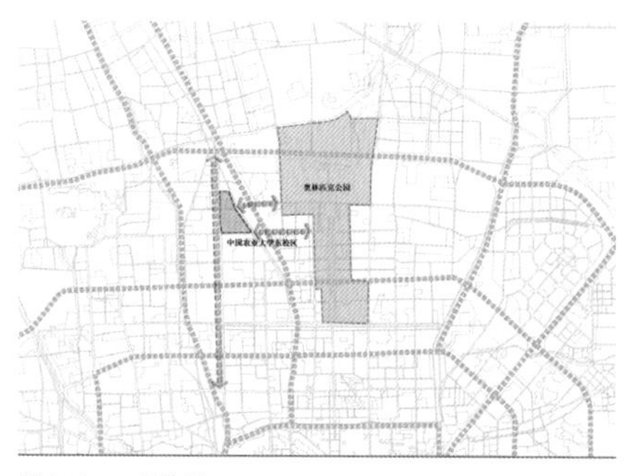

图（1）-1　区位图
Fig.（1）-1　Location map

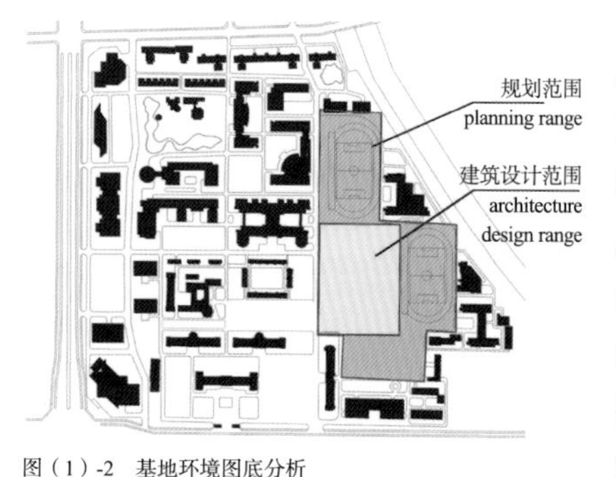

图（1）-2　基地环境图底分析
Fig.（1）-2　Ground analysis of site environment

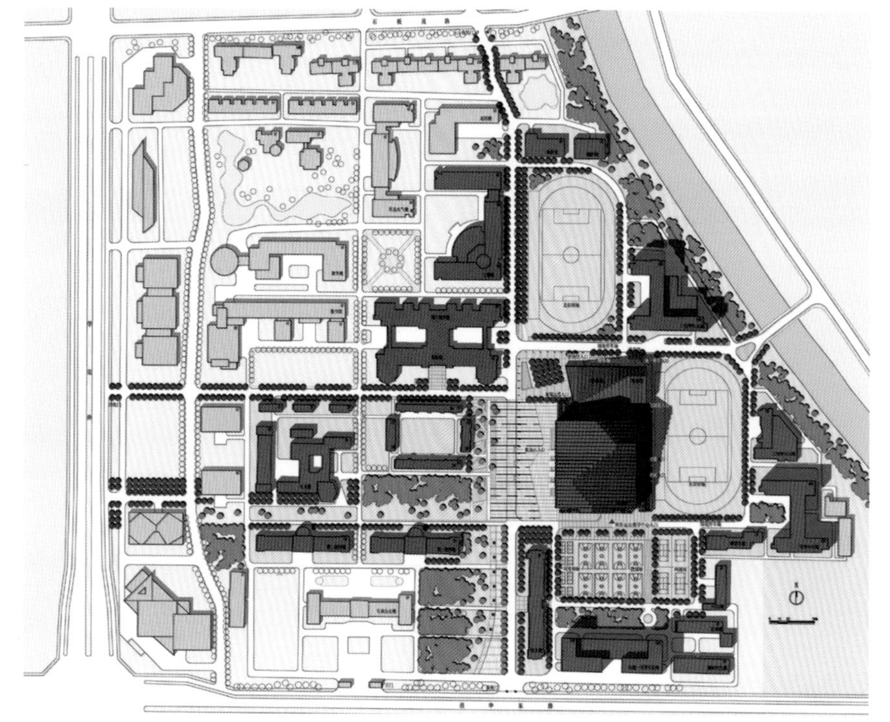

图（1）-3　最终总体布局方案
Fig.（1）-3　Final overall layout scheme

图（1）-4　总体鸟瞰图
Fig.（1）-4　Overall aerial view

设计探索

在 2008 年北京奥运会 11 座新建场馆中，位于大学校园里的体育馆具有奥运比赛用馆和高校体育馆的双重定位。作为奥运会体育比赛场馆，特别是位于大学校园内部的奥运体育馆，意味着奥运体育精神在年轻一代中的持续传播。奥运体育比赛馆的设计应该代表生生不息、永远奋斗的体育精神，使该体育馆真正成为 2008 年奥运会留给学校和城市的历史遗产和精神财富。而作为高校体育馆，体育馆的建设则意味着对校园实体环境质量的巨大改善。同时，赛后的场馆使用问题，体育和教学相结合等问题则是社会体育馆所不需面对的设计问题。此外，体育馆赛后功能的变化对校园和城市也存在长期的影响。

体育建筑，作为建筑类型中功能技术较为复杂的类型，长期的追踪研究是产生设计作品的前提。然而由于普遍的浮躁与急功近利使体育建筑的相关研究相对于快速的建设而言，显得缓慢而落后。更有甚者，由于违背体育建筑基本规律的决策误导，许多已定性的错误设计仍然重

Exploration

Among the 11 newly constructed venues for the Beijing Olympics, venues sited on university campuses targeted at both the Games and the schools. Olympic venues, especially those on campuses, carry the succession of the spiritual value of modern sports from generation to generation, symbolizing the unyielding soul of sports activities, thus become the Olympic legacy for the schools and the city. The completion of the gymnasium improves the overall environment of the campus extensively, while the transformation of its program and function post-game profoundly influences the campus and the city.

For sports facility, as a type of architecture with relatively complex technologies, persistent research is the premise of successful designs. In the context of social impetuosity and anxiety, flawed designs against the basic principle of sports architecture occur frequently, while referential researches drop significantly. In international tenders of domestic sports facility, said phenomenon have recurrently appeared.

图（1）-5　室外照片 1
Fig.（1）-5　Outdoor photo 1

复出现。在国内体育建筑的多次国际招标过程中，上述现象也不断在设计方案中出现。

2004 年 3 月，开始构思本方案的过程中，立足于我们对体育建筑、校园规划和城市设计的研究积累，试图通过求真的分析、自主的理念、务实的应对来回答这一国际招标设计题目。

Since March, 2004, throughout the whole process of out design, we endeavored to submit an honest, independent and pragmatic design for the international bidding based on our accumulated research and experience in sports architecture, campus planning and urban design.

1. 校园特色

中国农业大学东校区，是一座典型的源自 20 世纪 50 年代的北京校园，逻辑严整的校园建筑布局、亲切的毛主席塑像是国内为数不多的能留住历史记忆的大学校园。其中独具一格的是校园内几栋旧的砖混教学建筑，虽然造型与 20 世纪 50 年代的教学建筑一样普通，材料也是质量一般的红砖，然而，建筑师将砖体按照规律略有嵌出，改变了墙面的质感，几栋主要教学建筑虽细节有不同，但均采用了类似的手法，加上养护良好的树木绿化，使校园形成独特的氛围。这一景象令人印象深刻，也深深地影响了我们对建筑的构思。

1. Campus Features

China Agricultural University, where the Beijing Olympic Wrestling Hall is located, is a typical Chinese campus planned and built in the 1950s. The rigorous layout of the university and the amiable statue of Chairman Mao retain the transient sense of recent history. A few distinct brick-concrete structure academic buildings inside the campus, although as ordinary as any other campus building in style and material, have a special wall texture of pixelated bumpiness in red bricks. Although different in details, these academic buildings follow similar approaches, composing a scenery in concert with the well-maintained vegetation, instantly creating a unique campus atmosphere. The overall sights of the campus deeply influenced our design concept for the gymnasium.

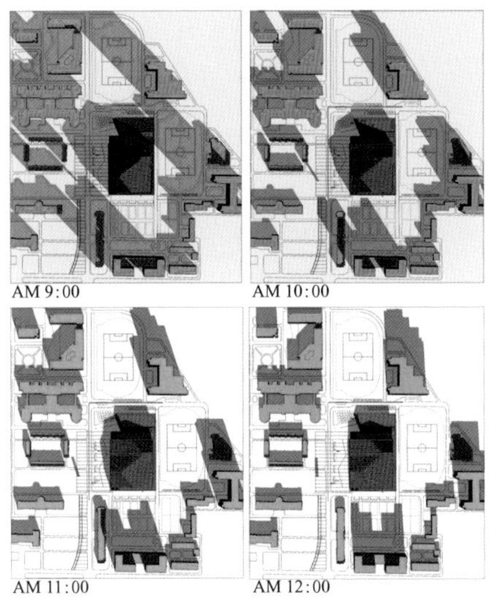

图（1）-6　周边建筑日照分析图
Fig.（1）-6　Shadow analysis of the surrounding building

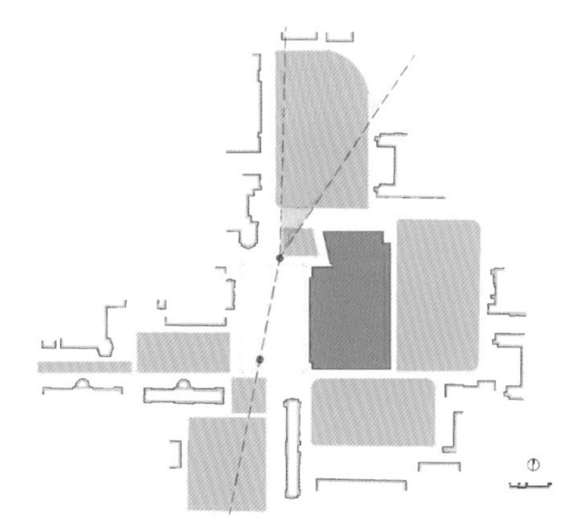
图（1）-7　校园开放空间及周边建筑界面分析
Fig.（1）-7　Analysis of the open space at the campus and the interface of surrounding building

2. 总图布局

为迎接奥运，中国农业大学确定了位于校园体育活动区内，150m×200m 的体育建设用地。尽管局促，却是校园内不可多得的地段。在陆续建设的校园新建筑的簇拥下，体育馆成为校园中心区最为突出的主体建筑。因此在建筑体量处理和景观处理上应突出体育馆主体建筑的标志性作用；另一方面，鉴于校园内建筑的紧凑布局方式，应该采用相应的手法，使体育馆的尺度能够与周边小尺度的建筑相衔接。把握好体育馆与校园环境的协调性和体育馆主体建筑的标志性，是体育馆设计的关键。

从校园总体规划的角度考虑体育馆的定位，将建筑单体与景观统一设计，塑造校园中心区。突出大体量体育馆主体比赛馆的标志性形象，将附属用房与周边建筑相协调。由于用地北面的食品试验楼和三号学生公寓大楼的存在，将体育馆主体建筑布置于用地北部或中部都将使布局较为局促。将体育馆主体建筑布置于用地南端则相对减弱了对周边环境产生不良的压迫感，使原先较为松散的校园开敞空间得以有效的界定和组织，有利于突出体育馆作为校园中心区主体建筑统筹全局的作用。通过综合优化比较，设计方案将体育馆主体布置在用地南端，游泳训练中心布置在用地北端，西侧形成集散广场，并和西侧原有的绿带一起形成校园中心区的主要空间节点。

在用地相对紧凑的情况下，合理的规划赛后用房改造是我们功能设计的基本出发点。作为奥运比赛馆，体育馆的形象必须考虑突出比赛馆的标志性和纪念性，也要求体育馆设计要考虑相应的协调性和相容性。

基于上述分析，我们确定了摔跤馆简洁、内敛的性格定位，以此作为结构选型、体量组合的原则。在经过多种结构形式的比较后，体育馆由比赛馆、游泳馆以及局部平台组成。

2. Site Plan

In preparation for the Olympics, the University wanted the sports gymnasium, which would serve as the Olympic wrestling hall and as the gymnasium after the event, to be established on the land sized 150m × 200m on the campus activity area. Despite the limited site, it was a precious section of the central campus. Amid a cluster of newly erected buildings, the gymnasium would become the most prominent building in the central area, therefore, the massing and the landscape design in the site plan should emphasize on the iconicity and visibility of the architecture. On the other hand, given the compact layout of the campus buildings, appropriate approaches should be adopted to keep the dimension of the gymnasium consistent with the surrounding buildings of smaller dimension. Coordination between the environment of the gymnasium and the campus, and adjustment to the iconicity of the gymnasium are the key to the architectural design.

In the design process, we pondered the proper targeting for the gymnasium from the perspective of the overall campus planning in attempt to shape the campus central area by unifying the building and the landscape, highlighting the iconic image of the gymnasium, and coordinating the auxiliary rooms with surrounding buildings. Because of the food laboratory building and the No.3 student apartment building on the north of the site, it would be unwise to compress the space further by putting the main gymnasium on the center or the north of the site. On the contrary, arranging the gymnasium in the lower south part of the site would weaken its oppression on the surrounding environment, re-structure the disorder open spaces, and highlight the dominating position of the gymnasium as the main building in the central campus. After comprehensive comparison, the main gymnasium was sited in the lower south; the swimming training center, the upper north; the distributed squares, the west. Compensated with the existing

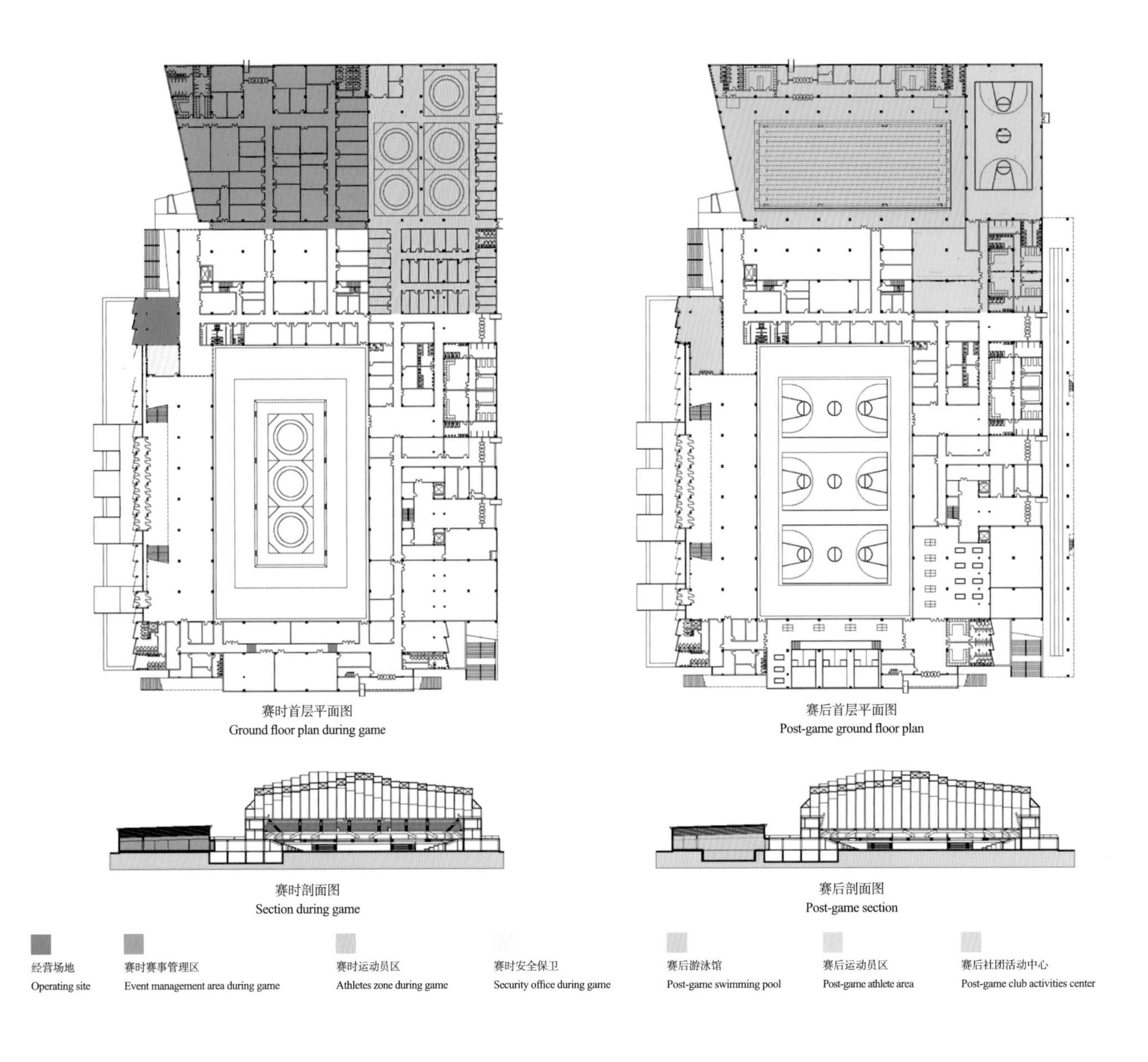

赛时首层平面图
Ground floor plan during game

赛后首层平面图
Post-game ground floor plan

赛时剖面图
Section during game

赛后剖面图
Post-game section

经营场地
Operating site

赛时赛事管理区
Event management area during game

赛时运动员区
Athletes zone during game

赛时安全保卫
Security office during game

赛后游泳馆
Post-game swimming pool

赛后运动员区
Post-game athlete area

赛后社团活动中心
Post-game club activities center

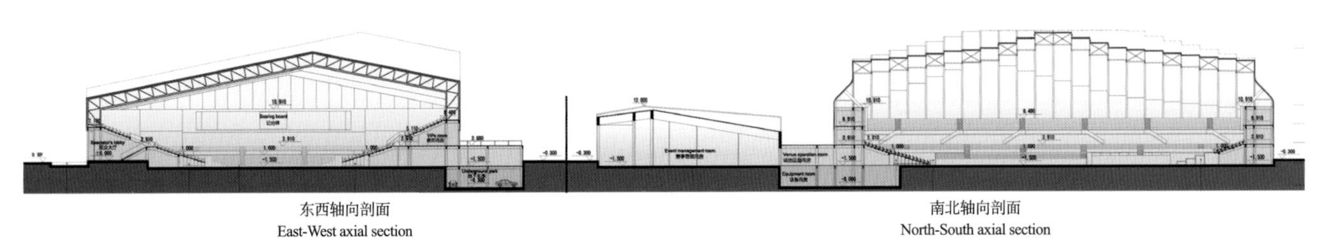

东西轴向剖面
East-West axial section

南北轴向剖面
North-South axial section

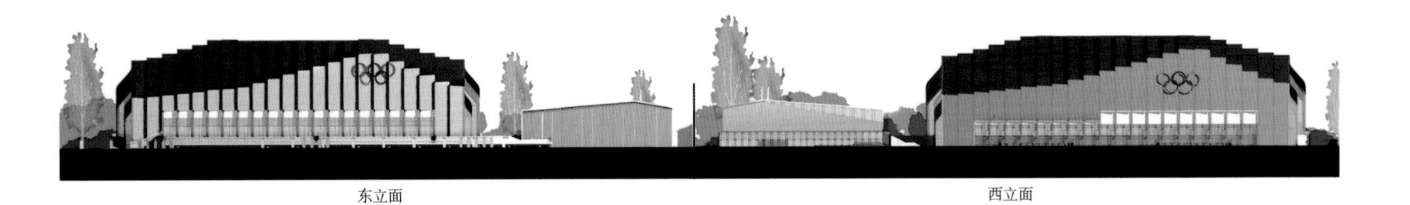

东立面
East elevation

西立面
West elevation

图（1）-8　主要技术图
Fig.（1）-8　Main technical diagram

图（1）-9　室内照片 1
Fig.（1）-9　Indoor photo 1

vegetation west to the main campus streets, aforementioned locations guide the spatial interaction among the campus architecture and landscape.

Due to the relatively compact site condition, the design of function and program relied on logical transformation of rooms post-game. Besides maintaining its iconicity and monumentality for the purpose of the Olympic Games, the gymnasium has to be provided with a certain extent of adaptability and compatibility.

比赛馆是主要体量，平面约 90m × 90m，屋面为反对称折面，采用巨型门式钢架结构，使建筑在规则的平面下富有韵律感。东西立面的檐口是反对称的起伏折线。通过硬朗的折线檐口和屋脊，丰富了天际轮廓，展现阳刚之美，突出纪念性。游泳馆处理同样弱化体量，关照附近其他建筑。

Henceforth, we based the massing and structural design on a concept of concise and introverted architecture. After several adaption, the gymnasium ended up with a competition hall, a natatorium and additional platforms, among which the 90m×90m competition hall is the major volume, with an asymmetrical poly-surfaced roof supported by a series of gigantic steel moment frames, creating a rhythmic profile. Cornices of the east and the west façades are of asymmetrical poly-lines. The sharp silhouettes of the cornices and the ridge enrich the skyline of the site in a monumental manner. The natatorium, rendered modest in terms of volume, coordinates with the surrounding buildings.

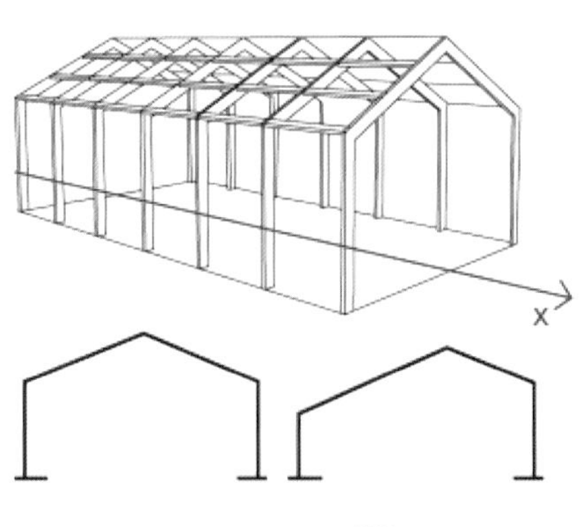

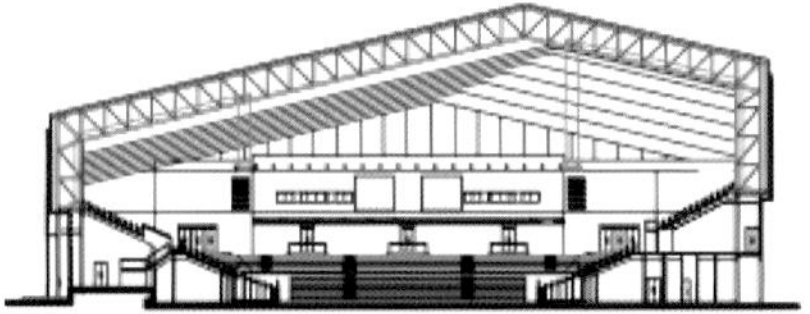

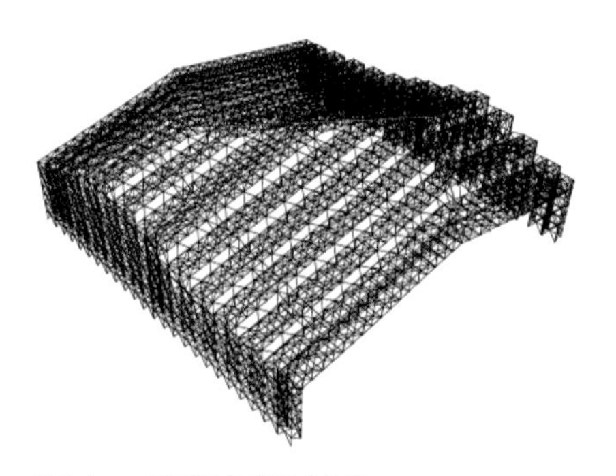

图（1）-10　巨型门式刚架结构体系
Fig.（1）-10　Giant portal frame structure system

可持续发展

1. 场地选型

作为新时期的高校体育馆，其场地大小的确定需要充分考虑日常高校教学与多种活动的需求。经比较，我们确定 40m × 70m 的场地尺寸最适合于本规模的高校体育场馆的使用。不仅满足奥运会的摔跤比赛，残奥会的排球比赛，会后教学场地的布置也具有极大的灵活性。当学校的重大典礼与集会、演出时结合活动座席布置，灵活调整场地大小与座位数量，馆内最大容量将可以超过万人，最大程度满足中国农业大学作为国内一流高校的使用。

Sustainable Development

1. Field size

As a modern gymnasium, its floor shall adapt to various activities besides daily teaching and training. After comparison, we determined the 40m×70m field most suitable for the gymnasium in terms of flexibly meeting the requirements of the wrestling competition of the Olympics, the sitting volleyball games of the Paralympics, and post-game daily teaching. Assisted with movable seats and adjustments to the field size, the maximal capacity of the gymnasium reaches 10,000, satisfying the demands of the university to the farthest extent.

图（1）-11　室内照片 2
Fig.（1）-11　Indoor photo 2

图（1）-12　室外照片 2
Fig.（1）-12　Outdoor photo 2

图（1）-13 室内照片 3
Fig.（1）-13 Indoor photo 3

2. 视点定位

视点的确定同样经过了多方案的比较，除奥运会摔跤比赛外还考虑到体操、篮球、手球、集会、演出等多种需求，视点确定为摔跤垫边线外 1.5m 处，视线差为 6cm，以便于确保舒适的视觉质量。

2. Vantage Point

After several adaptions concerning the demands for activities including wrestling, gymnastics, basketball, handball competitions, assemblies and performances, the vantage point is placed 1.5m outside the boundaries of the wrestling mat with a tolerance of 6cm to ensure comfortable visual experience.

3. 采光通风

本设计赛后将以学生使用为主，节能降耗成为高校体育馆能否真正为学生服务的关键。基于多年来的研究与体育场馆使用的调研，我们在设计中将自然采光通风的可能性作为十分重要的原则来遵守。经过艰苦努力，建筑造型与自然通风、采光很好地结合起来，层层错开的屋面与外墙便于引入自然光，也便于利用主导风向组织通风。初步建成的效果令人十分满意。

3. Natural Lighting and Ventilation

Since the design is developed mainly for student-use post-Olympics, energy saving becomes the key for the university gymnasium to truly serve the students. Complied with the principle, natural lighting and ventilation are imported to the design. From our experience of sports facilities design, we have successfully integrated natural lighting and ventilation with the architectural form. The scattering roof and exterior walls provide satisfying ventilation and lighting effects.

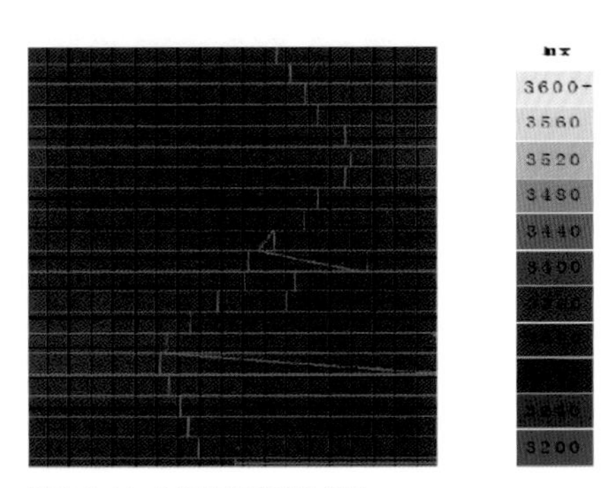

图（1）-14 自然采光和通风分析图
Fig.（1）-14 Natural lighting and ventilation analysis diagram

4. 赛后利用

奥运体育场馆的重要特点就是赛时所需要大量的用房，将面临赛后的处置问题。一味强调商业用途的转换，即便是社会场馆也不会轻易成功，本次奥运的赛后利用还要经过实践的严格考验。我们在本馆的设计中，始终注意空间的集约，在有限的体积内尽力完成最大灵活性的功能转换。结果是，在体积不变的情况下，主馆仍然能满足国际单项赛事要求，赛后馆内还将出现一个供学生使用的、具有一个标准游泳池的、设施完善的游泳馆。

以上是我们对北京奥运摔跤馆的工作小结。大型公共建筑的设计与研究在我国还有许多路要走，希望科学与理性成为永远的基点，以此与大家共勉。

（本案例的文字稿主要根据参考文献 [39]、[43]、[45]、[36] 整理）

4. Post-Game Use

One of the characteristics of Olympic venues is the large number of auxiliary rooms suspended for alternative function after the games. Functional conversion is already a difficult task for social venues, let alone for sports facilities. Therefore, we focused on designing compact spaces. Every effort was made to achieve upmost flexibility in conversions of functionality within the same volume. As a result, the building can still accommodate international competitions with an additional natatorium for students with a standard swimming pool and complete facilities without expanding in volume.

Above all is the summary on the design of the Beijing Olympics Wrestling Hall, open for further criticism and discussion. Considering the difficulties ahead for domestic design and research of large-scale public architecture, we hope that scientificity and rationality would become universal principles for designs.

图（1）-15　总体鸟瞰图

Fig.（1）-15　Overall aerial view

图（1）-16　室外照片 3（上）
Fig.（1）-16　Outdoor photo 3 (up)

图（1）-17　室外照片 4（下）
Fig.（1）-17　Outdoor photo 4 (down)

图（1）-18　室内照片 4（上）
Fig.（1）-18　Indoor photo 4 (up)

图（1）-19　室内照片 5（下）
Fig.（1）-19　Indoor photo 5 (down)

2008年奥运会羽毛球比赛馆（北京工业大学体育馆）中国，北京，2004

2008 OLYMPIC BADMINTON HALL (BEIJING UNIVERSITY OF TECHNOLOGY GYMNASIUM) BEIJING, CHINA, 2004

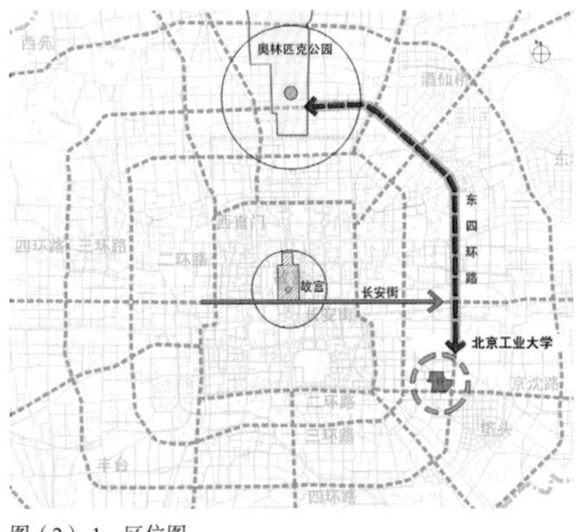

图（2）-1 区位图
Fig.（2）-1 Location map

图（2）-2 基地环境图底分析
Fig.（2）-2 Ground analysis of site environment

项目背景

北京工业大学体育馆位于北京东南部的北京工业大学校区内，2008年奥运会期间这里将作为羽毛球比赛用馆，奥运会后成为大学室内综合体育活动中心。

2004年3月，华南理工大学建筑设计研究院通过现场答辩，资格预审，在40个报名单位的竞争中取得了参加北京工业大学体育馆（2008年奥运会羽毛球比赛馆）方案国际竞赛的资格。11家国内著名设计单位参加的国际招标随即展开，2004年6月评选出华南理工大学建筑设计研究院、中国建筑设计研究院、英国奥雅纳与荷兰联合体3个优秀入围方案。2004年7月至12月，经过多轮优化比较，对规模、功能及技术细节进行了分析和比选，其中还经历了羽毛球馆项目从取消到恢复以及用地调整的过程。2005年元月确认华南理工大学建筑设计研究院为中标实施单位。经过3年努力，北京工业大学体育馆全面落成。

Background

Beijing University of Technology Gymnasium, located on the campus of Beijing University of Technology southeast of Beijing, was the badminton competition venue for the 2008 Olympics and currently serve as a multi-functional sports center of the university after the game.

In March, 2004, after preliminary presentation and audition of 40 architectural firms, Architectural Design & Research Institute of SCUT was qualified to participate the international bidding for the gymnasium. From June, 2004 to January, 2005, projects submitted by Architectural Design & Research Institute of SCUT, Architectural Design & Research Institute of China, and Union of Arup Limited and Netherlands went through rounds of conscientious evaluations on scale, function and technical details, as well as adjustments to the site, cancelation and resume of the project; eventually, the design of Architectural Design & Research Institute of SCUT was approved for execution. Three years later, the gymnasium was thoroughly completed.

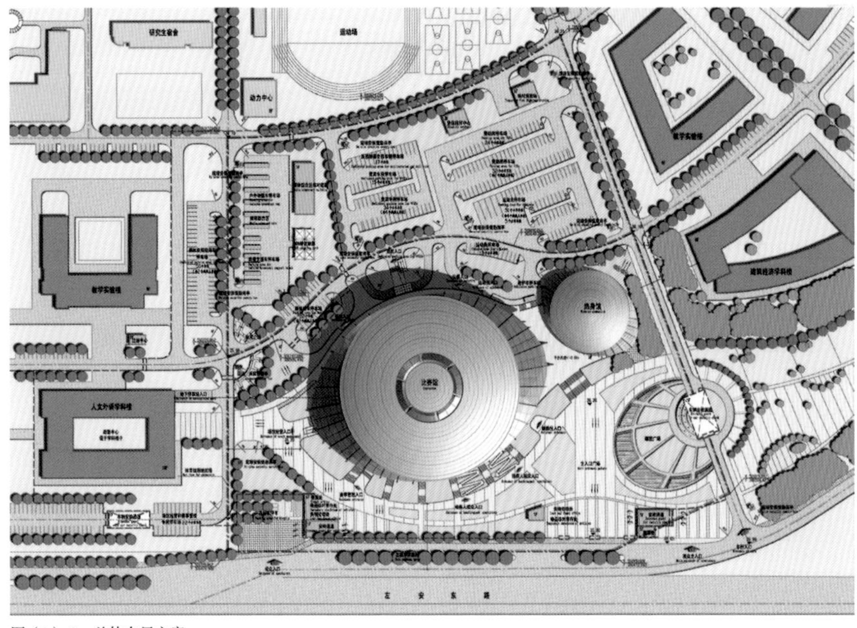

图（2）-3 总体布局方案
Fig.（2）-3 Overall layout scheme

图（2）-4　总体鸟瞰图 1
Fig.（2）-4　Overall aerial view 1

场馆定位

北京工业大学位于北京市东南部朝阳区。校区从西北向东南逐步发展，形成了 20 世纪 50 年代、60 年代的东西向轴线。北京工业大学体育馆位于校园东南正在拆迁的新规划用地。在规划为教学区和运动区的校园新区中，体育馆处于学校边缘，是校园空间和城市空间的临界地带，现状杂乱，城市空间破碎而不确定。同时，由于东临东四环路，南接京沈高速公路西延线，交通便利。

从城市区位看，北工大体育馆所处的位置及其重要，四环路与京沈高速公路在这里立体交叉，该区域为北京东南部的城市出入口，北工大体育馆的兴建也构成了北京东南部重要的景观节点，并为周边地区提供了重要的发展契机。

Targeting

Beijing University of Technology locates in Chaoyang District of southeast Beijing. The campus was developed from northwest to southeast, creating an east-west axis from the 1950s to the 1960s. The gymnasium was initially designated at the newly-planned area in the southeast of the campus, close to the border of the university in the campus planning. The site, at the intersection of urban space and campus space, was under chaotic conditions, whereas a main avenue and an extension of an expressway adjacent to the site provided convenient transportation opportunities.

From an urban perspective, the site of the project was ultimately significant, where the Beijing-Shenyang expressway and the 4th Loop intersect spatially. As an urban threshold in the southeast of the city, the

图（2）-5 投标方案总体鸟瞰图
Fig.（2）-5 Overall aerial view of the bidding scheme

gymnasium would become an importance urban landscape and offer an opportunity of development for the surrounding area.

图（2）-6 投标方案模型照片
Fig.（2）-6 Model photos of the bidding scheme

基于其特殊的位置，决定了羽毛球馆的设计首先应积极考虑城市设计问题，它建成的意义也不应该只是一个简单的体育场馆，而应该为城市提供明确的、有活力、有场所感的公共空间。北工大体育馆将在环境中扮演多重角色：作为校园建筑，它将是校园建筑群体的有机组成部分；它位于校园边缘区，也承担着界定校园边界的作用；在城市中它将是整个北京东南区的入口和全民健身中心，其形象应该有一定的标志性；而作为奥运会的羽毛球项目的比赛馆，又赋予了它另一层新的主题意义。

The particular location makes the design of the gymnasium/badminton hall not only an issue of sports facility, but an identifiable participator of the urban public space system with distinct and energetic characters. The gymnasium bears various responsibility, including the landmark of the university border, the organic constituent of campus building clusters, the threshold of the southeast of the city and the fitness center for the masses, thus its architectural image requires a certain extent of iconicity. Its identity as the badminton competition hall for the Olympics imbues the design another layer of thematic importance.

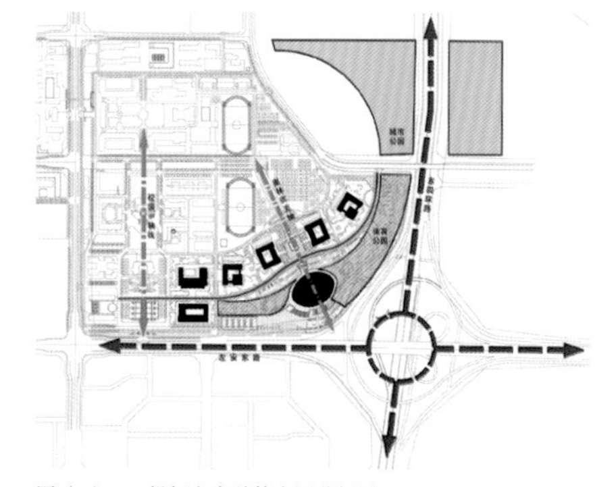

图（2）-7 投标方案总体布局分析图
Fig.（2）-7 General layout analysis of the bidding scheme

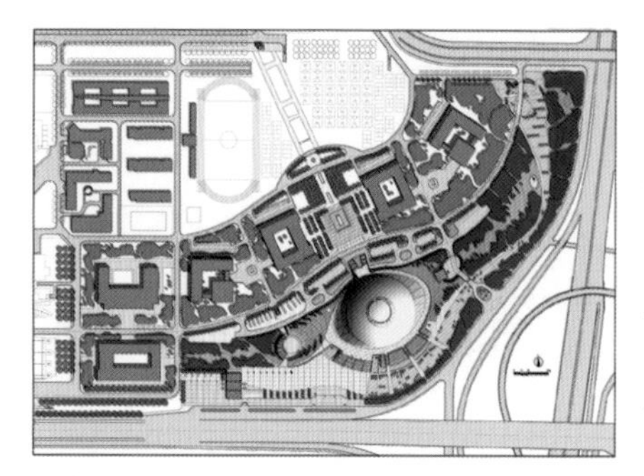

图（2）-8 投标方案总平面图
Fig.（2）-8 General layout of the bidding scheme

1. 总图布局

方案竞赛阶段：方案创作过程中，我们花了很多时间去研究不同体型建筑的各种总图的排布方式与周围环境的关系，在竞赛成果展览上几乎看到了推敲过程中被放弃的各种组合。从最后的结果看，这种投入是必要的。结果是把体育馆的主体建筑和邻近的两幢教学楼形成一个有机的建筑群体，三者之间围合成一个广场，呼应校园的“体育轴线”，体育馆附属的训练馆和热身馆顺应地形，分列两侧，这样的总图布置既考虑到与校园的结合，又呼应城市，使建筑和周围环境很好的融为一体。首先是确定了把用地范围包括100m的城市绿化带范围作为体育公园来设计，体育公园界定了学校和城市的边界，为市民提供一个健身活动公共场所。体育馆的主体和附属建筑作为公园上的节点，和公园一起组成连续的景观，构成北京东南区的城市入口景观。

1. Site Plan

During the design process, we went through a considerable amount of different site plans to investigate into the relationship between the massing distribution and the surroundings, which were presented at the exhibition of the bidding. Consequently, the main volume of the gymnasium formed an organic cluster with two adjacent academic buildings and outlines a plaza in between, corresponding to the axis of the site plan. The auxiliary training hall and warm-up hall were located at the two sides according to the terrain condition. The site plan coordinated with both the campus and the urban environment, allowing the gymnasium to merge with the surroundings. Within the given land area, a 100 square-meter urban green area was used as a sports park that divided that campus and the city where citizens could exercise. The park, with the gymnasium as a visual and structural punctuation, provided a coherent and harmonious urban landscape for the urban threshold.

图（2）-9　室内照片 1
Fig.（2）-9　Indoor photo 1

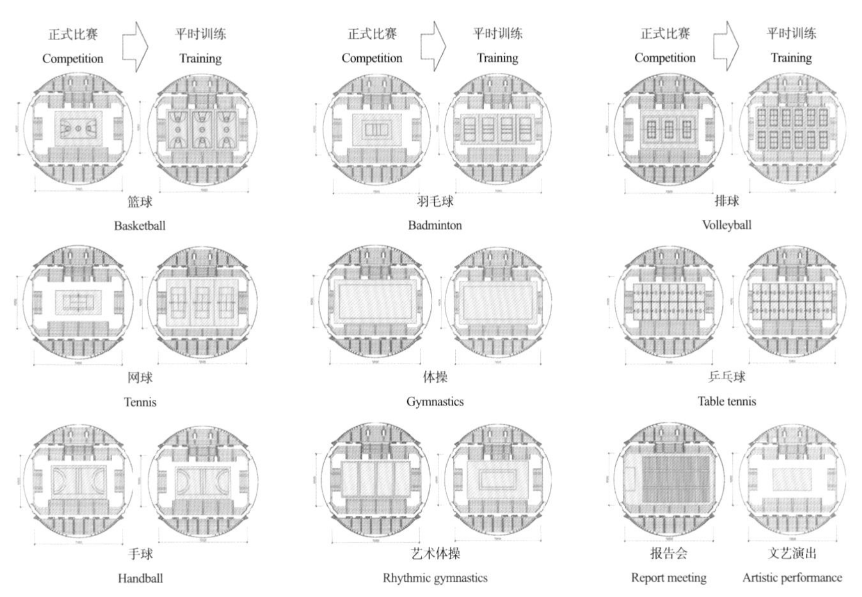

图（2）-10　场地多功能分析图
Fig.（2）-10　Analysis of the multi-functional floor

工程实施阶段：由于羽毛球馆建设决策过程中的一些变化，最终将羽毛球馆的建设用地西移，规划也进行了修改。球馆规模从万人减少到 8 000 席，同时将热身馆与训练中心合并设计。最终的总图有了较大的调整。

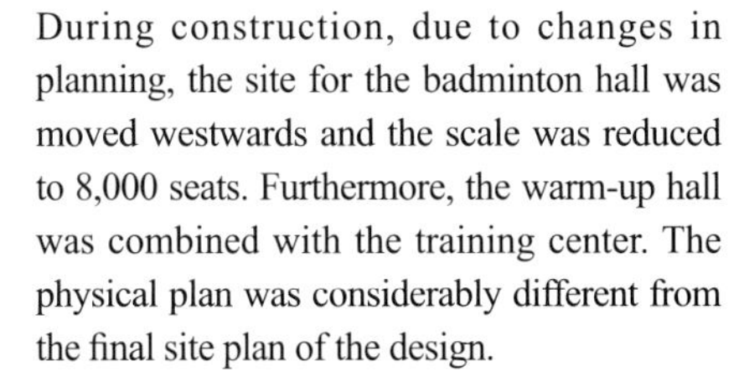

During construction, due to changes in planning, the site for the badminton hall was moved westwards and the scale was reduced to 8,000 seats. Furthermore, the warm-up hall was combined with the training center. The physical plan was considerably different from the final site plan of the design.

2. 功能配置

多功能体育馆主要体现在以下方面：

1）场地选择满足最大的使用可能性。

2）座席的布局与形式具有多种组合的可能性，适应各种场合的使用。

3）注重赛时和赛后的功能转换与用房的适用性。

北京工业大学体育馆，是高校体育、竞技体育、全民健身的一个结合点，对体育场馆的多功能设计提出了更有挑战性的要求。作为羽毛球馆，场地的确定较为简单。但考虑到作为大学体育馆以及赛后多种用途的要求，我们以确保体育场馆的最大灵活性、适应性为原则，确定了 40m × 70m 的比赛场地，以便在赛后满足包括国际体操赛事在内的各种比赛要求。

2. Function and Program

The multi-functionality of the gymnasium was achieved by the following measures:

1) The floor was designed to adapt to maximum utilization;

2) The seats could be composited differently for various occasions;

3) Be aware of the function conversion between intra-game and post-game uses and the adaptability of spaces.

Beijing University of Technology Gymnasium was venue for higher education institute sports, competitive sports and civil sports. The multi-functionality of the gymnasium was a challenge to designer. It would have been relatively easier to design the field if only for a badminton hall, yet for a venue responsible for various activities, it was required for the designers to adhere to the principles of maximum adaptability and compatibility. A 40m×70m field was chosen to meet the requirements of various post-Olympics Games including the International Gymnastics Tournaments.

图（2）-11　室内照片 2
Fig.（2）-11　Indoor photo 2

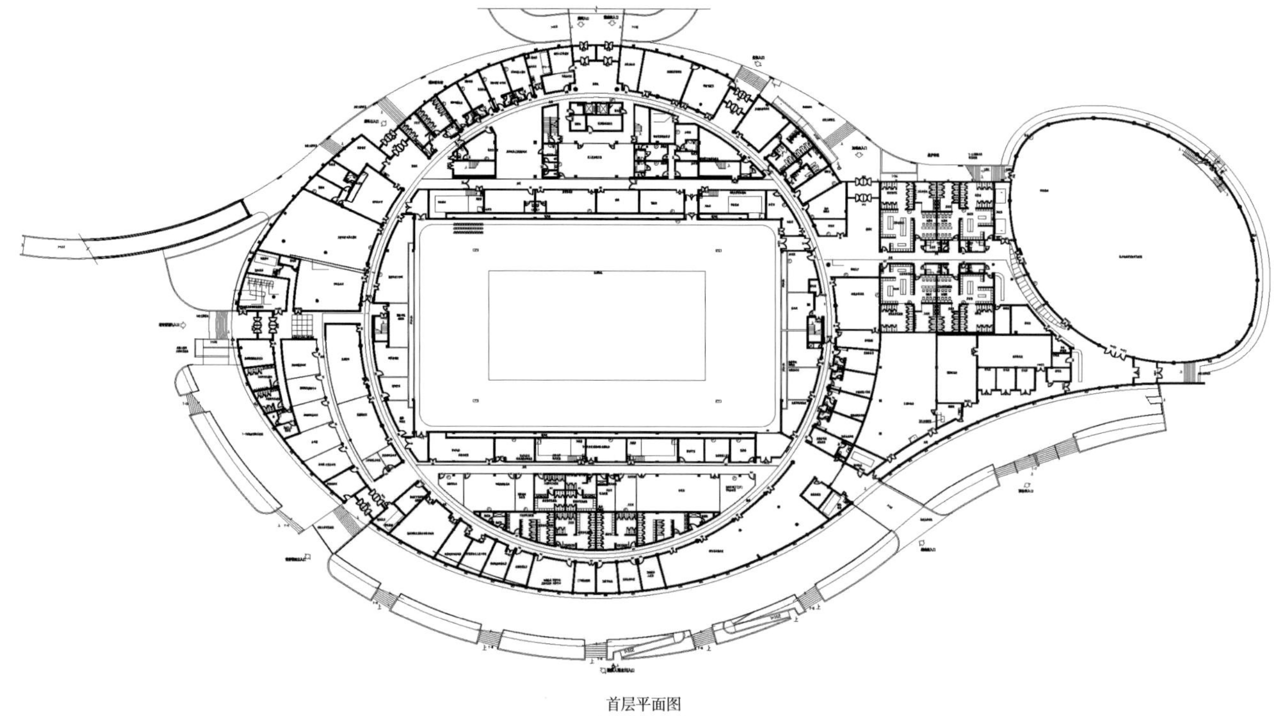

首层平面图
Ground floor plan

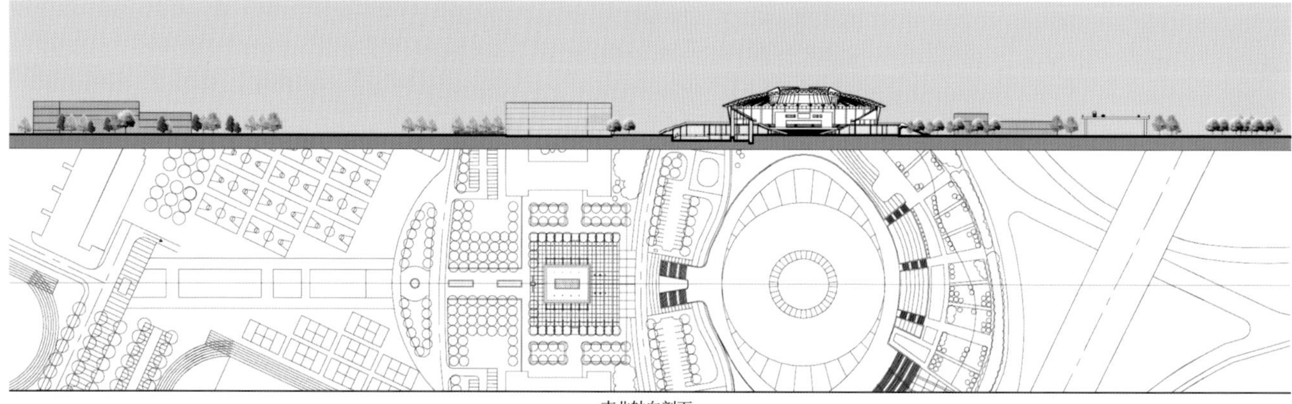

南北轴向剖面
South-North axial section

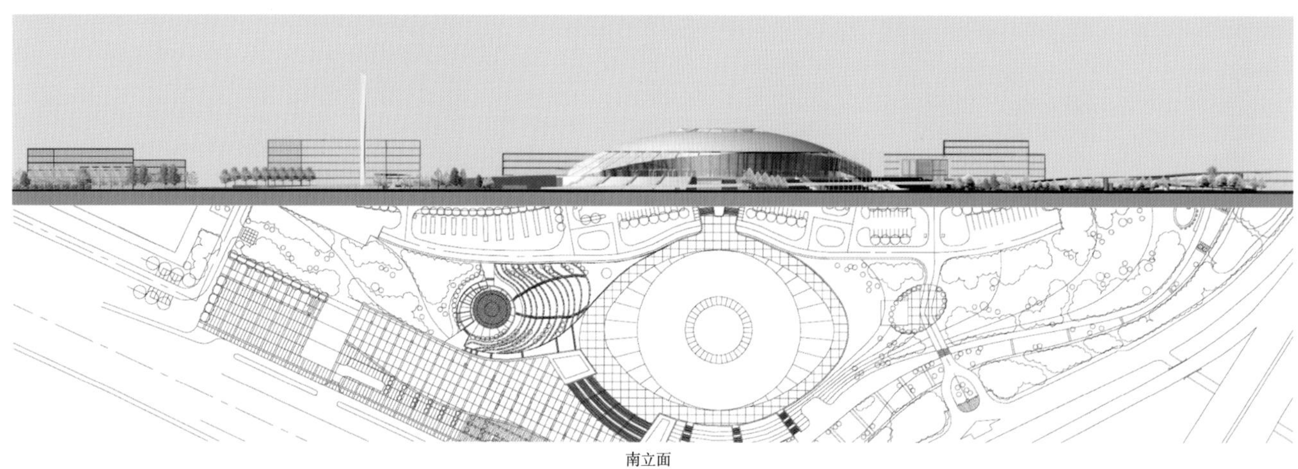

南立面
South elevation

图（2）-12　主要技术图
Fig.（2）-12　Main technical diagram

图（2）-13　室外照片 1
Fig.（2）-13　Outdoor photo 1

调整用地与规模后，使热身馆能够满足单独对外开放并达到羽毛球训练基地的功能要求。

The warm-up hall could be used independently as a public venue and a badminton training base after the scale and the site was adjusted.

3. 建筑形式

体育馆的造型设计，除了要反映体育馆本身的结构、功能形式之外还要满足各种不同的心理需求。

北京工业大学体育馆本身就是一个具有多层意义的建筑，所以它的造型设计也应该相应地满足多方面的心理需求。作为城市主入口的标志性建筑，必然要求建筑形体简洁、一目了然。作为奥运会的羽毛球比赛馆，又必然要求在建筑造型上给与一定的诠释。另外，顺应总体设计的理念，建筑体量之间的关系，也制约这建筑造型的选择。我们对建筑造型的定位是简洁、流畅、舒展、富有体育建筑气息、与公园完美地融合。

一直以来，体育建筑外形被要求了太多的意义，许多“俗成”的压力给体育建筑设计带来了严重干扰却又不得不面对。羽毛球馆要像羽毛球或球拍、摔跤馆要像摔跤、游泳馆要有“水”的理念，凡此种种不一而足。此外，体育建筑设计弥漫着追求表皮精致与概念，完全忽略体育建筑本质，造成建造成本增加，技术复杂，最终的建设效果却无法达到渲染图的夸张表达。

3. Architectural Form

The form of the gymnasium shall not only reflect physical aspects such as its structure and functions, but also satisfy psychological demands.

Beijing University of Technology Gymnasium, as a building of multiple thematic and physical significant, its form should satisfy various psychological demands. The form should desirably be clear and distinct due to its position as a landmark of an urban threshold; self-explanatory due to its identity as an Olympic venue. Additionally, the overall design concept and relationship between buildings restricted the design of the form. Our fundamental concept was to design a concise, coherent, relaxing and sports-related form that was consistent with the atmosphere of the park.

Since long, the architectural form of sports venues has been given too much meanings. Conservative concept brings inevitable pressure and interference to the design of sports venues. A commonly heard theory is that badminton halls should resemble badminton balls or rackets; wrestling halls, wrestling; and natatoriums, the water. Furthermore, the prevailing trend for sports venues to fall into surficial designs regardless of the spirits of sports venues has

图（2）-14　室外照片 2
Fig.（2）-14　Outdoor photo 2

图（2）-15 屋面结构施工照片
Fig.（2）-15 Construction photo of the roof structure

led to increased investment and complex techniques, which, however, still fail to materialize the exaggerated effect in the digital renderings.

北工大体育馆的造型设计中，并没有在造型上刻意模仿羽毛球，只是在建筑形式上追求一种和羽毛球运动相暗合的飘逸、空灵、轻松、洁净的效果。由于有这种定位，最后的造型设计既有羽毛球的感觉，但又不会刻意去模仿，较为含蓄。

Beijing University of Technology Gymnasium did not deliberately imitate the form of a badminton ball, but rather attempted to mimic the light, purified and relaxing atmosphere of badminton as a sport. Therefore, the final design subtly embodied the essence of badminton without overtly duplicating the form.

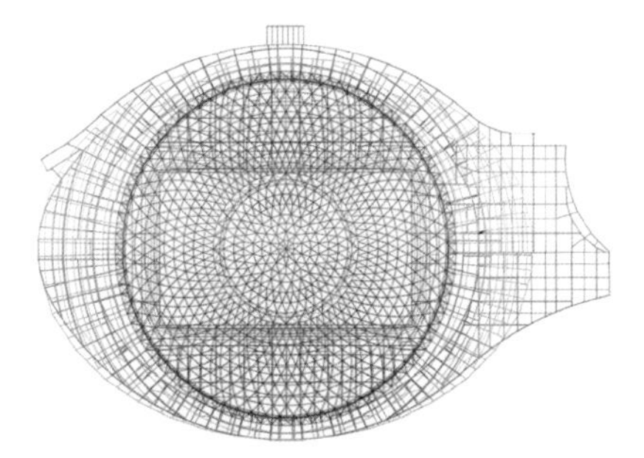

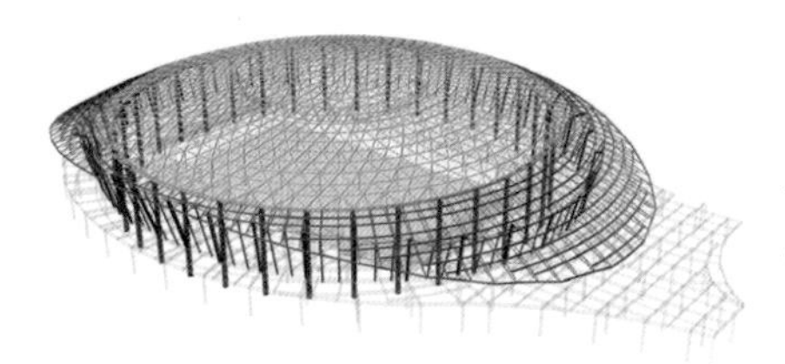

图（2）-16 屋盖结构体系分析
Fig.（2）-16 Analysis of the roof structure system

4. 结构选型

由于众多的限制条件，在确定了形状体量之后，结构选型应积极配合建筑构思。特别是考虑到造价的严格控制，本方案在投标阶段确定了采用球网壳的结构形式。力学性能良好的球网壳的结构经过切割，产生流畅的曲线，不仅满足了建筑功能的要求，同时利用网壳结构的灵活性，创造出流畅的曲线造型。在网壳的边缘再加以曲线型的玻璃雨棚，突出结构本身的美感和韵律感。

进入初步设计阶段，业主提出了采用预应力钢结构的设想，我们结合建筑室内外空间形式的要求，在初步设计阶段提出了三种结构方案。最终完成了预应力张悬结构方案使羽毛球馆屋盖结构成为本次奥运会体育场馆中最为先进的结构形式。

4. Structure Design

Due to a number of limitations, especially the strict cost control, the structure design was subject to the overall architectural concept. A reticulated shell structure was chosen at the tendering stage. The remarkable structural performance of reticulated shell allows the structure to be cut to create smooth curves that meet the internal function and program demands of the gymnasium. Curved glass rain-sheds were arranged along the edge of the shell to prompt the elegant rhythm of the shell structure.

As the project developed, the employer proposed to adopt a pre-stressed suspend-dome structure. Based on the spatial form required for the interior and exterior of the gymnasium, we designed three structural solution to the architecture form. Eventually,

图（2）-17 室外照片 3
Fig.（2）-17 Outdoor photo 3

羽毛球馆的设计，从构思、调整到深化、建成，经历了多方面的协调与互动。大到城市空间、校园环境，细微到活动座椅的伸缩变化，使我们经历了重要体育场馆建造的一次全程体验。我们逐渐感受到了成功的喜悦，也留下了许多思考：体育建筑适用、美观的原则如何确保；结构技术的先进性、适用性与易建造性的辩证关系；大型体育赛会对体育场馆建设的影响等，当 2008 年北京奥运会的欢呼响起的时候，我们的求真好像才刚刚开始。

（本案例的文字稿主要根据参考文献 [47]、[48] 整理）

the new structural system of the pre-stressed suspend-dome was selected as the main structure of the roof of the Badminton Hall. It became the most advanced structure among the 2008 Beijing Olympic venues.

The design of the badminton hall, from preliminary design to completion of construction, underwent a multi-level cooperation. Form urban structure to adjustable seats, we participated in the whole process of the establishment of a major sports venue. The success has led to subjects open to further discussions, i.e., methods to protect the aesthetics and functional adaptability of sports architecture, the critical relationship between advance structure and simple construction, the impact of major sporting events to the venues, etc. As the 2008 Beijing Olympics awoke, our exploration continued.

图（2）-18 设备系统施工照片
Fig.（2）-18 Construction photo of the equipment system

图（2）-19　总体鸟瞰 2（上）
Fig.（2）-19　Overall aerial view 2 (up)

图（2）-20　室外照片 4（下）
Fig.（2）-20　Outdoor photo 4 (down)

图（2）-21 室内照片 3（上）
Fig.（2）-21 Indoor photo 3 (up)

图（2）-22 室内照片 4（下）
Fig.（2）-22 Indoor photo 4 (down)

2010年亚运武术馆（南沙体育馆）中国，广州，2007

2010 ASIAN GAMES WUSHU HALL (NANSHA GYMNASIUM) GUANGZHOU, CHINA, 2007

图（3）-1　区位图
Fig.（3）-1　Location map

南沙体育馆位于广州市南沙区黄阁镇，建成后作为2010年广州亚运会武术比赛馆并承担亚运会卡巴迪项目的比赛，是本届亚运会的新建大型场馆之一。南沙体育馆总建筑面积3万多平方米，亚运期间总座席数为8 000多个，其中固定座席约6 000个，活动座席约2 000个，赛后如拆除新闻媒体席等临时工作席而恢复为普通观众座席，则总容量可达8 800多席。建筑主体采用了钢筋混凝土结构及钢结构，总高度为29m，其中比赛大厅主体钢结构采用了先进的双层环形张弦穹顶结构，主跨度达到了98m，处于亚洲同类结构体育馆的前列。

Nansha Gymnasium, located at Huangge Town, Nansha Development Zone, Guangzhou, as the Martial Arts and Kabaddi Gymnasium of 2010 Guangzhou Asian Games, featured a total construction area of 30,000m^2, and a total of 8,000 seats, including 6,000 fixed seats and 2,000 mobile seats. If temporary seats for the press and media removed, the maximum capacity of the building reaches 8,800. Nansha Gymnasium was structured in reinforced concrete and steel, with a total height of 29m. The main steel structure of the hall was an advanced circular suspend-dome with a span up to 98m, ahead of the majority of similar gymnasiums in Asia.

总图构思

广州南沙区体育中心用地约42.2ha，南面为珠江水系的江面—蕉门水道，十分靠近珠江的出海口。用地原为农田、池塘和蕉林等，周边丘陵连绵起伏，自然环境优美，具有典型的滨海地域特征。整块用地将分两期进行建设，一期工程即为亚运会期间投入使用的南沙体育馆以及整个用地范围内的道路停车、园林景观等设施；二期工程将在一期建设的基础上，在预留的用地内建设20 000座的体育场、露天游泳中心及其相关的配套设施。

Site Plan

The total land area for the Nansha Sports Center project was approximately 42.2 ha north to Jiaomen Waterway of the Pearl River. Originally used as farmland, including ponds and banana plantations, the site was surrounded by undulating mountains and pleasant nature, a typical environment of southern coastal areas. The construction of the Sports Center had two phases, one for the Asian Games related Nansha Gymnasium, its parking and landscape; the other for a 20,000-seat stadium and an open-air swimming center and auxiliary facilities.

图（3）-2　总体布局
Fig.（3）-2　Overall layout

图（3）-3　总体鸟瞰图 1
Fig.（3）-3　Overall aerial view 1

在对用地进行规划及体育馆单体进行设计的过程中，建筑与城市、自然环境的协调始终是设计灵感的源泉。只有对城市、环境做出认真思索，发扬建筑的场所精神，才能创作出优秀的建筑作品。方案中标后通过多次推敲和比较，并结合亚运期间及远期发展的要求进行综合考虑，在一期的规划中将体育馆布置在临近蕉门水道的用地，东南部体育场、游泳中心的预留用地，在亚运期间建设成为绿化景观、停车场等，以满足亚运期间的需求。通过对道路广场和园林景观的精心设计，突出了体育馆流畅的曲面造型，使之与周边连绵起伏的丘陵相互映衬，成为了蕉门水道边上的一道亮丽风景线。

From planning to architectural design, the coordination between the building, the city and the nature had always been the inspiration. Otherwise, without scrupulous research and analysis of the city and the environment, successful architecture would not have been produced. After we won the bidding, we placed the gymnasium of the first phase near Jiaomen Waterway and used the undeveloped area intended for the stadium and the swimming center in second phase for landscape and parking. By detailed design of the landscape and the circulation, the curved exterior of the gymnasium was emphasis among the background of mountains, water, and vegetation.

1. 海天一色的基地环境

南沙，广州的最南端，珠江三角洲的咽喉之地。体育馆临水而建，水波浩渺的景观环境令人遐想无边。南沙体育馆作为 2010 年广州亚运会武术及体育舞蹈比赛馆，是本届亚

1. The environment among the mountains and rivers

Nansha is the southernmost of Guangzhou City, the heart of the Pearl River delta area. Nansha Gymnasium, built by the boundless water, lights up the visitors’ imagination. As

图（3）-4　总体鸟瞰图 2
Fig.（3）-4　Overall aerial view 2

运会新建大型场馆之一，总建筑面积 3 万多平方米。

从接标开始，海天一色场景就始终徘徊于脑际。尽管从规划布局到单体功能分析，需要严肃的理性分析；尽管复杂的结构体系与流线分区常常禁锢浪漫飞扬的概念，在对城市、环境的条件做出认真思索，努力发扬建筑的场所精神，并据此进行精心设计的基础上，我们通过多次的方案推敲和比较，并结合远期建设要求进行综合考虑，完成了总图设计。规划中将体育馆布置在了临近蕉门水道的东南部，体育场、游泳中心的预留用地，则通过道路广场和园林景观的设计，突出体育馆流畅的造型设计，使之成为蕉门河水道上一道亮丽的风景线。

the Martial Arts and Rhythmic Gymnastics Hall for the 2010 Guangzhou Asian Games, the gymnasium was one of the largest construction of the Games with a total construction area of over 30,000m^2.

Though rational analysis form master-planning to internal programs, complex structure and intricate circulation often circumscribe creativity and imagination, we had been adhere to the inspiring environment in our design since the bidding. After rounds of adjustments to the site plan, concerning the long-term utilization of the venue, we placed the phase one gymnasium near Jiaomen Waterway and used the undeveloped area intended for the stadium and the swimming center in phase two for landscape and parking. By detailed design of the landscape and the circulation, the curved exterior of the gymnasium was emphasis among the background of mountains, water, and vegetation.

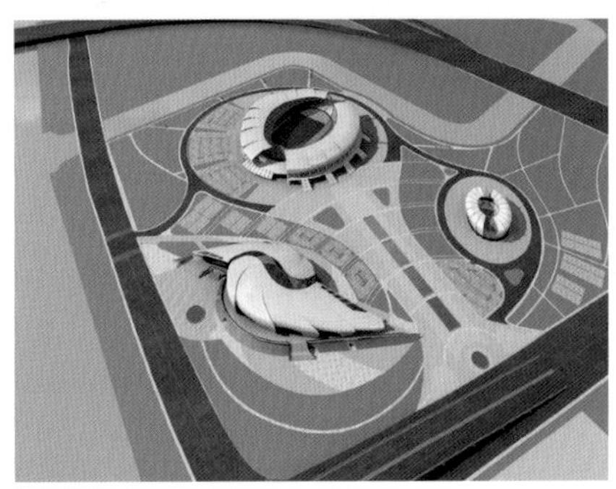

图（3）-5　投标方案设想的体育中心布局
Fig.（3）-5　Sports center layout envisaged in the bidding scheme

2. 经典平面与特色形体

建筑发展中产生了许多经典的平面布局，如体育馆建筑中，最令人难忘的代代木体育馆。圆形的比赛大厅，沿切线方向外延，形成两个宽敞而方向感明确的休息厅。南沙体育馆则在理解和学习这一处理手法的基础上，在一侧安排了热身训练馆，并结合二层平台，形成了适合亚热带气候特色的半开敞休息空间。

2. Classic plan and special features

With understanding of and reference to the classic layout of Yoyogi National Gymnasium, Nansha Gymnasium was designed with a round competition hall tangentially extended to two spacious lounges with clear senses of direction. A warm-up training center was arranged on one side of the gymnasium, forming a semi-open resting and activity space in combination with a two-storey platform that suited the subtropical climate in Guangzhou.

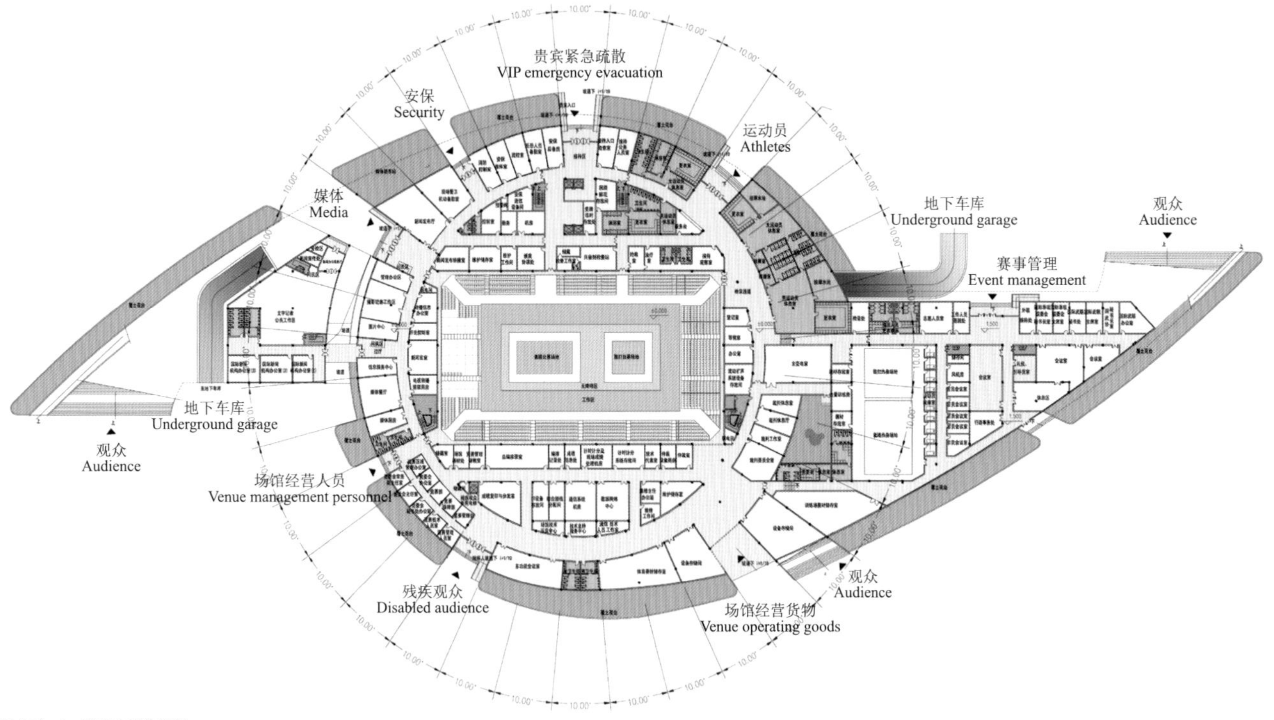

图（3）-6　平面功能分析图

Fig.（3）-6　Analysis of the plan function

图（3）-7　平面意向图

Fig.（3）-7　Intention map of the plan

同时，作为亚运会武术比赛的主赛场，武术特点与相关议题让我们不得不关注。建筑平面形态与太极图的巧合，给了我们巧妙利用“武术”概念的机会。设计中，将组成体育馆外壳的9个曲面单元片片层叠，并分为南北两组以比赛大厅圆心为中心呈螺旋放射状展开。这样处理的目的是将单一的建筑体量一分为二，并以一种富有动感的方式将两者紧密联系。运用近似太极阴阳图的构成方式，隐喻中国武术的最高境界——“阴阳俱合，天人合一”，成为设计的一个特色。

此外，结合广东地区独特的地域文化——海洋文化的特征，我们又借鉴了富有肌理变化的“海螺”外壳作为造型设计的意向，通过金属屋面的片片层叠处理及颜色深浅的变化，力求创造出一种蕴含了地域文化特征的建筑形态，将中华传统文化和地域文化巧妙结合是我们的设计目标。

上述三个方面的融合，逐步形成了我们最初的设计构思，并使设计创意为评委和业主接受，最终成为实施方案。应该强调的是，尽管

This classic plan, resembling the Taiji Diagram, subtly implicated a ‘Wushu (Chinese martial arts)’ concept. In the design, nine curved parts of the exterior shell were layered in spiral order in two groups from the south and the north, with the competition hall as the oculus. The single mass was divided into two parts that were closely connected with each other in a dynamic way. The form of Taiji Diagram implied the highest level of Chinese Wushu– the combination of Yin and Yang, the unity of heaven and human.

Additionally, combined with the unique local culture of Guangdong – the marine culture, with reference to the changing texture of sea snails, the design of the layered metal plates of varying colors of the roof had evoked reminiscence of Guangdong’s regional and cultural features, which corresponds with our goal to integrate Chinese traditional culture with regional peculiarity.

The abovementioned measures consisted our essential concept, which was approved by the employers and the judges. It was noteworthy that, although the conch

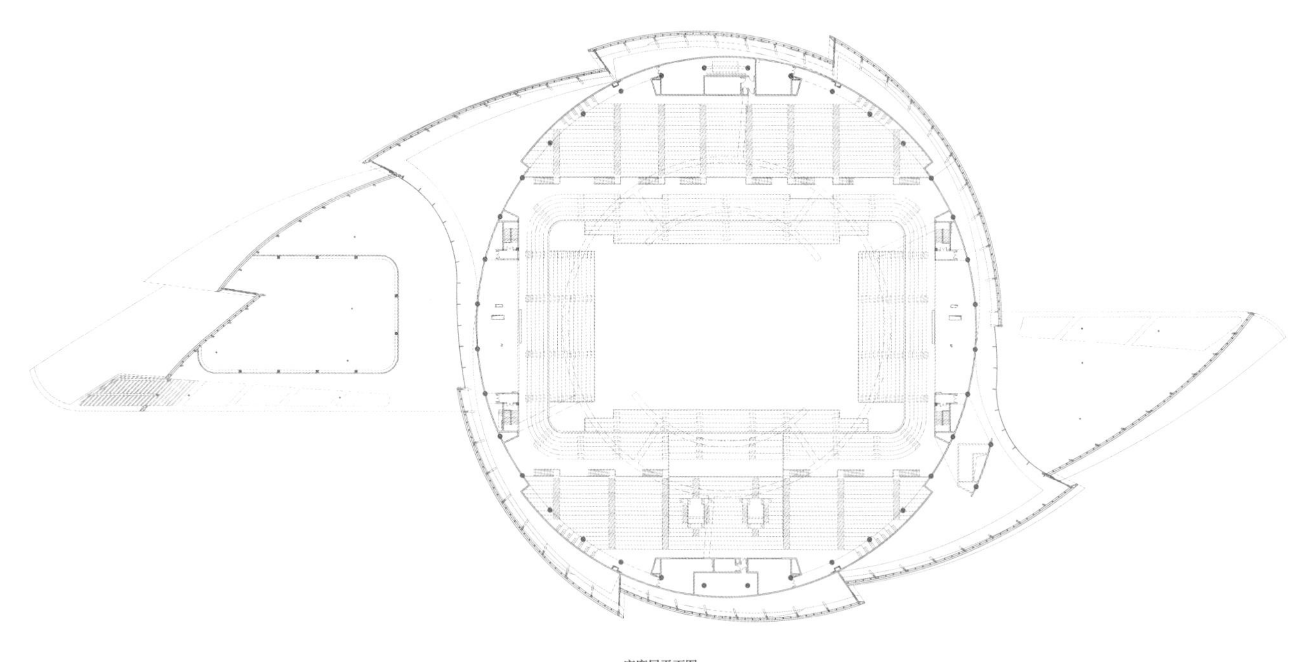

座席层平面图

Floor plan of the stands level

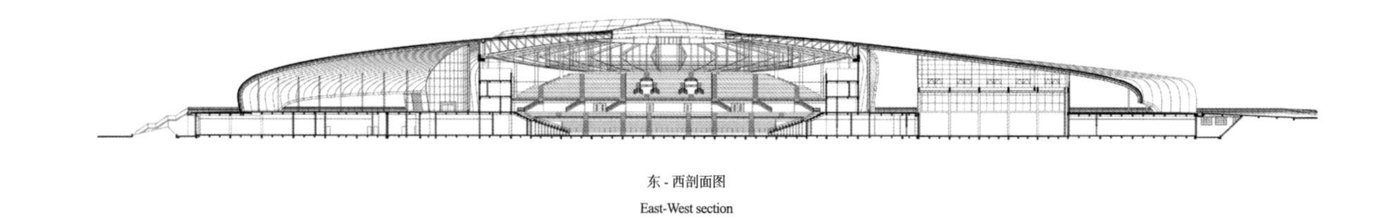

东 - 西剖面图

East-West section

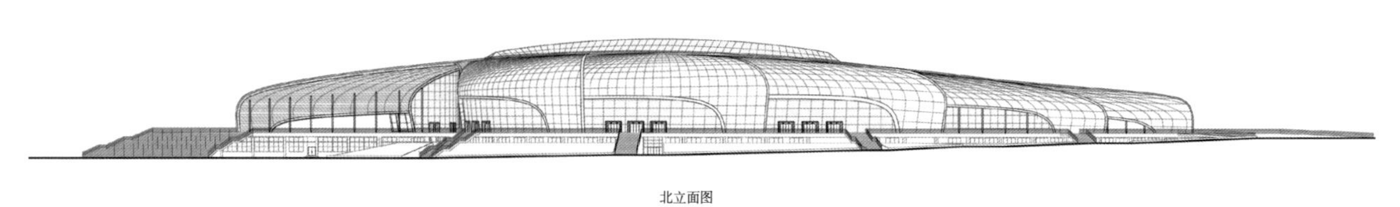

北立面图

North elevation

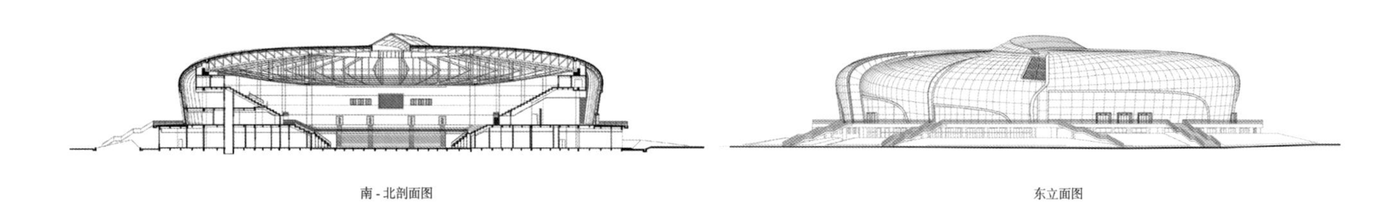

南 - 北剖面图

South-North section

东立面图

East elevation

图（3）-8　主要技术图

Fig.（3）-8　Main technical diagram

图（3）-9 室外照片 1
Fig.（3）-9 Outdoor photo 1

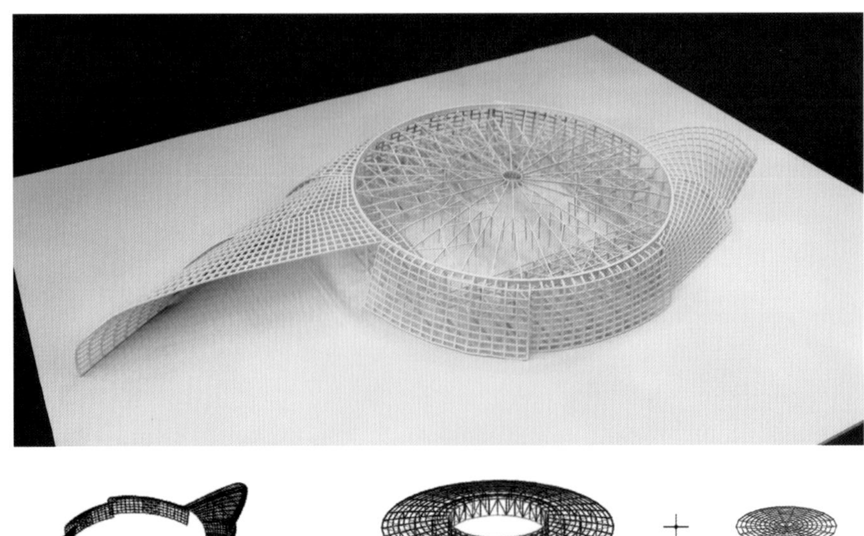

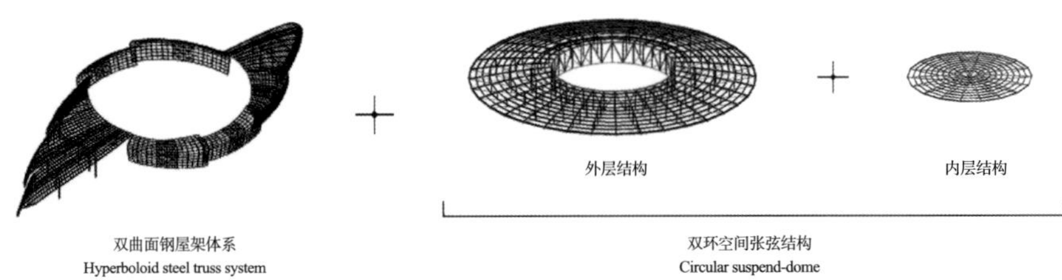

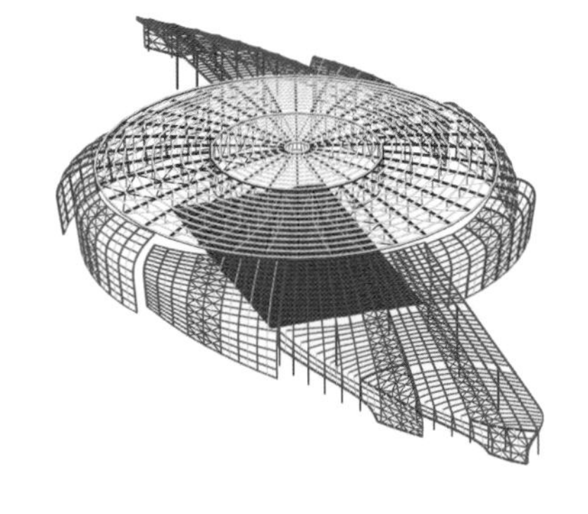

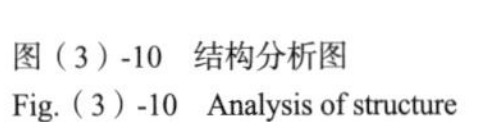

图（3）-10 结构分析图
Fig.（3）-10 Analysis of structure

在投标文本与媒体报道中，海螺的造型最为引人注意，但设计最根本的还是对经典平面的理解与升华，特别是结合功能特点的全新组合。正是有了这样的理性根基，设计才历经不断完善与修改，以科学理性的方式最终完成。

shape won most of the press exposure, the fundamental design was the comprehensive improvement to the classic plan, especially in terms of the functional configuration. The design principle laid a solid foundation for further modifications, and pushed the project to its final success in a rational way.

3. 浪漫解读与理性构成

设计工作进入工程阶段，面临着一系列理性构建的合理化过程。浪漫动感的屋盖造型即经历两个方面的推敲：

首先是运用计算机技术推敲确定屋盖的几何形状与参数。大跨度结构的特点决定了屋盖必须遵守基本的结构受力特点和屋面材料特性。脱离结构基本受力特点,将使构件复杂，工艺繁琐。无视屋面材料特性与国内工程技术条件，则将严重影响建筑建成的完成度。许多国外建筑师在中国的作品正是由于上述两方面原因，建筑建成后与原设计效果相去甚远。为此我们尽可能运用计算机技术，结合材料特性与工程技术条件，使建筑体量的确定与材料技术条件相适应。

3. Romanticized analysis and logical composition

During construction, to logically structuralize the romanticized form of the roof, two ways of development was processed:

Firstly, digital computation was used to parametrically design the roof. The large span restricted the roof structure to follow fundamental mechanisms and the characteristics of the material; otherwise, the design would be resolved in an unnecessarily complex way with inadequate techniques incapable with the current technology level of China. Some domestic buildings designed by foreign architects, unfamiliar with the said constrains, were not able to realize the original design in physical construction. Henceforth, we mostly relied on computer technologies, material performances and available mechanicals to

图（3）-11　结构模型照片
Fig.（3）-11　Photo of the structure model

其次，体育馆复杂体型的构建，必然需要大空间结构的组合适应。巧妙的组合结构设计，不仅可以实现建筑构思，还使合理、健康的结构技术集成。南沙体育馆比赛大厅主体钢结构部分采用了先进的环形张弦穹顶，主跨度达到了 98m。外围结构则采用渐变的钢桁架组合而成。两套结构的共同工作，形成了具有非线性曲面效果的外形，又坚持了结构施工技术的理性原则。

fit the venue into the appropriate technical conditions.

Secondly, a smart configuration of parts not only met the requirements of the architectural concept, but also represented a logical management of techniques. The main steel structure of the hall was an advanced circular suspend-dome with a span up to 98m, with steel trusses peripheral support. The system constructed the doubly-curved exterior as well as adhered to the principle of rational use of technology.

4. 多功能设计及赛后利用

南沙体育馆作为亚运比赛场馆仅仅是一个短期需要，而其更主要的角色是在赛后作为南沙城市中心区重要文体活动中心。南沙区拥有港口、汽车等多种工业企业，利税高，贡献大。但由于是新建区，公共建筑缺少。体育馆是该区最重要的新建公共设施。因此，在功能分区、流线组织合理的前提下，以多功能性来适应赛时赛后不同需求的变化，也成为了本项目功能设计中的重要出发点。

体育馆比赛大厅的场地我们选择了 40m × 70m 的尺寸，同时利用活动看台组合变化来适应各种比赛对于场地的不同要求，力求达到场地适应性的最大化。在首先满足亚运比赛要求

4. Multi-functional design and game use

Nansha Gymnasium would, after the games, serve as an important center of cultural and sports activities in the urban center, an excellent place to organize sports, art performances and large gatherings. Thus, on the premise of reasonable functional zoning and flow organization, the versatility and adaptability to different needs during and after the game was the key to our design. In the design of Nansha Gymnasium, various design methods had been applied for different nature of functional zoning.

The main field sized 40m×70m, while mobile stand combinations would be applicable to different requirements of the competitions for maximum adaptability. Besides meeting the demands of the

图（3）-12　室外照片 2
Fig.（3）-12　Outdoor photo 2

图（3）-13　天窗照片

Fig.（3）-13　Photo of the skylight

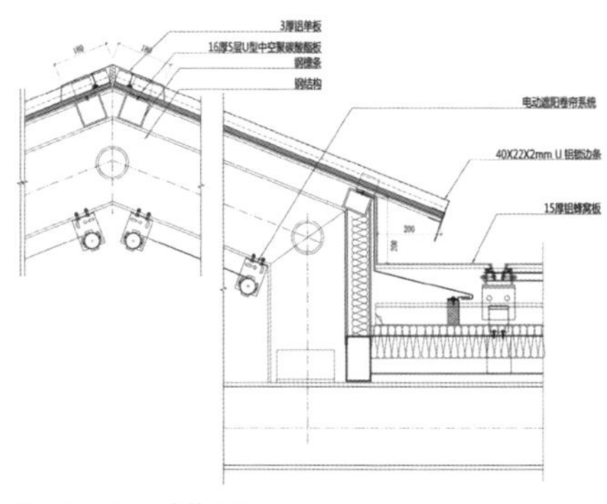

图（3）-14　天窗构造图

Fig.（3）-14　Construction of the skylight

的前提下，更多地考虑了赛后各色企事业单位的使用，便于举办各种比赛和企业文化活动。

在辅助用房的设计方面，我们也尽可能预留改造的空间，同时也考虑了设备使用上的灵活性，满足赛后使用的多种需求。功能上的合理配置和对赛后利用的重视，为实现体育场馆的自负盈亏，达到以馆养馆的目的奠定了良好的基础。热身馆、运动员餐厅等可以改造为对外开放的休闲餐饮、健身中心；新闻媒体及部分赛事管理使用的区域将具有会议功能；主场地可以沿坡道直接进入，便于重型设备与货物的搬运，有助于发挥展览中心的功能。半开敞的休息平台，营造出通风良好、符合亚热带气候特点的公共空间，有利于场馆赛后的利用。

（本案例的文字稿主要根据参考文献 [32]、[34] 整理）

Asian Games, the venue focus more on other post-game function demands from business or enterprise units for their various competitions and cultural activities.

In the design for auxiliary rooms, we reserved as much space as possible for future transformation, as well as for flexibility of equipment use and other post-game use. Rational function configuration and awareness of post-game utilization laid a good foundation for the self-sustainability of the gymnasium. The warm-up hall and the athlete's restaurant could be transformed for public dinning and civil fitness; rooms for the press and other management purposes could be used for conference; the main fields could be accessed from the ramp, facilitating transportations for heavy equipment and supplies; the semi-enclosed resting platform creates a ventilation-friendly public space applicable to post-game utilization.

图（3）-15　室内照片 1

Fig.（3）-15　Indoor photo 1

图（3）-16　总体鸟瞰图 3
Fig.（3）-16　Overall aerial view 3

图（3）-17　室外照片 3
Fig.（3）-17　Outdoor photo 3

图（3）-18　室外照片 4
Fig.（3）-18　Outdoor photo 4

图（3）-19　室内照片 2
Fig.（3）-19　Indoor photo 2

图（3）-20　室内照片 3
Fig.（3）-20　Indoor photo 3

2010年亚运会省属游泳跳水馆 中国，广州，2007

2010 ASIAN GAMES IN PROVINCIAL SWIMMING AND DIVING HALL GUANGZHOU, CHINA, 2007

图（4）-1 区位图
Fig.（4）-1 Location map

2010年广州第16届亚运会，广东奥林匹克体育中心承担多个项目的比赛和训练任务。位于主体育场以北的游泳跳水馆是亚运会游泳跳水项目的主要比赛场馆之一，在此进行游泳、跳水和现代五项游泳比赛及亚残会的游泳比赛场馆建筑倍受各界关注。场馆赛后组建成为IOC/OCA国际训练中心、国家南方训练基地，满足国家队冬训及亚运会之后举行其他重大赛事的要求。

During the 2010 Guangzhou Asian Games, Guangdong Olympic Sports Center accounted for aplenty of competitions and pre-game training. The swimming and diving hall north to the main stadium was a major venue for swimming and diving competition during the Games. Afterwards, the hall serves as the IOC/OCA international training center and southern China training base, meeting the demands of other major events after the Games.

游泳跳水馆用地9.9ha。总建筑面积约33 331m^2，总座席数为4 584座。馆内分别设置一个51.5m × 25m × 3m的标准比赛池（含移动池岸）、一个50m × 25m × 1.4~1.8m的训练池和一个25m × 25m × 5.5m的跳水池，建筑总高度约29m。

The total land area of the project was 9.9ha, with a construction area of 33,331m^2 and a total seats of 4,584. The hall included a 51.5m×25m×3m standard pool (inclusive of the mobile deck), a 50m×25m×1.4~1.8 m training pool and a 25m×25m×5.5m diving pool.

1. 基地解读

广东奥林匹克体育中心，规划建设于1999年，其中8万人体育场是第九届全国运动会开闭幕式场馆与主比赛场。游泳跳水馆规划位置处于基地的边缘，由于单侧临市政干道，无辅助街道沟通，人员流线组织困难，对于安保要求高的国际赛事场馆而言，矛盾尤为突出。同

1. Site analysis

Guangdong Olympic Sports Center, planned in 1999 and designed for the 9th Naitonal Games, had an 80,000-seat stadium. The Swimming and Diving Hall located at the border of the center, where urban traffics occupied one side of the site without subordinate streets to coordinate. For a major venue in need of high level security,

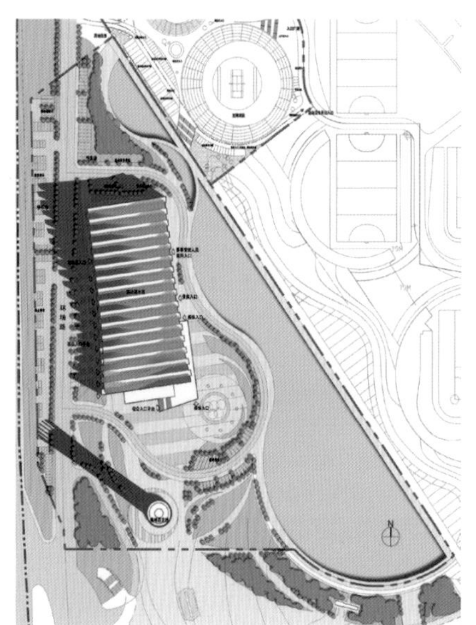

图（4）-2 总体布局
Fig.（4）-2 Overall layout

图（4）-3 室外照片1
Fig.（4）-3 Outdoor photo 1

图（4）-4　总体鸟瞰图
Fig.（4）-4　Overall aerial view

时规划位置还临近形态张扬的奥林匹克体育场，与规划中的奥林匹克塔形成犄角关系，使建筑设计的限定因素增加。紧邻游泳馆用地的，还有同时建设的广州亚运会网球中心，该建筑与游泳馆同时招标，同时建设，让建筑群体空间的不定性更加突出。经过分析与比较，我们采用了最简约的体量处理办法，用完整的矩形来应对局促的基地与复杂的周边环境要求。

the issues of circulation was prominent. Meanwhile, the site was close to the main stadium of an exaggerated exterior. Lastly, a tennis center planned to be built simultaneously with the swimming hall was stuck beside the site for the hall, which exacerbated the uncertainty of the spatial relationship within the site. Through careful analysis, we adopted a simple and pure massing of cube to deal with the compact area and complicated surroundings.

2. 空间组合

近年来，体育建筑的标志性被片面夸大，“高、大、空”的设计倾向重新抬头。加之体育建筑科学研究的屡被忽略，体育建筑的相关科学研究问题在我国体育场馆建设的策划、可研、设计、评审等环节普遍存在缺位。特别是在大型场馆的建设中，对基本空间尺寸控制失误的实例比比皆是。连体育建筑的大空间应该怎样与功能空间科学吻合，都很难确保实现了。

2. Spatial relationship

Lately, the iconicity of sports facility is unfairly exaggerated. Spacious and superficial designs are back into fashion. Additionally, the lack of research relevant to sports facility has led to issues in planning, feasibility studies, design, examination, and various other aspects of sports facility designs, especially of large-scale sports venues, where improper scale frequently occurs. Even the correlation between the

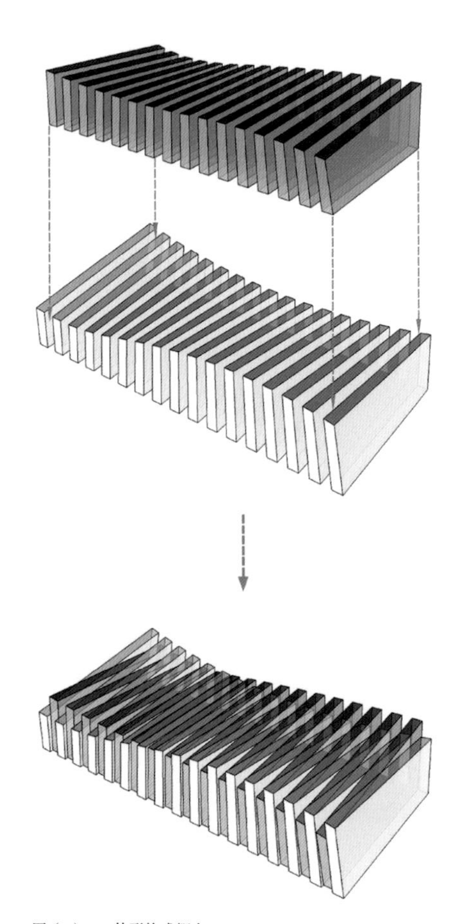

图（4）-5 体形构成概念
Fig.（4）-5 Concept of the form

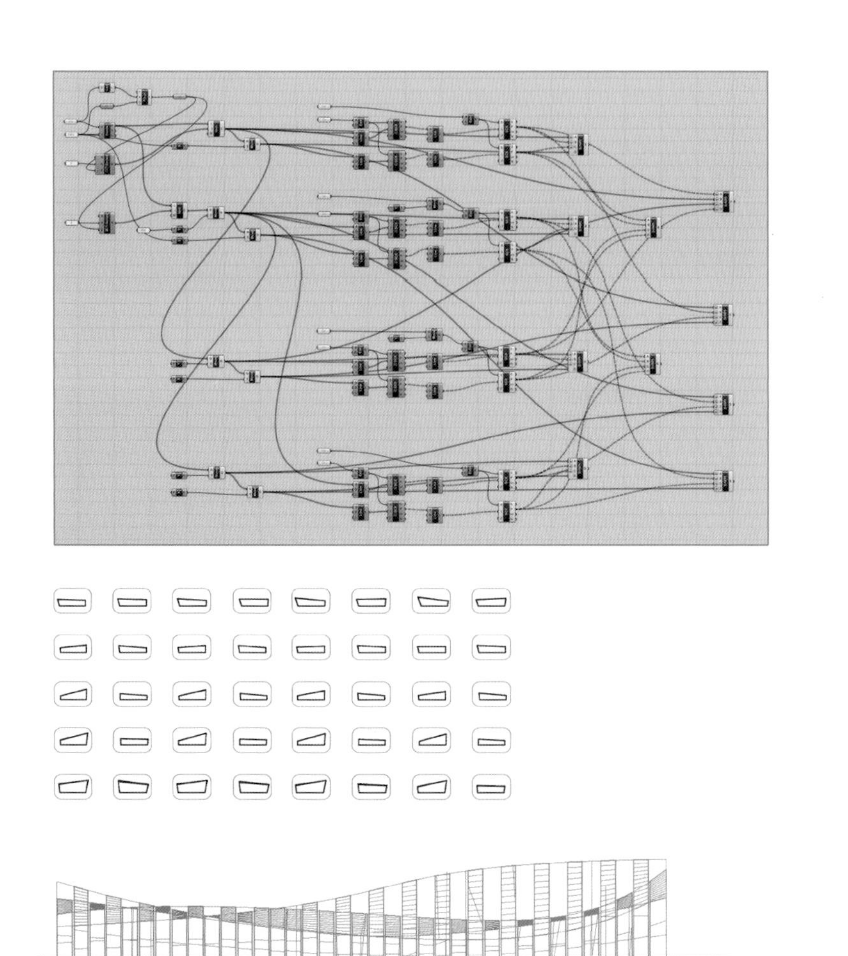

图（4）-6 体形参数化设计
Fig.（4）-6 Form parametric design

游泳馆是体育建筑中对室内温度控制要求最高的类型。因此，室内空间高大将增加室温控制的难度，也使赛后运营的能源消耗成本加大。同时，游泳馆空间组合的重要一点，就是处理好跳水区域与游泳区域的高度差。为此，1990年代国内曾经有过将泳池空间与跳水空间分设的尝试，以期使不同空间区别对待，如：汕头游泳馆、珠海游泳馆。进入新世纪以来，体育部门大都明确提出，希望两池合并同一空间内建设。这样，如何使游泳馆空间紧凑合体就重新成为设计的难点。就游泳馆高度而言，游泳馆室内空间的高度不应过大，其最大高度，以能满足跳水区域的最大高度即可。

设计中，我们将建筑分为3个区段：跳水区、游泳比赛区、训练区。分别按照不同功能要求确定室内空间高度。跳水区域最高，比赛区其次。训练区域则尽可能降低高度，将最浅

large volume and the functional space of sports facility becomes an uncertain task.

Natatoriums, among sports facility, require the most precise temperature control. Therefore, the larger the volume is, the more difficult the temperature control is, thus the higher the expenses are. In the 1990s, some domestic venues had attempted to separate the swimming and the diving areas for different technical treatments, e.g., Shantou Natatorium and Zhuhai Natatorium. Since the 21st century, however, most authorities have specified that the two area should be built within the same mass, which makes the compression of volume a difficult task. In terms of heights, the maximum height of natatoriums shall not exceed the minimum height for diving.

In our design, we divided the architecture into three parts: the diving area, the swimming competition area, and the training area, and determined the height

图（4）-7　室外照片 2
Fig.（4）-7　Outdoor photo 2

的训练池置于首层。赛后，这里将对市民开放，而紧凑的空间环境，将减少使用能耗，便于对外开放，增加社会效益。

of each part accordingly. The diving area ranked the highest, the competition were the next. Since the height of the training area was reduced, the shallowest training pool was placed on the ground level. After the Games, the hall would be open to the public. The relatively compact space was preferable for energy saving and social benefits.

3. 诗意表达

体育建筑的大体量、大跨度，使其不可避免地成为建设成就的象征。决策者希望得到宏伟壮观的业绩体现，市民各界也陶醉于品味易懂有趣的建筑形象，对于体育建筑的本质内涵则不甚了解。这使我们的体育建筑创作，在体现理性精神的同时，既要努力创新，也必须考虑社会大众适应的能力。

游泳跳水馆所处地理位置特殊，曾经承办过第九届全国运动会的主体育场在亚运会中也会担当主会场的角色。因此，游泳馆的设计力求与原基地环境相协调，总体布局上延续奥林匹克体育中心的总体规划设计理念。

根据游泳跳水馆的结构处理形成的层级渐变、和缓起伏的总体形态，我们大胆运用蓝白两种颜色，间隔布置，形成穿插对比的动态造型，使矩形的体量展现出丰富的动感。主体造型采用双色螺旋流动造型，主体建筑白色和蓝色相间，既巧妙地隐喻了广州“云山珠水”的城市地理特征，又是对主体育场“飘带”曲线的延

3. Poetic expression

Sport venues with large mass and span are inevitably considered as the symbol for construction achievements. Policy makers hope to have impressive results, while the public also revel in contemplating unpretentious and interesting architectural images, though often neglect the nature of sports facility. Therefore, for sport facility design, we need to take into consideration the spirit of rationality and creativity alongside with public acceptance.

The Swimming and Diving Hall, locating within the Guangdong Olympic Center, where the Asian Games had also taken place, had a special geographic significance. Therefore, the design for the natatorium shall be compatible with the original site condition and consistent with the overall planning concept of the Center.

Based on the general form of the swimming and diving halls, which are gradually changing and gently undulating, we boldly used the color blue and white alternatingly in dynamic shapes to create dynamic movements on the rectangular massing.

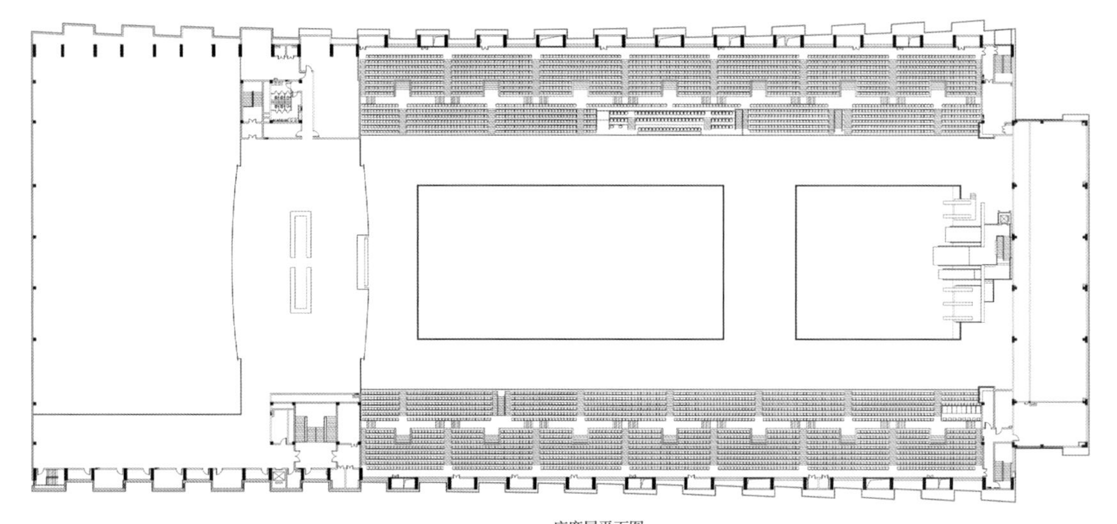

座席层平面图
Floor plan of the stands level

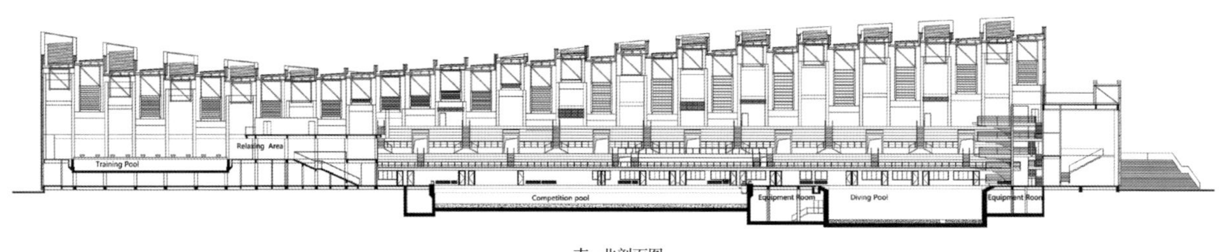

南 - 北剖面图
South-North section

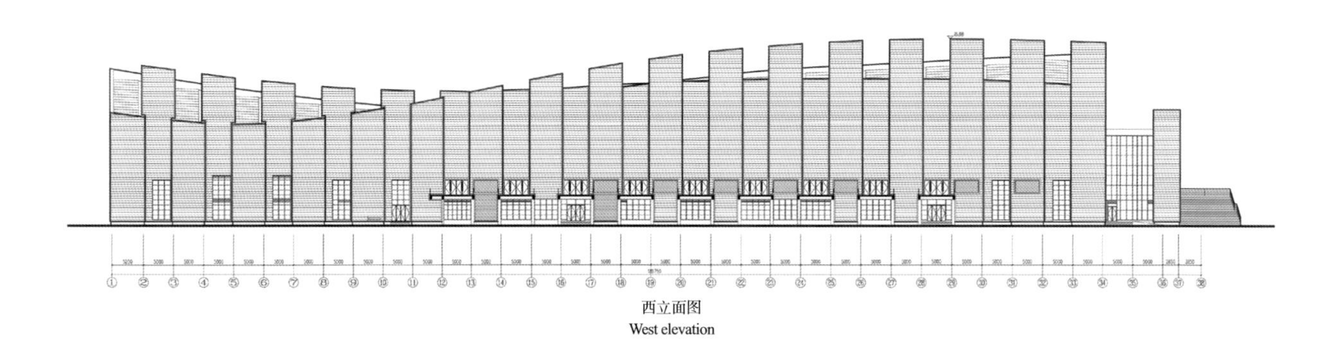

西立面图
West elevation

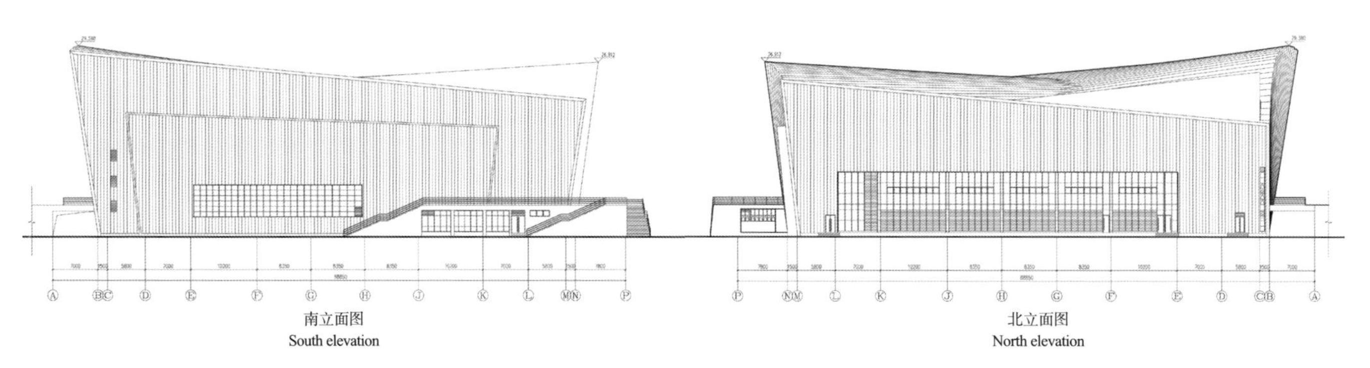

南立面图
South elevation

北立面图
North elevation

图（4）-8　主要技术图
Fig.（4）-8　Main technical diagram

图（4）-9　室外照片 3
Fig.（4）-9　Outdoor photo 3

续。同时通过相互穿插流动造型，结合建筑朝向，很好地满足了建筑内部空间高度、采光通风、建筑节能以及合理布置设备管道的需求。

The flowing double-color spiral of blue and white of the building subtly implied the geographical features of Guangzhou's 'Baiyun Mountain and Pearl River' and continues with the 'streamer' curve of the main venue. Additionally, by fluidly intertwining the volumes and carefully orienting the building, we had sufficiently met the requirements of ceiling height, lighting, ventilation, energy efficiency and piping equipment planning.

游泳跳水馆在外观起伏变化，充满动感的同时，力求本质内敛，理性朴实。从喧嚣纷争的都市环境进入场馆，淡淡的天光从简单明确的屋架间隐隐泄下，一泓蔚蓝的清池，碧波荡漾。“云水”秀外，“禅心”慧中的设计理念经受了低造价、差施工的考验，度过了无数次无奈而焦虑的抗争，用简单的材料构建出基本符合设计构思的建筑。

Besides the undulating and dynamic appearance of the swimming and diving halls, reserved, rational and simple characteristics are also preferable. Coming from the busy city into the venue, you can see a light from the sky coming between the clear and simple roof trusses, with a rippling pool of water. The concept of 'Outer landscape, inner Zen' has once undergone low budget and low construction quality. Yet after countless anxious struggles, it is eventually realized with simple materials.

4. 节能降耗

作为大型国际体育赛事比赛场馆，在设计中除了要满足亚运会复杂的使用要求，还要充分考虑体育设施赛后其他可能的比赛，以及全民健身的需求。因此，我们将节能降耗作为赛后利用成败的关键环节看待。主体空间的紧凑布局，成为减少能耗的核心。建筑构件错列布局形成了天窗采光，使训练与对外开放时能耗大大降低。

4. Energy saving

As a large-scale international sports venue, besides satisfying the complex demands of the Asian Gama, the natatorium shall also be adaptable to post-game uses and civil fitness. Therefore, energy and cost saving have also become the key to success. The compact space of the main volume facilitates the reduction of consumption; the scattered arrangement of structural members lowers the energy waste.

在满足亚运比赛要求的基础上，在比赛池中设置移动池岸，可以满足赛后训练和进行短池比赛等多方面要求；功能房间、建筑空间的布局也结合赛后利用进行了充分合理的考虑。由于受到极端局促的投资限制，在

Besides meeting the demands of the Asian Games, mobile decks were provided in the competition pool, which could be useful for post-game training and short course competitions; the correlation of functional rooms and the overall architectural space was designed with thorough consideration

图（4）-10　屋面采光天窗照片
Fig.（4）-10　Photo of the roof skylights

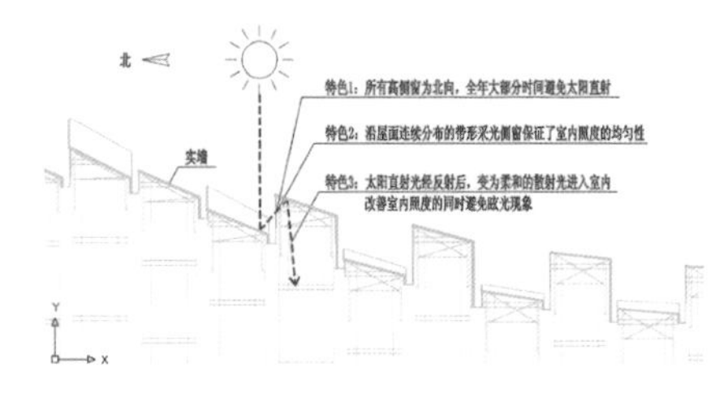

图（4）-11　屋面采光设计剖面
Fig.（4）-11　Section of the roof skylights design

设计中不得不综合考虑初期投资和赛后运营费用的矛盾，设计中提供了赛后为方便比赛、大众健身空间的不同使用要求，而对建筑空间灵活分割的可能。

运动员赛时检录用房及其顶层，赛后转换成餐饮休闲设施，与训练池组成大众健身娱乐区，对外开放并不影响比赛训练区域的专业运动员正常使用。

5. 技术措施

亚运游泳跳水馆的建筑形体变换复杂，结构单元又清晰简单，使形式与逻辑之间的关系颇为重要。传统设计手段无法胜任，我们使用了计算机参数化辅助设计的手段。这不仅体现

on post-game functions. Due to the extremely strict investment control, the design also proposed several flexible division and configuration of the overall space convenient for civil fitness and other post-game uses.

Athlete's check-in rooms and their top floors were converted into dining and entertainment facilities, and became the civil fitness and entertainment area along with the training pool. Public-area would not interfere with professional trainings within the private training area.

5. Technical measures

The overall form of the Swimming and Diving Hall was complicated, whereas the structural units were straightforward and simple. The relationship between the form and the logic was important to a certain

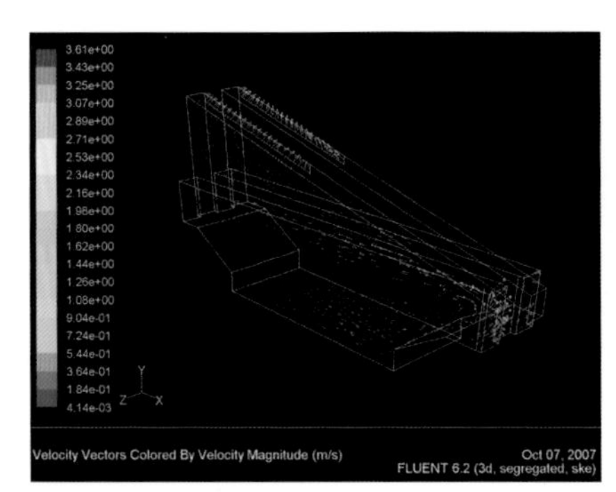

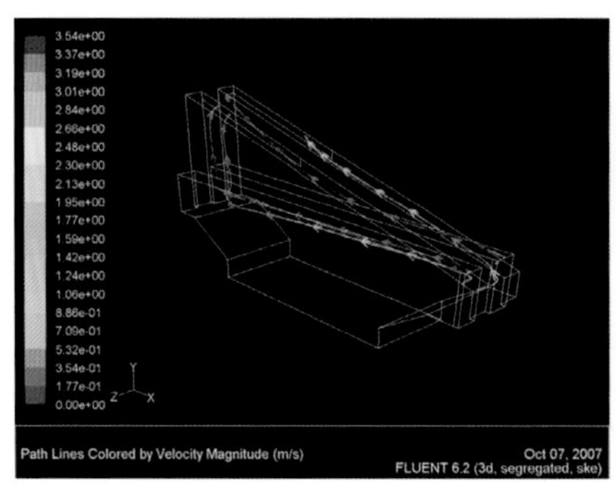

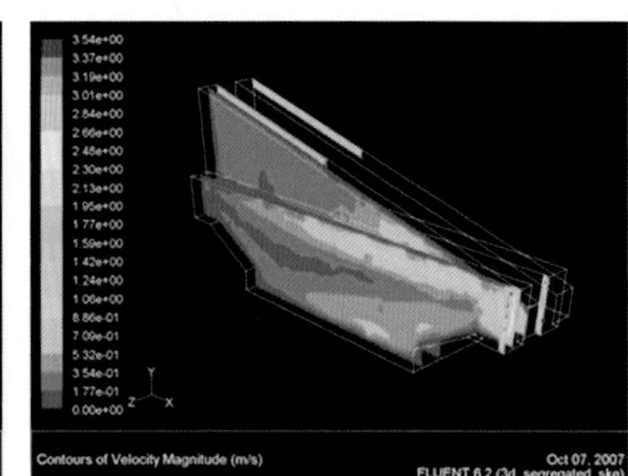

图（4）-12　Fluent 模拟
Fig.（4）-12　Simulation by Fluent

图（4）-13　屋面结构照片
Fig.（4）-13　Photo of the roof structure

在方案阶段的体型推敲上，在后期技术深化阶段也尤为重要。例如，为了节约造价，我们对建筑跨度严格控制。投标方案的跨度控制在75m，中标后的修改曾压缩为66m；后来继续深化过程中在造价和功能空间要求之间取得折中，确定为72m。如此反复推敲，计算机模型修改50余次，由于有参数化模型和相关技术保证，对于如此复杂的建筑形式，使我们有了做深入、细致推敲的可能。

又如，为了在建筑形体不断深化修改与变化的同时，控制自然采光的面积，设计中采用参数化方法，弄清采光面积与形体错落之间的原则，将数学逻辑贯彻入参数化程序，从而在深化修改形体的同时及时知道自然采光面积

extent. Traditional design method would not be able to suffice, therefore, parametric design was adopted as a digital aid. The measure was evident throughout the entire process, especially at the technical design stage. For example, as a means to lower the construction cost, we strictly controlled the span of the building. The span was once reduced to 66m from the original 75m; after mutual compromise between function demands and overall cost, the span was eventually determined 72m. Parametric measures and appropriate technical support guaranteed the opportunities for us to carefully tailoring the design for over 50 adaptions of digital model.

Other utilizations of parametric design were digitalized controls and real-time displays of natural lighting and acoustic performance throughout the ever-changing design, which provided reliable data for each step of development in terms of the relationship between the form and the correspondent amount of natural light or internal acoustic quality.

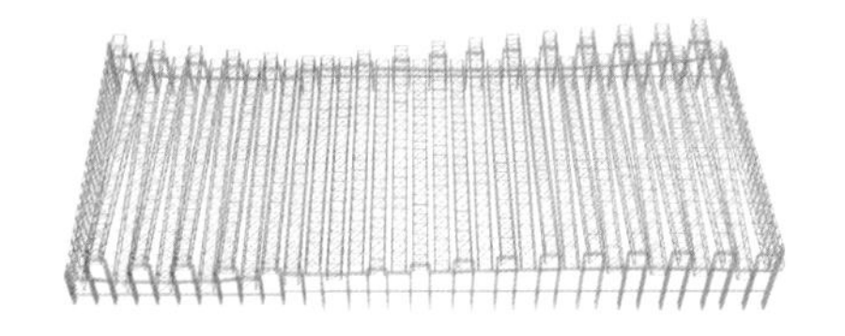

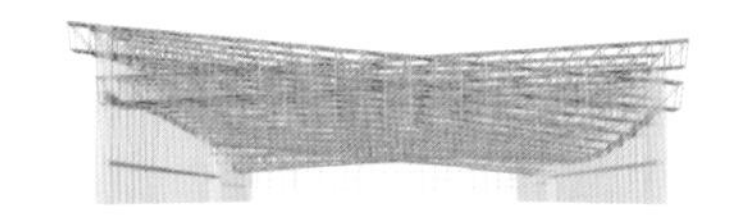

图（4）-14　结构体系设计
Fig.（4）-14　Design of the structure system

的变化，并加以控制。再如，游泳馆的体积与场馆的声学质量有很大关系，因此体积的控制也是重要指标，同样用传统方法很难统计的数据，在参数化的模型里，可以实时显示。

在结构选型方面，为了与复杂的建筑造型相协调，本工程屋盖钢结构采用 32 榀方型空间钢管桁架结构作为竖向承重结构，桁架宽 3.3m，高 3.2m。桁架支座一端为固定铰接支座，另一端为可以沿桁架长度方向（y 向）滑动的滑动支座，以释放温差引起的变形。该方案的优点在于屋面钢结构具有较好的刚度，整体性相对较好，用钢量相对较少。

（本案例的文字稿主要根据参考文献 [31]、[33] 整理）

For the structure, in order to visually support the complicated form, 32 rectangular space-frames were used in the vertical system. Each space frame measures 3.3m in width and 3.2m in height, had one pinned connection and one roller connection with the columns to adapt to thermal expansion. The advantage of the structure was its relatively high stiffness, level of completeness and low steel consumption.

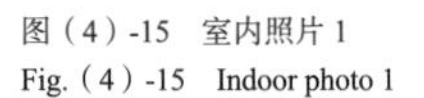

图（4）-15　室内照片 1
Fig.（4）-15　Indoor photo 1

图（4）-16　室外照片 4（上）
Fig.（4）-16　Outdoor photo 4 (up)

图（4）-17　室外照片 5（下）
Fig.（4）-17　Outdoor photo 5 (down)

图（4）-18　室内照片 2（上）
Fig.（4）-18　Indoor photo 2 (up)

图（4）-19　室内照片 3（下）
Fig.（4）-19　Indoor photo 3 (down)

图（4）-20　室内照片 4（上）
Fig.（4）-20　Indoor photo 4 (up)

图（4）-21　室内照片 5（下）
Fig.（4）-21　Indoor photo 5 (down)

2010 年亚运柔道摔跤馆（广州大学城华工体育馆）中国，广州，2005

2010 ASIAN GAMES JUDO WRESTLING HALL (GUANGZHOU UNIVERSITY CITY SCUT GYMNASIUM) GUANGZHOU, CHINA, 2005

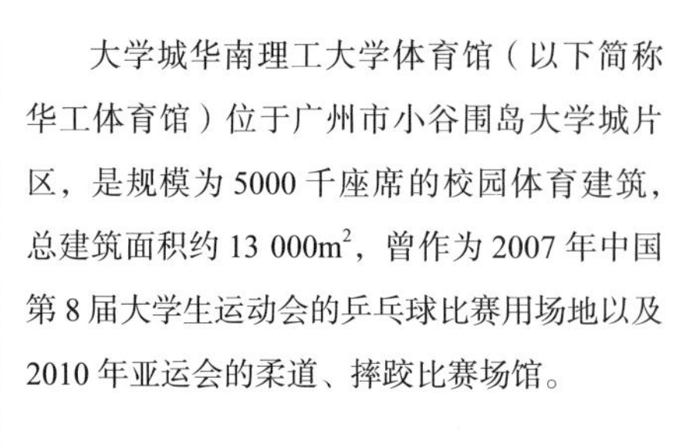

图（5）-1　区位图
Fig.（5）-1　Location map

大学城华南理工大学体育馆（以下简称华工体育馆）位于广州市小谷围岛大学城片区，是规模为 5000 千座席的校园体育建筑，总建筑面积约 13 000m²，曾作为 2007 年中国第 8 届大学生运动会的乒乓球比赛用场地以及 2010 年亚运会的柔道、摔跤比赛场馆。

建筑方案构思希望新校区的体育馆形式能与华工北校区的岭南风格的历史建筑有所呼应，同时力图将可持续设计策略运用其中，探讨湿热气候下机械辅助式自然采光、通风在大型公建运用的可能性。

The Gymnasium of South China University of Technology, Higher Education Mega Center is located at Xiaoguwei Island, Guangzhou, with a scale of 13,000m² and 5,000 seats. The Gymnasium served as a table tennis competition venue of the 8th National Game of College Students in 2007 and Guangzhou Asian Games Judo and Wrestling Venue in 2010.

The project requirements proposed that the gymnasium should be formally consistent with the southeast Chinese styled historical buildings in the north campus, as well as compliant with sustainable strategies by using mechanic-assisted natural lighting and ventilation systems.

1. 体现基地特色的规划布局

体育馆建设基地整体呈长条型土台，东面与校内道路高差约为 2m，西面与校际道路高差约为 4m，根据基地的地形特点，我们在方案初期就决定做一个半覆土的建筑。狭长地形中，利用地形的高差，不仅将建筑体量减小，营造出亲切怡人的建筑形象，通过立体交通，自然有效的解决了二层观众与首层体育馆内部人员分流的问题，同时半地下的建筑降低了底层的空气温度，为后续的通风设计埋下伏笔。

1. Site plan featuring local conditions

The site for the project was a rectangular terrain approximately 2m higher than the campus paths at the east side and 4m at the west. Based on the existing condition of the topography, our early design already aimed at semi-underground architecture. The narrow land and the elevational drop not only reduced the building mass to a more amiable size, but also effectively ordered the division of second-floor audience and ground-floor staff. Meanwhile, the underground structure helped cooling the temperature of the ground floor, creating excellent condition for ventilation design.

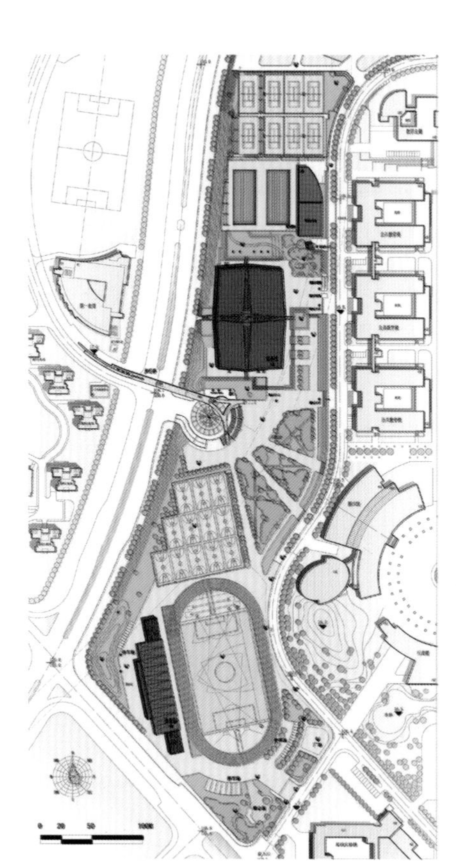

图（5）-2　总体布局
Fig.（5）-2　Overall layout

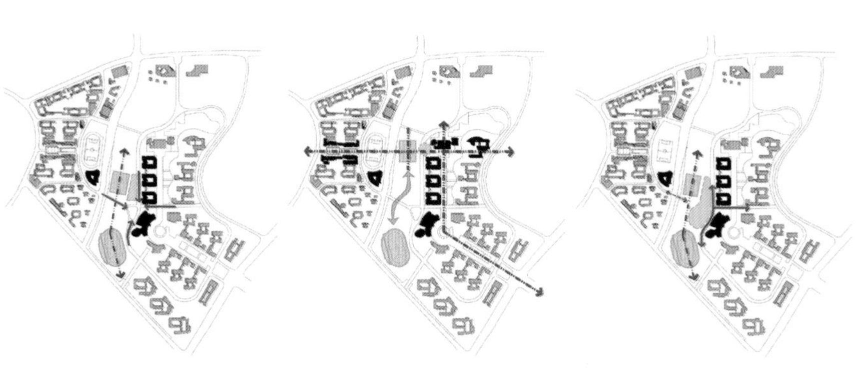

图（5）-3　总体布局多方案比较
Fig.（5）-3　Comparison of the overall layout scheme

图（5）-4　总体鸟瞰图
Fig.（5）-4　Overall aerial view

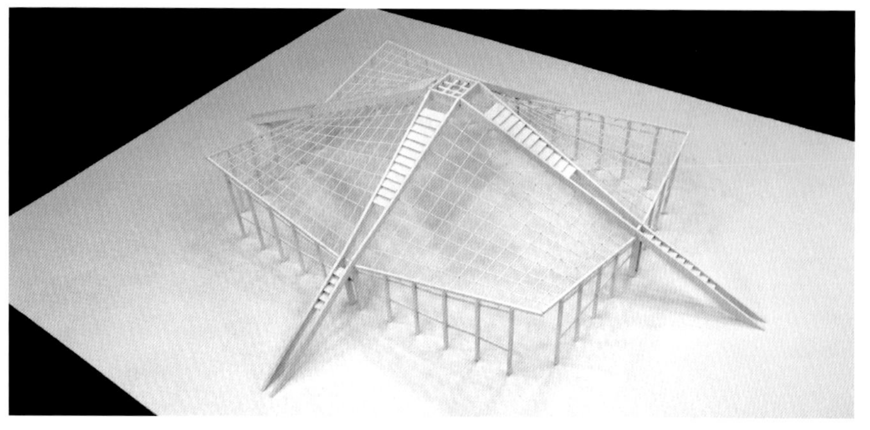

图（5）-5　结构模型照片
Fig.（5）-5　Photo of the structure

2. 契合设计构思的结构选型

华工体育馆在方案设计的初期，便将可持续设计策略考虑进来，使之与建筑设计有机结合，并从结构体系出发，创作能够体现设计构思的结构形式，使之成为建筑的特质。

根据体育馆的规模，5 千人的座席数不算大，我们倾向于采用非对称的看台布局；体育馆的比赛大厅进深大，无法利用风压通风，我们希望找到利用热压通风的形式；考虑到校园体育馆的赛后运营成本，比赛大厅要是能利用自然采光，必定对降低成本有所帮助。

2. Structure Selection consistent with design concept

Since the beginning, the design of Gymnasium of SCUT had included the principle of sustainability organically into the architecture. The structure of the building became the reflection of the design concepts and the feature of the architecture.

Since the 5,000-seat capacity was a relatively modest scale, we inclined to adopt an asymmetrical arrangement for the grandstand. The large depth of the competition hall hindered the design of wind pressure ventilation, therefore, we were seeking a way to implement a thermal pressure ventilation system in the building. Additionally, natural lighting system was desirable for its ability to help reduce operation cost.

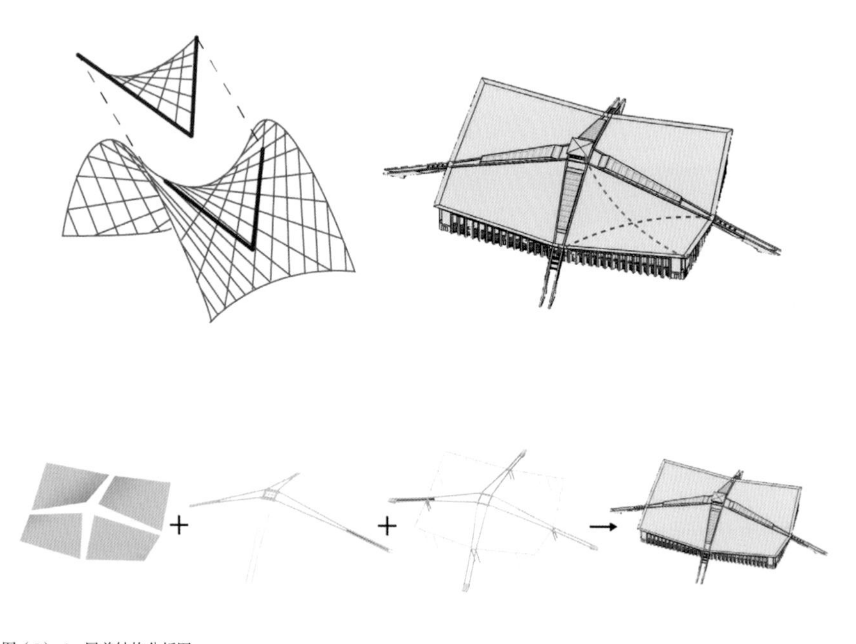

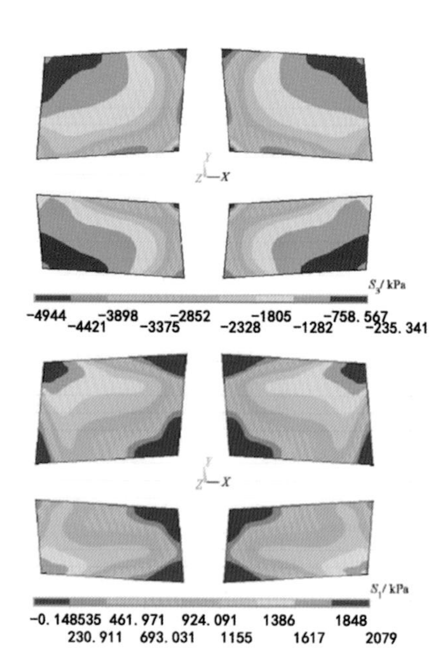

图（5）-6　屋盖结构分析图
Fig.（5）-6　Analysis of the roof structure

图（5）-7 施工过程外部照片
Fig.（5）-7 External photo of the construction process

图（5）-8 施工过程内部照片
Fig.（5）-8 Internal photo of the construction process

从这些基本的设计构思出发，在概念性方案阶段，建筑师便与结构工程师进行了深入的探讨。我们最终形成了以组合钢筋混凝土双曲抛物面扭壳结构作为屋面结构形式的构思。

建筑屋面由 4 片非对称的扭壳（扭壳的最终厚度定为 130mm）组合而成，支承于周边边缘构件和两榀正交拱架之上，在两向正交拱架落地点间设置预应力水平拉杆，使结构成为自平衡体系。这样的结构形式与非对称的看台布局相适应，形成比赛大厅四周低、中央高的空间格局，也造就了屋顶十字形采光天窗的雏形，为后续的自然通风及自然采光的设计提供了前提条件。

屋面结构形式成熟而又新颖：屋面预应力钢筋混凝土双曲抛物面扭壳结构方式经济节约，这类壳体由两个方向的直纹组成，支模、扎筋和预应力筋铺设等都很方便。华工体育馆屋面结构的创新之处在于，我们设计的是非对称的组合扭壳，通过壳面与支撑拱架的组合，屋盖的平面投影达到了长轴 99.8m，短轴 70.0m 的大跨度，屋盖的水平投影面积约 6 568m^2。在这一项目中与设计理念相契合的结构形式给建筑带来了独特的生命力。

Based on these premises, we had cooperated with structural engineers since the preliminary phase of the project, and reached a disagreement on using a hyperbolic paraboloid torsional reticulated shell made of reinforced concrete for the main structure.

The roof was structured with four asymmetrical torsional shells whose thickness measures 130mm. The shells were supported by two perpendicularly intersected arches, which were connected by pre-tensioned levers at their feet, creating a structural equilibrium. The design suited the asymmetrical grandstand and provided a large cambered indoor space, which laid the foundation for desired ventilation and natural lighting design.

The hyperbolic paraboloid torsional reticulated shell of reinforced concrete was a cost-effective structural member. The shells, which could be de-composited into two sets of rotating straight lines perpendicular to each other, was relatively easy to construct. The innovation of this particular design was the impressive span. The plan projection of the entire structure reached 99.8m in the long axis and 70.0m in the short axis, while the horizontal projection area was about 6,568m^2. The structure, consistent with the design concept, gave the architecture a special energy.

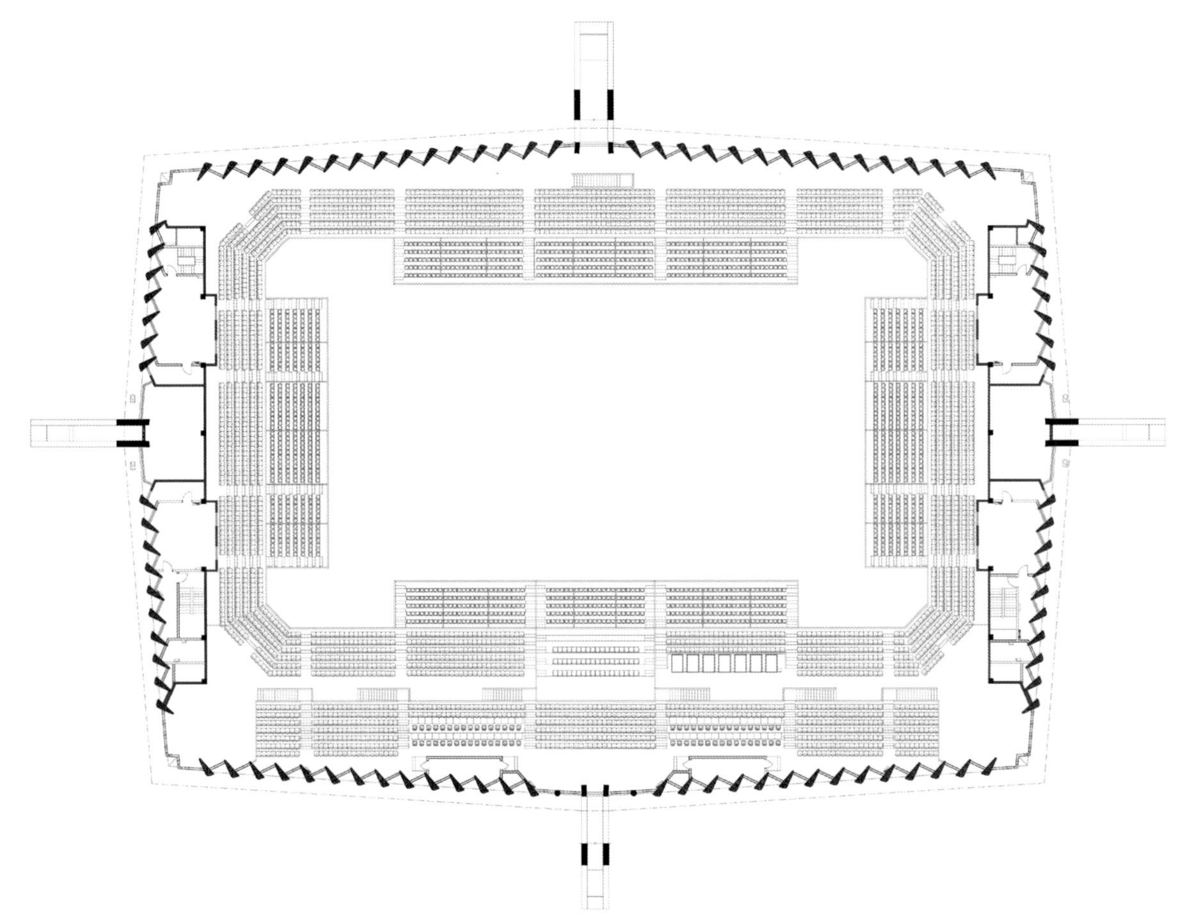

座席层平面图
Floor plan of the stands level

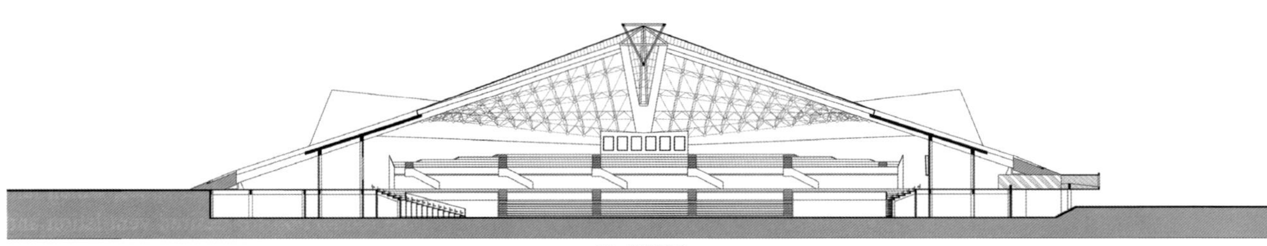

南 - 北剖面图
South-North section

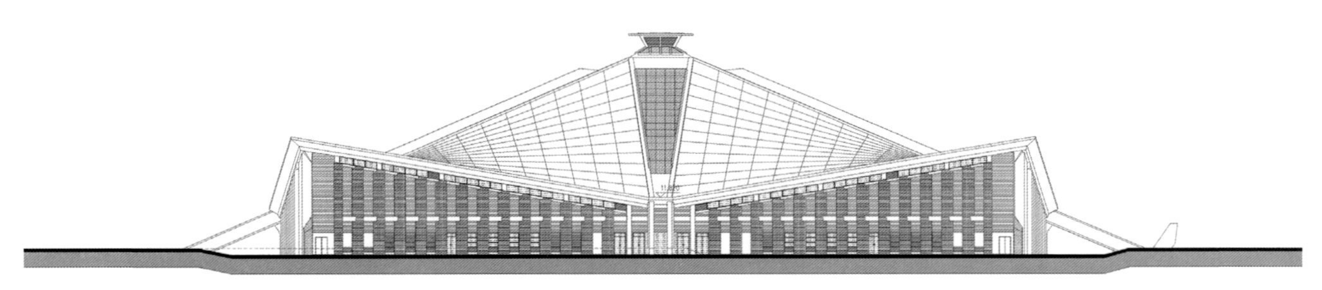

西立面图
West elevation

图（5）-9　主要技术图
Fig.（5）-9　Main technical diagram

图（5）-10　室外照片 1
Fig.（5）-10　Outdoor photo 1

3. 多功能使用的场地设置

高校体育场馆在满足学校正常教学要求的基础上，兼顾大型运动会的比赛要求，赛后作为学校的体育设施加以利用，这是当前高校体育场馆发展的重要模式之一。体育馆的场地选型为 40m × 70m，具备举办国际性乒乓球、篮球、排球、羽毛球、武术、手球等项目比赛的功能，并满足平时训练、集会、演出等功能。两千多个可自由伸缩的活动座席使体育馆可以适应不同比赛对于场地的要求。

3. Multi-purpose use of the site

According to the demands of the university, with extra considerations for large-scale games, the floor of the gymnasium was set to 40m×70m, capable of hosting competitions of international scale for table tennis, basketball, volleyball, badminton, martial arts, handball, etc., as well as daily training, assembly and performance. Over 2,000 retractable seats enable the gymnasium to switch between different functions.

4. 机械辅助式热压通风设计

华工体育馆的外观造型与它所运用的机械辅助式自然采光、通风策略相一致，通风空调专业与照明专业在方案设计的前期就与建筑专业紧密协作，不同一般的建筑设计，其他专业介入的时间深度都超越常规的程序。

4 Mechanic-assisted thermal pressure driven ventilation

Professional ventilation and lighting teams had been cooperated with our architectural design team since the beginning of the design. The form of the building was consistent with the adopted strategies for ventilation and lighting. The level of participation of these professionals into the gymnasium project was higher than average.

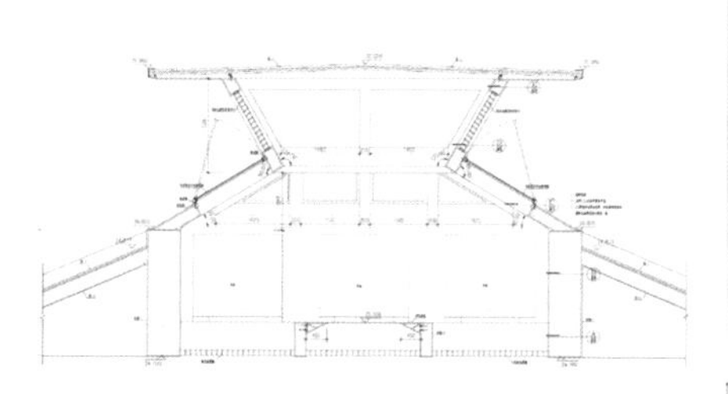

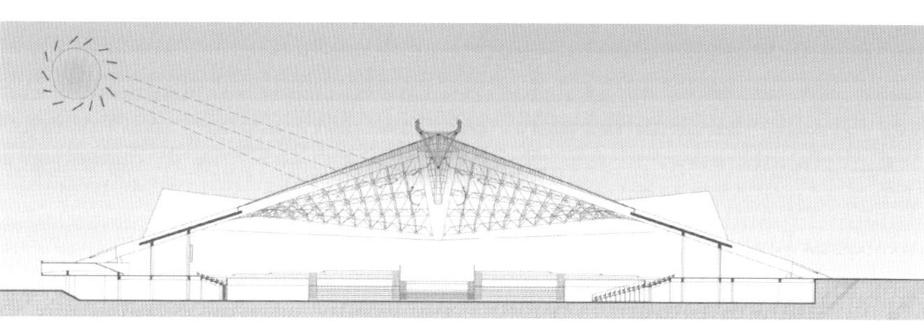

图（5）-11　屋顶出风口设置及通风示意图
Fig.（5）-11　Roof outlet setting and ventilation diagram

图（5）-12　通风系统细部照片
Fig.（5）-12　Detailed photo of the ventilation system

广州属于我国夏热冬暖地区，受典型的南方湿热气候影响。该地区的通风潜力以风压通风为主，体育馆大空间利用热压通风需要依靠风机辅助，增强驱动力。传统南方民居建筑设计中有运用小型拔风构件的先例，但是现代大空间建筑在顶部拔风口设置排风机的属创新设计。

体育馆首先利用基地的地形高差，将首层功能用房埋入地下，为热压通风创造了底层温度较低的空气（在通风模拟中，暖通工程师认为地下室具有 2℃的自然冷却效应）。前面已经提到，组合扭壳结构形成了由四边向中部升起的室内大空间造型，比赛场地与屋盖顶部的高差达 32m，为热压通风创造了优良的气流高差条件。最终确定的机械辅助热压通风系统包括进风口、风道组织、风扇送风及屋顶排风几个部分，是一个完整而科学的通风体系。新风通过地下室的冷却，经由风道引向赛场，再通过设于活动座席后方辅助用房内的风扇，将新风送往赛场，辅助用房靠走廊的一面均设百叶墙，以利于风的流通。冷空气送入赛场后，吸收内部的热量，温度升高，往顶部上升，位于屋顶拔风构件中的 8 台风机将上空的热空气排出室外，形成负压，继而吸引下部的热空气持续上升，从而达到通风的效果。

对于机械辅助热压通风效果模拟显示，在过渡季节以及冬季能够有效地改善场地中心的热状况，绝大部分空间的温度都能控制在27℃以下。可以达到完全不使用空调的节能目的，同时场内风速均不超过 0.2m/s，符合体育馆建设初期举办乒乓球比赛的体育工艺要求。

5. 比赛大厅自然采光设计

比赛大厅的自然采光利用，非常有利于校园体育建筑的日常运营，同时由于体育建筑的特殊性，利用自然采光最需要解决的是眩光的问题。

Guangzhou's weather is under the control of the typical humid and hot climate of subtropical area. The ventilation system of the region largely relies on wind pressure, while thermal pressure ventilation would normally request mechanical assistance. Small-scale vents had been used in traditional residences of this region, yet application of vents in modern large-scale buildings was unprecedented.

Based on the regional climate, we arranged innovatively exhaust fans at the top air dispensation outlet to greatly enhance the thermal pressure ventilation for the large space of the gymnasium. Moreover, taking advantage of the elevation difference of the site, we put the functional rooms at the first floor underground to create an air source of low temperature at the bottom for thermal pressure ventilation, resulting in a temperature drop of about 2°C. The gymnasium was of a hexagonal shape in plan, with a major axis of 97m and minor axis of 67m. The roof of the competition hall is structured in 4 torsional shells of reinforced concrete. Such an architectural form created a large interior space where the four sides rise towards the center with a height difference of 32m between the floor and the roof top, creating excellent conditions the height difference for thermal pressure ventilation. The mechanically assisted thermal pressure ventilation system was composed of the air inlets, duct system, fanned airflow and roof exhaust, which formed a complete ventilation system. Fresh air was cooled by the basement, led through the duct toward the gymnasium, and blown by the fans in the auxiliary rooms behind the retractable seats to the gymnasium. A louvered wall was provided on the corridor side of the auxiliary room to facilitate air flows. The cold air sent to the gymnaisum absorbed the heat inside and rose to the top, which was then discharged by the 8 wind turbines inside the air dispensation members of the roof. A resultant negative pressure vacuum kept the hot air below rising to feed the ventilation system.

图（5）-13　室内照片 1
Fig.（5）-13　Indoor photo 1

图（5）-14　室外照片 2
Fig.（5）-14　Outdoor photo 2

华工体育馆比赛大厅通过四个梯形的天窗以及东西屋檐下的高侧窗获取自然光照明。天窗的遮阳板与支撑结构的连接件结合于一体，并且四个朝向的遮阳板尺寸及遮阳角度各不相同，这是我们通过模拟太阳的运行轨迹发现的。如果要在日间场馆可能使用的时间内完全避免直射光进入赛场空间，朝向四个不同方向的天窗，其遮阳板都需要根据各自所处方位在白天经常使用的时段内，全年日照时间最不利点来确定其形式。我们在合理确定全年日照时间最不利点的基础上，通过模拟该时间点的日照情况，调整各个方向遮阳板的折板方式，板间密度，保证在该不利点遮阳板之间没有直射阳光透过，也就是地面阴影无光斑，从而确定在全年其他使用时间段内，均无直射阳光透过。

The simulation for the ventilation system showed that in transitional seasons and winters, the heat within the building can be effectively dispersed. Temperature of the majority of the space can be lowered to at most 27°C without using air conditioning. Meanwhile, the average wind speed was below 0.2m/s, suitable for hosting table tennis games.

5. Natural lighting of the competition hall

Natural lighting is extremely beneficial to the daily operation of the gymnasium. Due to the particularity of sports architecture, the ultimate goal for natural lighting design of sports architecture is to deal with sun glares.

The competition hall of the Gymnasium of SCUT received natural light from four trapezoidal skylights and the clerestories under the east and west eaves. The sun-shields, different in dimension and angled towards four directions determined by the local sun trajectory, were merged into the connections supporting the structure. In order to avoid direct sunlight, the shields should be distributed and oriented according to the least light-penetrable position throughout the day, so that no direct light is introduced onto the floor any time of the year.

图（5）-15　室外照片 3
Fig.（5）-15　Outdoor photo 3

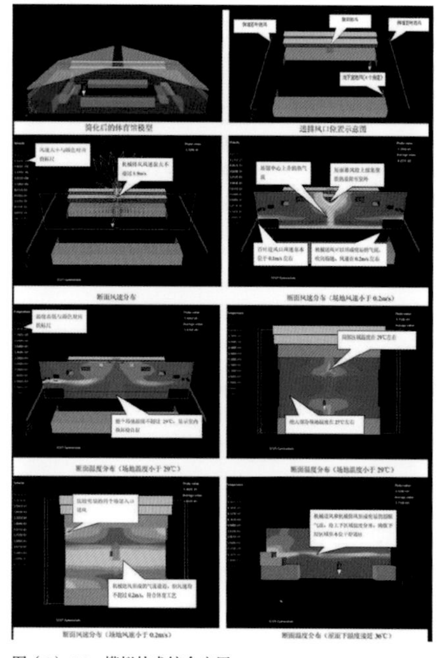

图（5）-16　模拟技术综合应用
Fig.（5）-16　Comprehensive application of simulation technology

6. 先进技术手段的运用

在体育馆的方案深化过程中，多个专业都采用了先进的技术手段辅助设计。暖通工程师对室内机械辅助的自然通风进行了计算流体力学（Computational Fluid Dynamics, CFD）模拟，通过计算得出建筑所需通风量，进风口的面积和位置、顶部排风口风机的总功率、台数（与噪音限制有关）以及安装风机需要占用的面积。建筑师运用 SketchUp 软件对四个梯形的天窗以及东西屋檐下的高侧窗获取自然光照明进行了日照模拟，以确定遮阳板尺寸及控制室内防眩光的自然采光效果。结构工程师对屋盖结构采用大型通用有限元程序 ANSYS 进行计算分析，主要进行了竖向荷载、风荷载、地震作用、温度作用、结构整体稳定计算和施工过程模拟分析，着重分析了屋盖壳体结构（包括支承壳体的边缘构件）的内力与变形，并对看台框架部分进行校核。

6. Use of advanced technological means

In the deepening process of the gymnasium, a number of professional had adopted advanced technology aided design. HVAC engineers, indoor mechanically assisted natural ventilation were CFD (Computational Fluid Dynamics, CFD) simulation, by calculating the building ventilation rate requirements under the premise of the inlet area and the location of the top of the exhaust port fan total power, number of units (with noise restrictions related) and the area to install the fan. Architects used SketchUp software skylights and four trapezoidal side windows under the eaves of things. Natural light was simulated to determine the size of the sun visor and interior glare control natural lighting effects. Structural engineer for a large common roof structure using finite element program ANSYS calculation and analysis, mainly for the vertical loads, wind loads, earthquake, temperature effects, the structure of the overall stability calculation and construction process simulation analysis focuses on the roof of the housing structure (including the edges of the support casing member) of the internal forces and deformation, and the stands were checked frame portion.

7. 亚运改造

华工体育馆建成之后，为了迎接 2010 年的广州亚运会，进行了翻新改造。除了根据亚运会竞赛需求对房间功能的调整外，我们也听取了使用者的意见。在使用过程中，业主给设

7. Transformation for 2010 Asian Games

After the project completion, the venue was facing the 2010 Guangzhou Asian Games. Besides transforming the rooms for the Games function, we also made adjustments according to user demands, i.e., introducing

计师反馈的意见是，他们希望改善位于地下室的运动员休息套间的自然采光以及贵宾室的自然通风条件。因此，在这次改造中，我们在运动员休息套间靠近通道的部分增加了采光高窗，这样，从通道顶部新风口进来的光线也能进入到房间中，使房间的舒适度得以提高。而贵宾区原来的顶部采光口是密闭的，现在改为带百叶的采光口，改造完之后，贵宾室的通风得到了很大的改善。建筑在使用的过程中还能不断更新完善，以变化求得可持续发展，也是我们设计师最想要实现的。

（本案例的文字稿主要根据参考文献 [52] 整理）

natural lighting into the basement resting rooms for athletes through aisle air vents; replacing the closed windows at the top of the VIP area with shutter to enhance the ventilation performance. The building was open to any future improvements that advance the level of sustainable operation of the venue.

图（5）-17　室内照片 2
Fig.（5）-17　Indoor photo 2

图（5）-18　室外照片 4（上）
Fig.（5）-18　Outdoor photo 4 (up)

图（5）-19　室外照片 5（下）
Fig.（5）-19　Outdoor photo 5 (down)

图（5）-20　室内照片 3（上）
Fig.（5）-20　Indoor photo 3 (up)

图（5）-21　室内照片 4（下）
Fig.（5）-21　Indoor photo 4 (down)

江苏淮安市体育中心
中国，江苏，2009

HUAIAN CITY, JIANGSU PROVINCE SPORTS CENTER JIANGSU, CHINA, 2009

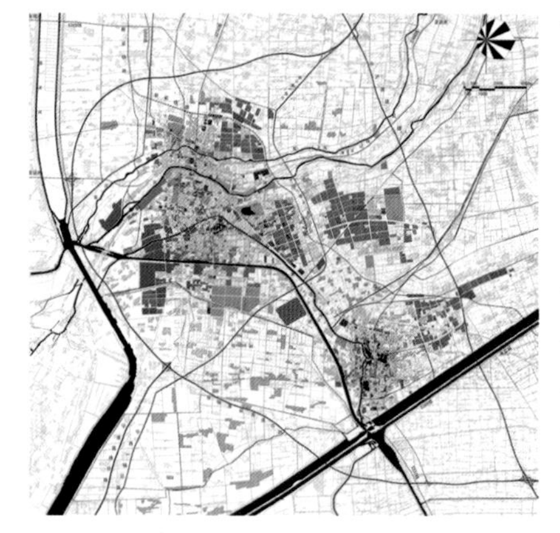

图（6）-1　区位图
Fig.（6）-1　Location map

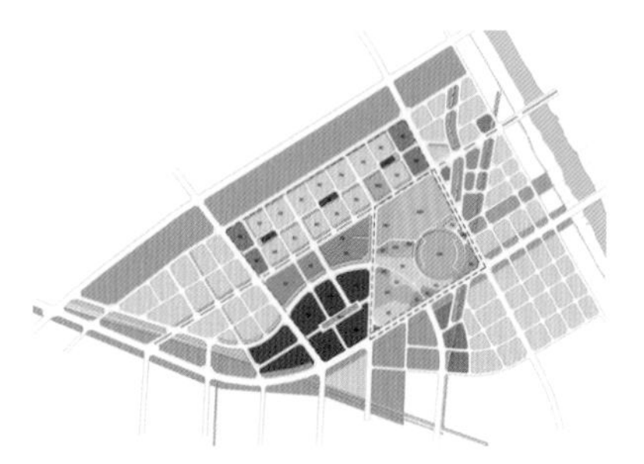

图（6）-2　片区控制性详细规划
Fig.（6）-2　Regulatory plan of the area

1. 总体规划

2008 年北京奥运会之后，中国城市体育场馆建设缓慢，但仍在继续。淮安体育中心是为 2014 年在淮安市举办的江苏省第十八届省运会而建造的。基地选址于淮安城市规划中的新城西区，希望通过兴建这一体育中心，带动新城区的发展。项目用地面积约 428 949m^2，建设包括一个体育场，一个体育馆、游泳馆综合体，一个运动员宾馆，以及远期规划的综合训练馆及网球中心，和一系列的室外运动场地。

在中国，传统的体育中心在规划和建设中，往往过于强调对体育中心自身标志性特征的凸显，希望将体育中心对城市土地运营和城市纪念性景观营造方面所发挥的推动作用最大化。这样容易忽视考虑体育中心与城市肌理的融入，与未来社区生活的衔接，从而常常产生体育设施面向城市生活社区的可达性差、日常使用不便、人气不足等孤岛效应，进而导致后期闲置与运营困难的困境。这不仅成为政府财政的负担，也难以发挥体育设施服务大众的积极作用。

1. Master plan

After the 2008 Beijing Olympic Games, domestic constructions of sports venues continued slowly. Among them, Huan'an Sports Center was constructed for the 18th Provincial Games held in Huan'an, Jiangsu. The center, sited in the West District of the new town, targeted to become the pioneer of the development of the new town. The land area measured 428,949m^2, upon which a stadium, a gymnasium, a natatorium, a hotel, a multi-function training center, a tennis hall and a series of outdoor sports fields would be designed and built.

Domestically, urban planning for sports facilities has focused more on the effect of their iconicity than of their integration with the urban fabric and community activities. Therefore, most sports facilities ended up with poor community service, inconvenient daily usage, low popularity, etc. The problematic and ineffective post-game operation has resulted in heavy burden on government funds and insufficient service for civil fitness.

图（6）-3　片区与城市整体发展
Fig.（6）-3　Integrated development of the area and the city

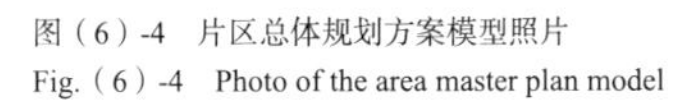

图（6）-4　片区总体规划方案模型照片
Fig.（6）-4　Photo of the area master plan model

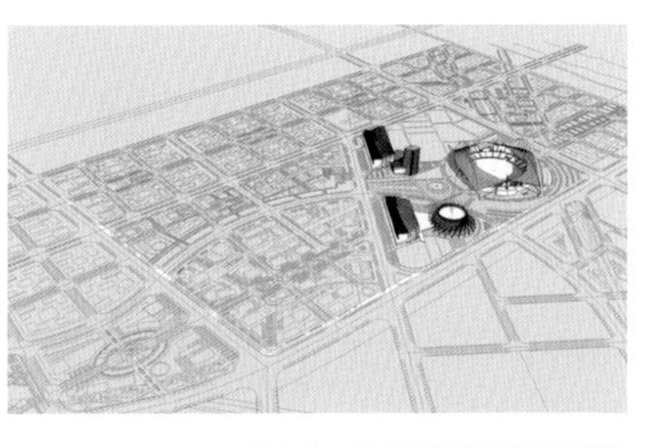

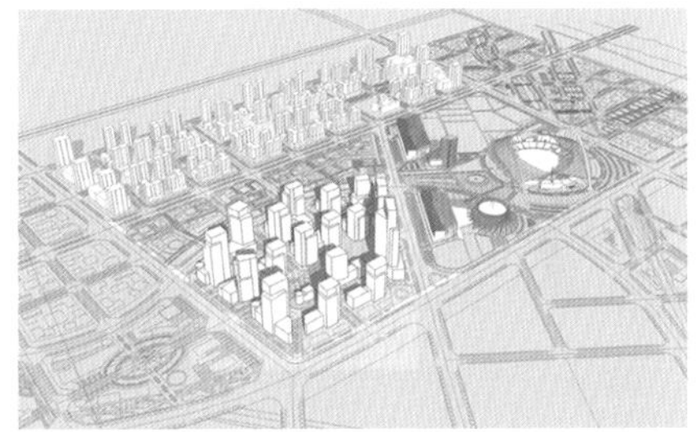

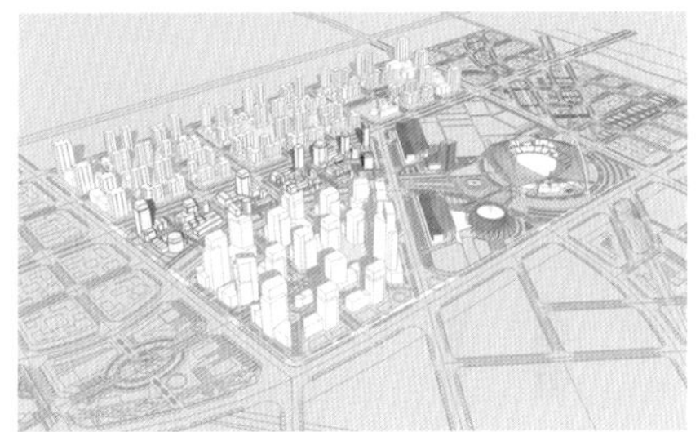

图（6）-5　片区分区开发设想
Fig.（6）-5　Development plan of the area

因此，规划将淮安体育中心定位为运动型城市活力中心，规划通过功能策划、规划布局、建筑设计，整体实现体育设施的可持续性营运、体育场馆的低造价建设与使用，以及市民使用的便利性。

Henceforth, the planning authority identified Huai'an Sports Center as a core of activities of the city. The planning specifics included guidelines on function configuration, masterplan and architectural design for the purposes of sustainable operation, low-cost construction and public convenience.

在做体育中心的规划布局之前，我们根据这一片区的城市总体规划，为该片区做了城市设计。通过功能策划，我们将该区域定位为：以体育产业为主导，以运河文化为纽带，集竞技比赛、休闲健身、商务会展、创意产业、生态体验、文化娱乐、体育旅游、健康宜居等多功能于一体的运动型城市活力中心。

Before laying out the site plan for the center, we looked into the overall urban planning of the district and finished the adjacent urban design. Eventually, the center would majorly target at the sports industries, assisted by entertainment, conferences, exhibitions, creative industries, cultural activities, ecological exploration, sports tourism, and healthcare.

图（6）-6　地块总体室外照片
Fig.（6）-6　Outdoor photo of the overall site

在规划布局上，我们的设计试图改变将体育场馆作为“纪念物”的单一作法。我们区别对待体育场与体育馆和游泳馆。体育场设计力图简单、完形，有一定的纪念性，满足市民与决策者的需求。将体育馆、游泳馆作为城市肌理的一个部分，强调与街道融合。主要建筑沿街布置，面向街区使用，以创造积极多元的生活界面，更好的满足社区化的服务需求。

In the masterplan, we attempted to avoid the stereotypical monumentality of the buildings, and aimed for a simple and integrated stadium which was given a certain extent of iconicity in order to satisfy the public and the policy-makers. The gymnasium, along with a natatorium designed differently, merged into the blocks of the urban fabric. The main buildings were oriented towards the community along the streets, creating a positively engaging channel for various activities and providing better service for the cultural development of the neighborhoods.

In terms of architectural design, besides meeting basic requirements for competitions and games, the center was provided with maximum flexibility in operation, so that the sports facilities could contribute to more social benefits and to the venue’s self-sustain.

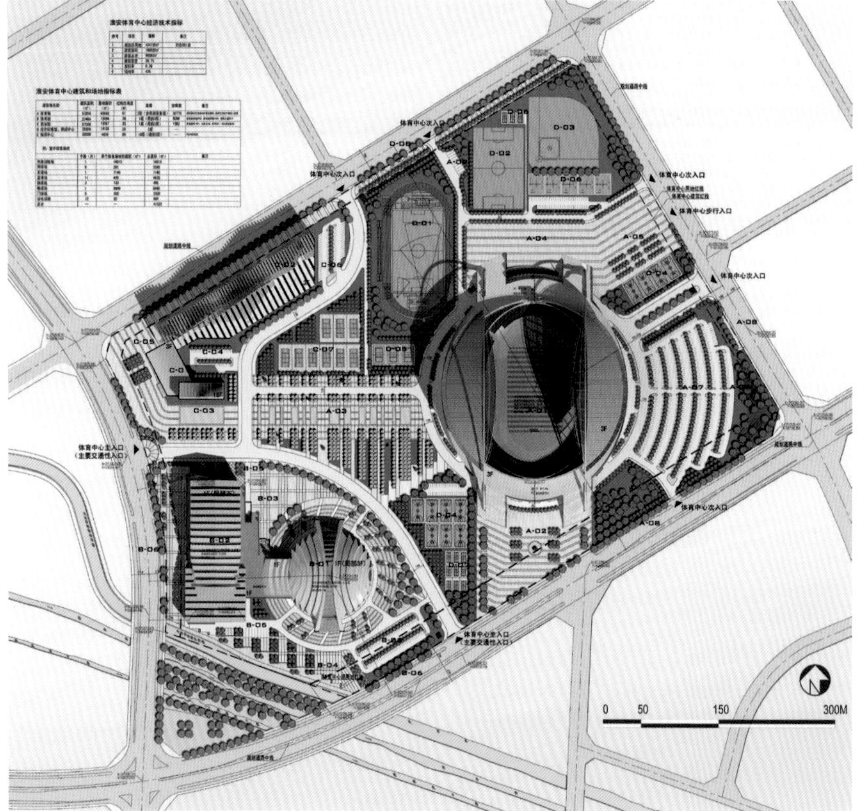

图（6）-7　体育中心地块总平面图
Fig.（6）-7　General layout of the sports center

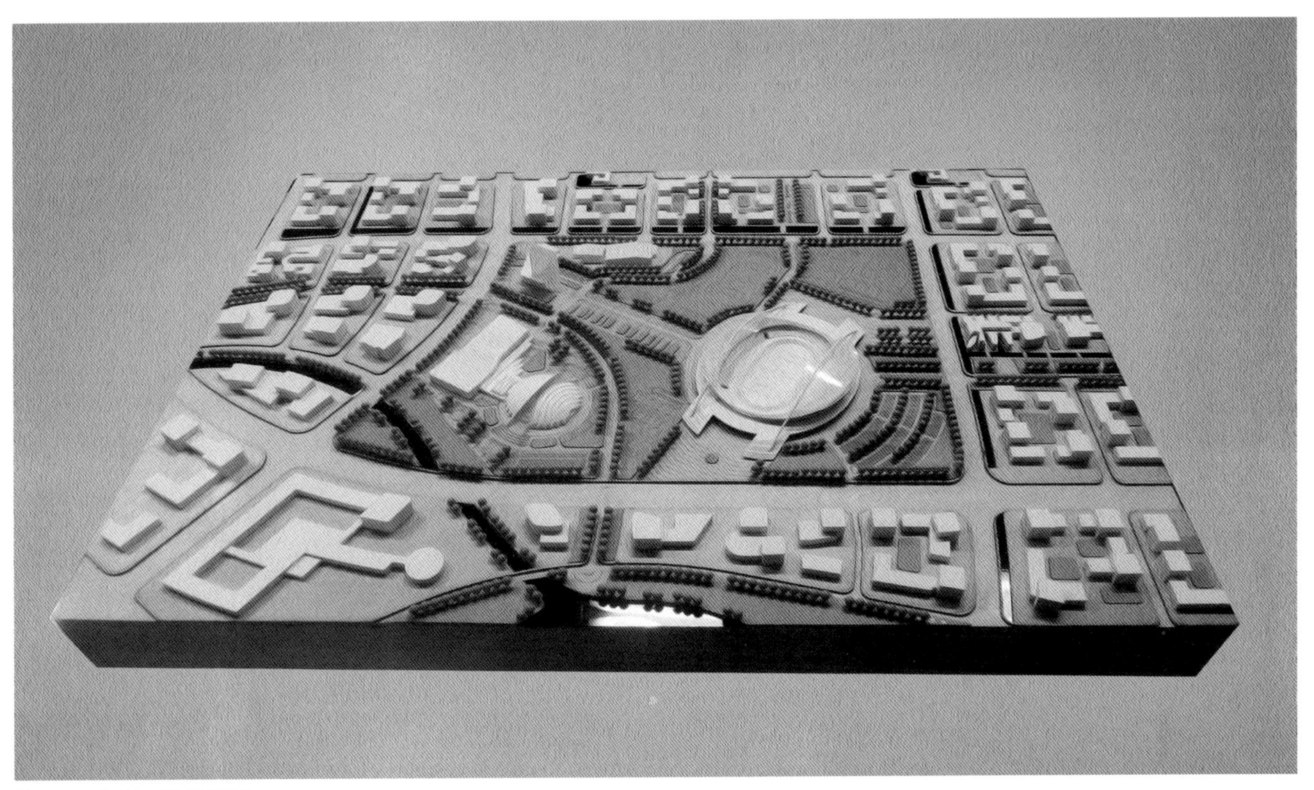

图（6）-8 体育中心地块模型照片
Fig.（6）-8 Photo of the sports center model

在建筑设计上，我们通过功能上的设置，在满足赛事基本需求的基础上，让体育中心具备最大的对外运营的灵活性，让更多的人能使用我们的设施，也使其在经济上能尽量自我维持。

基于以上功能策划、规划布局、建筑设计的成功，体育中心实际运营一年，功能和实用性良好，受使用者欢迎，给运营商利好。共接待了大约 1 万名运动员，19.8 万普通使用者以及 23.8 万观众。是淮安市承办 2014 年江苏省第十八届运动会的主要赛事场馆。

Thanks to the success we had achieved in function configuration, masterplan and architectural design, the center had survived a year of remarkable operation with high profits and popularity. Over the course of the year, the venue had hosted over 10,000 athletes, 198,000 amateur players and 238,000 spectators. It was the main venue for the 18th Provincial Games of Jiangsu in 2014.

图（6）-9 体育中心地块鸟瞰图
Fig.（6）-9 Aerial view of the sports center

图（6）-10 体育场室内照片 1
Fig.（6）-10 Indoor photo of the stadium 1

2. 体育场建筑设计

淮安在中国古运河文化时代，是一个重要的交通枢纽。在体育场建筑形象构思上，我们抽象出“运河文化”中最重要的运输工具“船”的一些形象要素，例如拉索、帆等，以此对城市历史有所回应。

不同于使用复杂表皮将体育场层层包裹的设计理念，我们体育场的设计回归体育场的基本需求，在满足大部分看台得到遮盖的基础上，最大限度的将看台后方的功能空间敞开，一方面避免了过度复杂的表皮带来的高能耗，另一方面，为了充分塑造建筑简洁、优美的造型，屋盖钢结构体系采用了创新的双拱支承结构体系。上部的单管拱与下部的三角形拱形折架组成的双拱体系具有传力明确、传力路径直接等优点，充分发挥了拱结构的受压性能，拱结构水平投影跨度达到 301.2m。

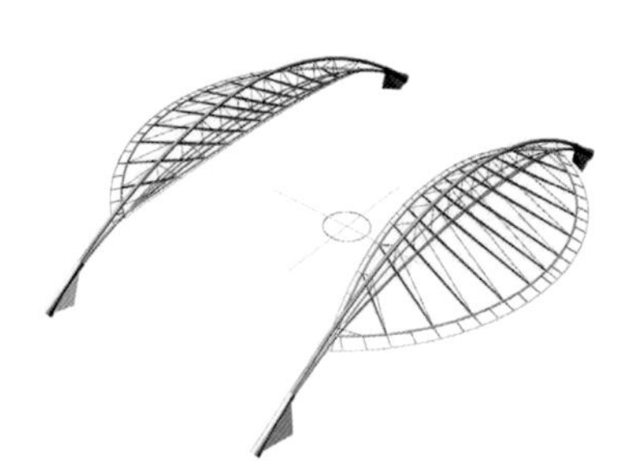

图（6）-11 体育场屋盖结构图
Fig.（6）-11 Roof structure of the stadium

2. Stadium design

Huai’an is an important traffic hub in the age of ancient canal culture era. Therefore, in the architectural image design, we used elements such as dragline and sail extracted from ship designs to respond to the history of the city.

Unlike the prevailing design concept of wrapping the stadium in layers of flamboyant shells or surfaces, our team, with respect to the fundamental function need of the stadium, open up the functional space on the back of the grandstand as much as possible given that the majority of the grandstand could be shaded, so as to reduce cost and energy consumption accompanying complex surface structures. Meanwhile, to shape the simple and elegant form, the stadium roof used an innovative long-span truss arches as its main perpendicular bearing member. Each truss arch was connected through horizontal trusses to produce the spatial pitch structure. The force flow within the structure was direct and straightforward. Ground projection of the roof structure reached 301.2m in the direction of the arches.

图（6）-12 体育场室外照片 1
Fig.（6）-12 Outdoor photo of the stadium 1

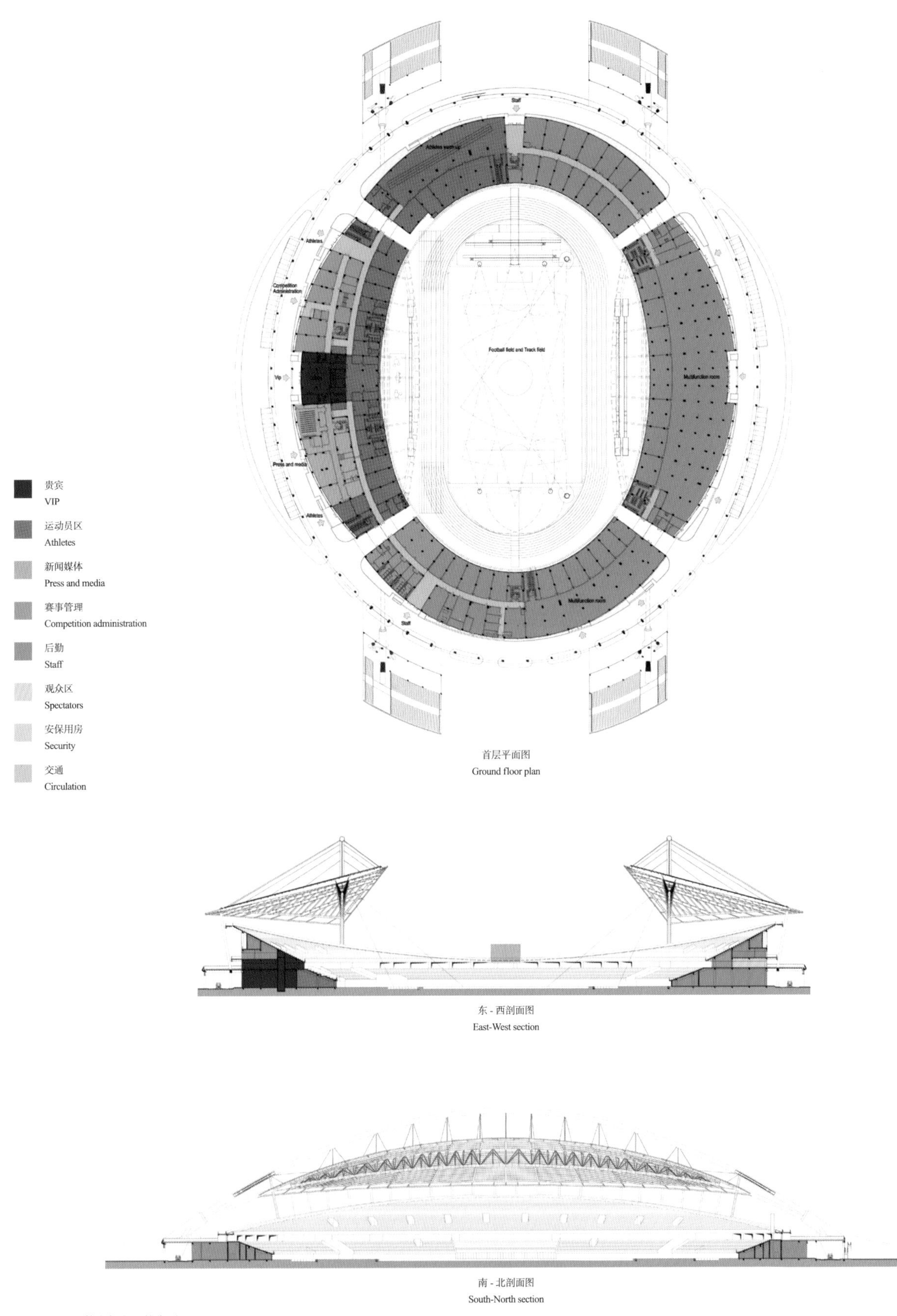

图（6）-13　体育场主要技术图

Fig.（6）-13　Main technical diagram of the stadium

图（6）-14 体育场室外照片 2（上）
Fig.（6）-14 Outdoor photo of the stadium 2 (up)

图（6）-15 体育场室外照片 3（中）
Fig.（6）-15 Outdoor photo of the stadium 3 (middle)

图（6）-16 体育场室内照片 2（下）
Fig.（6）-16 Indoor photo of the stadium 2 (down)

图（6）-17　体育场室外照片 4
Fig.（6）-17　Outdoor photo of the stadium 4

图（6）-18　体育馆与游泳馆综合体鸟瞰图
Fig.（6）-18　Aerial view of the complex of gymnasium and natatorium

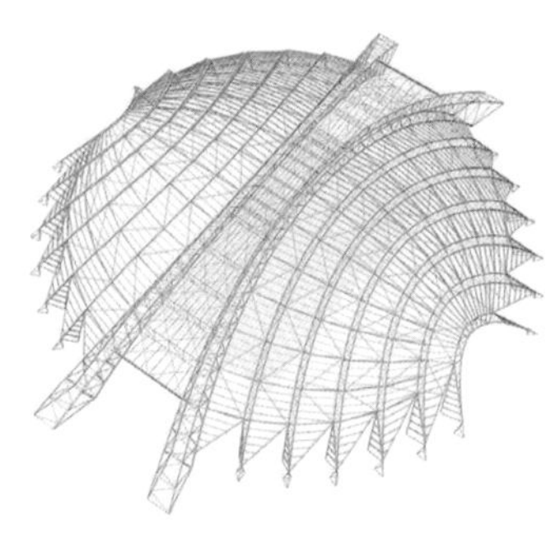

图（6）-19　体育馆屋盖结构图
Fig.（6）-19　Roof structure of the gymnasium

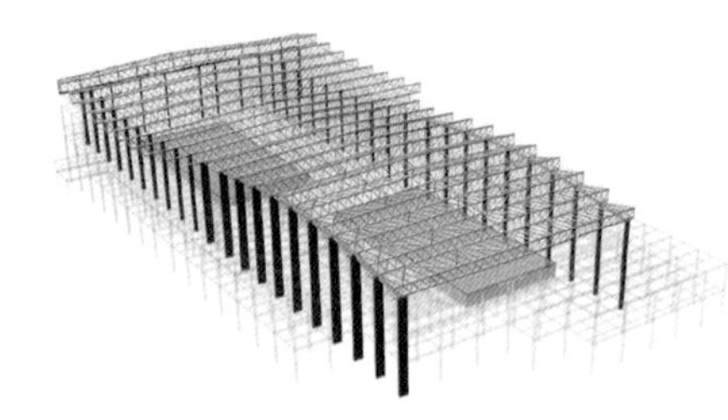

图（6）-20　游泳馆屋盖结构图
Fig.（6）-20　Roof structure of the natatorium

3. 体育馆与游泳馆综合体建筑设计

在体育馆、游泳馆综合体建筑形象构思上，取“水波”之意。创新性的将这两个功能结合在一起。电气主机房、消防水池以及新闻发布厅，这些能共用的功能用房都进行了整合，减少了总体的投资，在建设面积一定的情况下，将用于体育活动的设施面积最大化。

体育馆的屋盖采用大跨度桁架钢管拱作为主要竖向承重构件，并通过水平方向桁架将各榀桁架拱连结起来，形成空间折板结构。游泳馆采用了经济的空间门式钢架。两者的结构，都与建筑的外形相契合，塑造了具有韵律感的造型。同时我们还结合屋面的层叠造型，设置了天窗，为日常使用提供自然采光通风，节约建筑运营能耗。

3. Design for the complex of gymnasium and natatorium

The representational image for the gymnasium and the natatorium was ripple. Shared functional rooms such as the electrical rooms, the fire cistern and the press conference hall reduced the sum cost as well as maximized the area for sports facilities.

The gymnasium roof used long-span truss arches as its main perpendicular bearing member. Each truss arch was connected through horizontal trusses to produce the spatial pitch structure. The natatorium employed economical space portal frame. Both structures matched the exterior appearance and created elegant rhythms. We also designed skylights according to the stacked form of the roof to provide natural lighting and ventilation to reduce energy consumption.

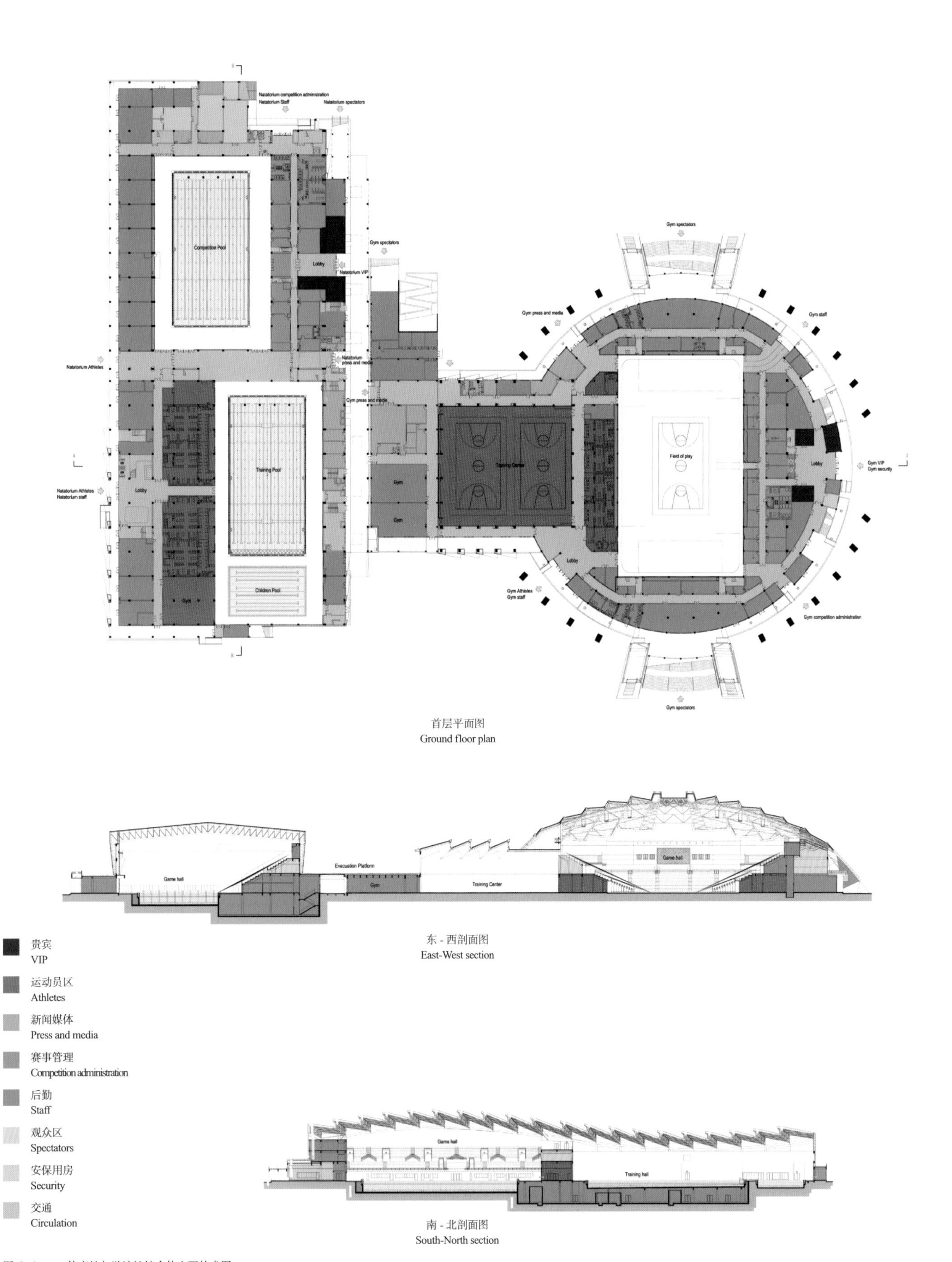

首层平面图
Ground floor plan

东 - 西剖面图
East-West section

南 - 北剖面图
South-North section

图（6）-21　体育馆与游泳馆综合体主要技术图
Fig.（6）-21　Main technical diagram of the complex of gymnasium and natatorium

图（6）-22　体育馆室外照片 5（上）
Fig.（6）-22　Outdoor photo of the gymnasium (up)

图（6）-23　游泳馆室外照片（下）
Fig.（6）-23　Outdoor photo of the natatorium (down)

图（6）-24　体育馆室内照片 3（上）
Fig.（6）-24　Indoor photo of the gymnasium (up)

图（6）-25　游泳馆室内照片（下）
Fig.（6）-25　Indoor photo of the natatorium (down)

广东梅县体育中心
中国，广东，2011

GUANGDONG MEIXIAN SPORTS CENTER GUANGDONG, CHINA, 2011

图（7）-1　区位图
Fig.（7）-1　Location map

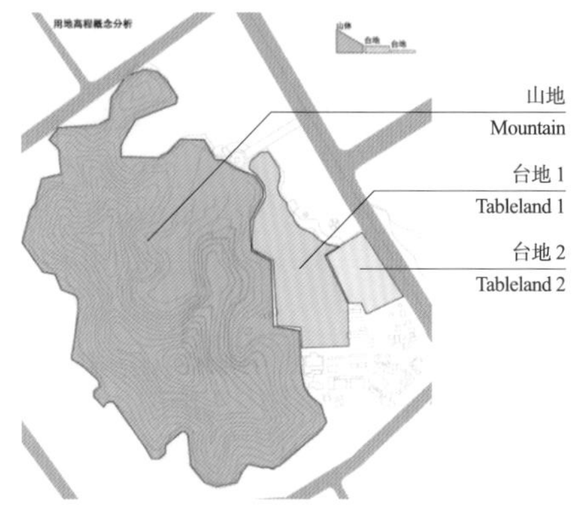

图（7）-2　基地高程分析
Fig.（7）-2　Base elevation analysis

为打造“世界客都”的城市形象，提升梅县的城市文化品质，梅县县政府于 2009 年始兴建梅县文体中心和体育场，项目于 2012 年初竣工。梅县体育中心位于梅县梅花山下，人民广场旁，项目建成后与梅花山、人民广场一起成为梅县市民休闲活动的中心场所。

梅县文体中心项目包括体育馆和体育场，设计利用地形，沿山体高度由西向东布置体育场和体育馆。和大多数城市体育场馆建设强调自身体量的做法不同，梅县体育场看台设置依山而建，以融入自然山体的手法，最大限度减少建设项目对原有山体的影响，确保城市与梅花山体之间的视线联系。体育场的外墙采用兼顾乡土和现代气息的石笼墙，石笼墙石块取自当地石材，突出了融入自然山体的主题。

通过巧妙的看台设置和场地竖向设计，成功处理了山体、场馆和城市的关系，并很好地解决了交通人流疏散的问题，实现了整体环境的协调。

The project of Meixian Cultural and Sports Center, including a gymnasium and a stadium, is located next to the People's Square at the foot of Meihua Hill, Meixian. The completed project, with the hill and the square, shapes the urban park of Meixian.

The design, based on the terrain, arranges the stadium and gymnasium from west to east following the hillside. Unlike most egocentric urban sports center, Meixian stadium has its grandstand merged with the natural mountains by the hillside. The environmentalist design principle protect the integrity of the natural landscape for the city and minimize constructional damages to the nature. The exterior wall of the stadium is made of rustic yet modernized gabion wall of local stones for the stadium to fit into the hill.

The position of the grandstand and rich sectional details successfully negotiate among nature, building and the urban conditions, smoothly directe the public circulation, and harmonize the general environment.

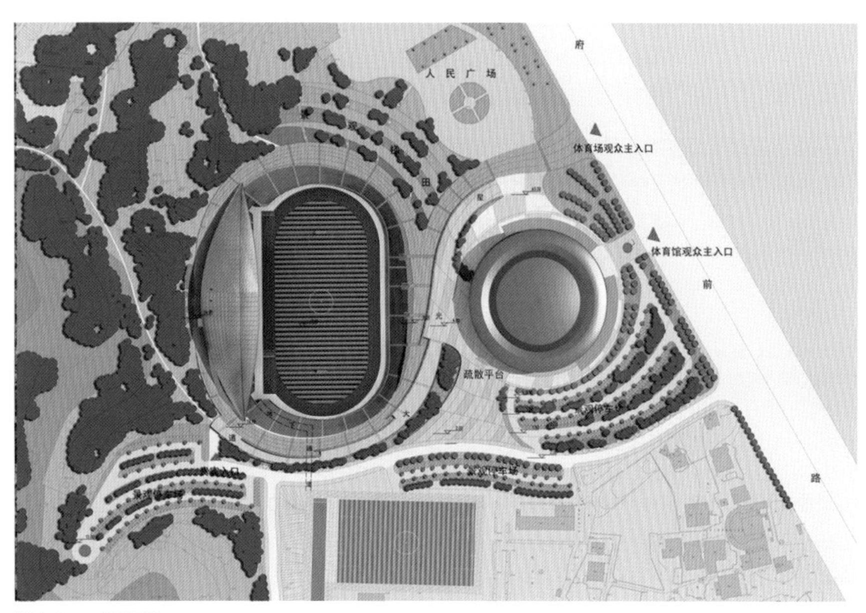

图（7）-3　总平面图
Fig.（7）-3　General layout

图（7）-4　总体鸟瞰图 1
Fig.（7）-4　Overall aerial view 1

梅县文体中心项目位于体育场与城市道路之间，拥有一个可容纳 7 000 人的多功能体育比赛厅和训练热身馆，可举办体育比赛、文艺汇演以及会展活动。各类人流出入口分布于一、二层。由于基地位于山体与城市道路之间的距离并不富裕，同时又要容纳体育场和文体中心两个公共建筑，按常规的布局和流线安排容易造成集散广场面积不足，散场时集中人流对基地和城市形成交通压力过大的问题。因此设计将大量观众的出入口布置于面向山体的一面，这样一方面可以利用地形高差，将观众人流通过坡道自然带到二层平台，且适当延长

The Meixian gymnasium is located between the stadium and urban streets, consists of a 7,000-seat multi-purpose sports hall and a training hall. The gymnasium holds sports competitions, art shows and exhibitions. The entrances and exits for pedestrians are distributed on the first and the second floors. Due to the limited space for the stadium and the cultural center between the hills and the city, regular solution for circulation inclines to reduce gathering space and aggravate urban traffics. By orienting the entrance and exit for audience towards the hill, the terrain will naturally lead the flow of audience to the second floor platform and suitably

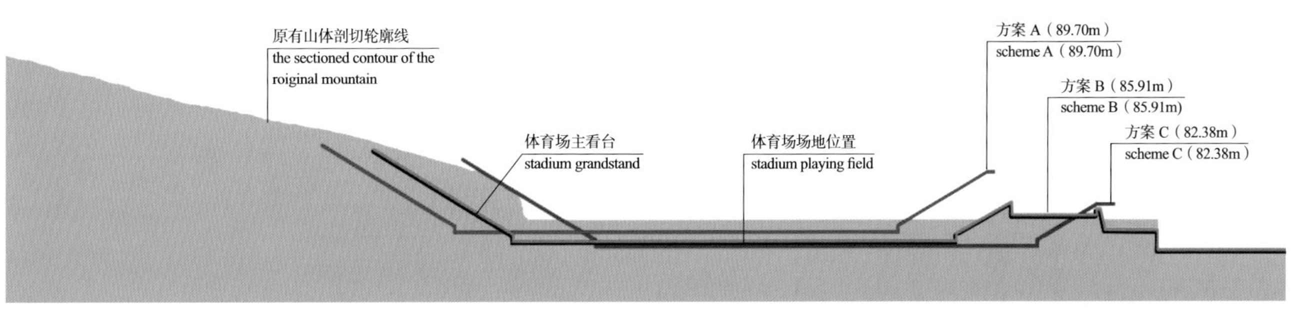

图（7）-5　高程设计多方案比选
Fig.（7）-5　Comparison of elevation design schemes

高程设计多方案比选　　表（7）-1

方案	体育场地高程取值	土方工程量	节约用地效果	场馆体量与山体的协调性	最终采用方案
方案 A	89.70 米	较大	最好	不协调	×
方案 B	85.91 米	中等	较好	较协调	√
方案 C	82.38 米	较小	不佳	协调	×

Comparison of elevation design schemes　　Tab.（7）-1

Scheme	Sports ground elevation	Earthwork	Land conservation	Coordination of the stadium mass and the mountain	Final schemes
Scheme A	89.70 m	Large	Maximum	Uncoordinated	×
Scheme B	85.91 m	Moderate	Good	Quite coordinated	√
Scheme C	82.38 m	Minimum	Poor	Coordinated	×

图（7）-6　总体鸟瞰图 2
Fig.（7）-6　Overall aerial view 2

图（7）-7　立面材料照片
Fig.（7）-7　Photo of the facade material

了观众散场时的流线长度，保证人员安全疏散的同时减缓了散场时集中人流对城市的交通压力。

梅县体育场（后更名为“曾宪梓体育场”）建设规模可容纳 2 万人，功能以满足作为“足球之乡”的梅县举办足球比赛为主。看台设置依山而建，以融入自然山体的手法，最大限度减少建设项目对原有山体的影响，确保城市与梅花山体之间的视线联系。体育场的外墙采用兼顾乡土和现代气息的石笼墙，石笼墙石块取自当地石材，突出了融入自然山体的主题。

extend the circulation, which enhances the safety level of evacuation and mitigates the impact of visitor traffics on the city. The position of the grandstand and rich sectional details successfully negotiate among nature, architecture and urban conditions, smoothly direct public circulation, and harmonize the general environment.

The stadium (once named Zeng Xianzi Stadium) hosts 20,000 people; its basic function is to serve football matches of the region, which has been praised as a ‘hometown of football’. Meixian stadium has its grandstand merged with the natural mountains by the hillside. The environmentalist design principle protects the integrity of the natural landscape for the city and minimizes constructional damages to the nature. The exterior wall of the stadium is made of rustic yet modernized gabion wall of local stones for the stadium to fit into the hill.

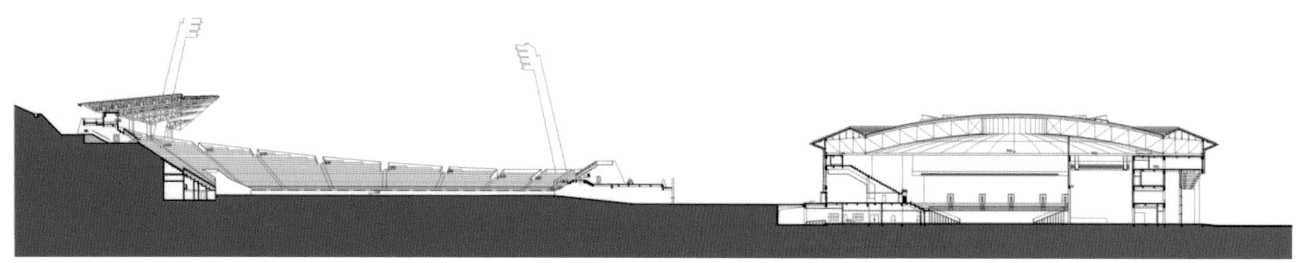

图（7）-8　最终确定的剖面
Fig.（7）-8　Finalized section

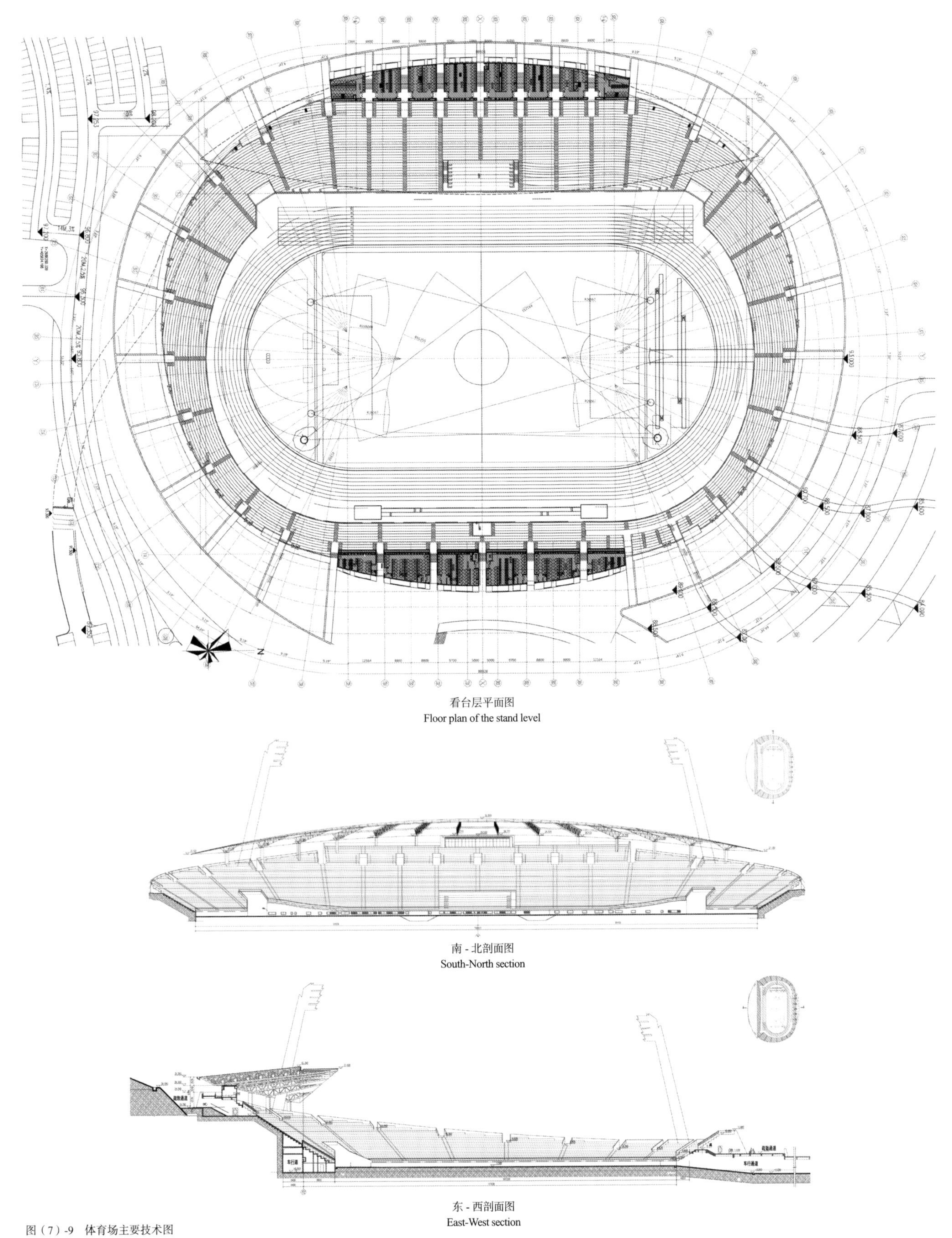

看台层平面图
Floor plan of the stand level

南 - 北剖面图
South-North section

东 - 西剖面图
East-West section

图（7）-9　体育场主要技术图
Fig.（7）-9　Main technical diagram of the stadium

图（7）-10　体育场室外照片 1（上）
Fig.（7）-10　Outdoor photo of the stadium 1 (up)

图（7）-11　体育场室内照片 1（下）
Fig.（7）-11　Indoor photo of the stadium 1 (down)

图（7）-12　体育馆室外照片 1
Fig.（7）-12　Outdoor photo of the gymnasium 1

体育馆建筑平面形式为圆形，座席按不对称方式布置，以利于未来举办演出和会议，提高场馆的空间利用率。训练馆和比赛厅使用可分可合，体育比赛时，通过活动分隔，训练区和比赛厅分为两个区域；文艺演出时，训练馆可作为演出的舞台或后台；举办会展活动时，两个空间区域可合二为一。

结构设计和设备系统设计均考虑到建筑的多功能使用需求，以灵活适应为原则，满足未来演出和各类活动的需要，并预留结构荷载裕度以及设备容量裕度。考虑未来举办更高级别体育比赛的可能，电力照明系统充分考虑了可扩展性和可更新性。

结合当地经济发展水平和造价情况，设计采用适宜技术，最大限度实现建筑自身的节能降耗，屋盖设置环形天窗，可为比赛厅和热身馆提供自然采光；天窗上设置自动开启窗扇，满足平时自然通风的要求。

图（7）-13　圆形土楼意向图
Fig.（7）-13　Circular earth building intention image

The asymmetric grandstand serves for future performances and conferences, enhancing the annual utilization rate. The training hall and the competition hall can be used separately or inter-dependently, e.g., for sporting events, temporarily partitions separate the training area from the competition hall; for theatrical performances, the training hall can serve as a stage or backstage; for exhibition activities, the two may be combined.

In structural and equipment system design, the multi-functional requirements of the building are taken into account. Flexible adaptation principle is adopted to meet the needs of future performances and various activities, and reserve structural load and equipment capacity margin. Considering the possibility of holding higher level sports competitions in the future, the electric lighting system takes sufficient account of the scalability and renewability.

The roof is provided with circular skylight to let in natural lighting to the competition and warm-up halls; the window sashes can be opened automatically to satisfy the ventilation needs. According to the local economic level and cost, appropriate technology is applied to save energy and reduce cost.

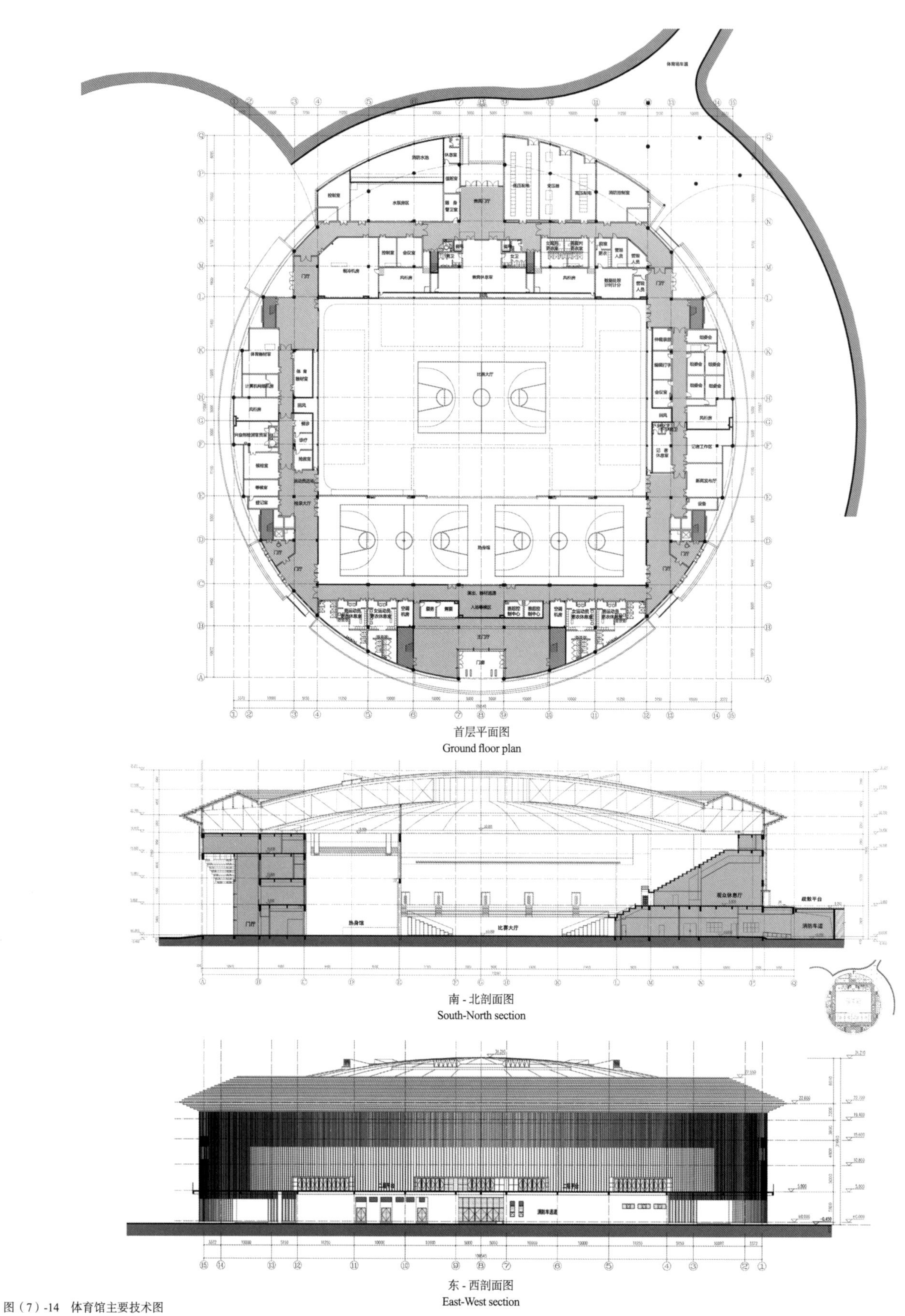

图（7）-14　体育馆主要技术图

Fig.（7）-14　Main technical diagram of the gymnasium

图（7）-15 体育馆室外照片 2（上）
Fig.（7）-15 Outdoor photo of the gymnasium 2 (up)

图（7）-16 体育馆室内照片 1（下）
Fig.（7）-16 Indoor photo of the gymnasium 1 (down)

图（7）-17　体育馆室外照片 3
Fig.（7）-17　Outdoor photo of the gymnasium 3

图（7）-18　体育馆室内屋盖照片
Fig.（7）-18　Indoor photo of the gymnasium roof

图（7）-19　体育馆入口照片
Fig.（7）-19　Photo of the gymnasium entrance

建筑屋顶直径长达 112m，使人联想到客家传统的围龙屋。文体中心的外墙采用竖向排列陶土棍，并选用接近客家传统民居墙体颜色的土黄作为主体基调，通过疏密有致的排列，形成独特的具有客家文化气息的外立面效果，同时满足了文体中心内部用房的采光和遮阳需求。

梅县体育场及文体中心的建成，翻开了梅县人民的文化体育事业建设的新篇章，不仅为梅县人民提供了重要的文化、体育活动的场所，也为素有“文化之乡”“华侨之乡”“足球之乡”美誉的梅县又增添了一道靓丽的风景。

（本案例的文字稿主要根据参考文献 [53] 整理）

The roof is 112m in diameter; its form reminds people of Hakka’s traditional Lung Wai house. The external wall of the culture and sports center is made of vertically arranged clay sticks using yellowish brown (wall color of traditional Hakka houses) as main color. With careful arrangement, the wall forms a unique exterior façade reflecting Hakka culture, and at the same time, provides daylight and shades.

The establishment of Meixian Sports Center marks a new chapter of the development of cultural and sports business of Meixian. It not only becomes an important place for cultural and sports activities for the local, but also brings a new scene into Mei County, which has long enjoyed being the ‘home of culture’, the ‘hometown of overseas Chinese’ and the ‘home of football’.

图（7）-20　体育馆室外照片 4（上）
Fig.（7）-20　Outdoor photo of the gymnasium 4 (up)

图（7）-21　体育馆室内照片 2（下）
Fig.（7）-21　Indoor photo of the gymnasium 2 (down)

图（7）-22　体育馆室内照片 3（上）
Fig.（7）-22　Indoor photo of the gymnasium 3 (up)

图（7）-23　体育馆室内照片 4（下）
Fig.（7）-23　Indoor photo of the gymnasium 4 (down)

华中科技大学体育馆
中国，湖北，2002

HUAZHONG UNIVERSITY OF TECHNOLOGY GYMNASIUM HUBEI, CHINA, 2002

图（8）-1　区位图
Fig.（8）-1　Location map

总平面布置

业主对该项目主要期望在于两点：1. 体育馆沿城市干道布置；2. 设置跨湖道路与体育馆入口广场、城市干道连接。

我们发现基地有一横跨湖面的埂，这成为方案中跨湖道路的选线；体育馆稍靠东面布置，在临湖面形成开阔广场；同时，体育馆主轴线与城市干道平行，并沿道路红线后退45m，形成良好的观赏角度；体育馆的定位与东北面校园运动区取得直接联系，方便使用；形成校园对景。

对于道路、流线组织，沿现阶段用地的东侧规划有供运动员、贵宾、记者、办公管理人员及器材用的道路和停车位，并设置不同的入口；沿用地的西侧规划有供观众用的道路及停车位。流线区分明确，避免了不同人流的相互交叉、干扰。对于广场、绿地，在体育馆及校园道路和水塘之间规划有供学生休息的广场、绿地，形成独特的校门前区空间。空调机组设在体育馆的平台下，开敞布置，以利于通风、散热，降低造价。

General layout

The owner expected two phenomenon in the final design: the gymnasium oriented towards the city streets; the gymnasium entrance connected to the city streets by a path over the lake.

During the course of investigation and design, we discovered a ridge across the lake where the path could be located. The gymnasium was placed to the east to form square by the lake; meanwhile, the main axis of the gym lay parallel to the city streets 45m away from the red line, so that vantage points for views could be created. The gymnasium drew relationship with the sports activity area northeast of the campus for convenient use and visual connection.

The circulation system divided the site for two parts, in which the east one provided circulation and parking for athletes, VIPs, journalists, officers, management crews. The west for audience. The clear division prevented interference between the two flows of visitors. Between the gymnasium and the lake, squares and vegetation for student activities and recreations created a special frontal area for the campus gate. The HVAC system was installed beneath the platform in an open space for better ventilation, faster cooling and lower constructional cost.

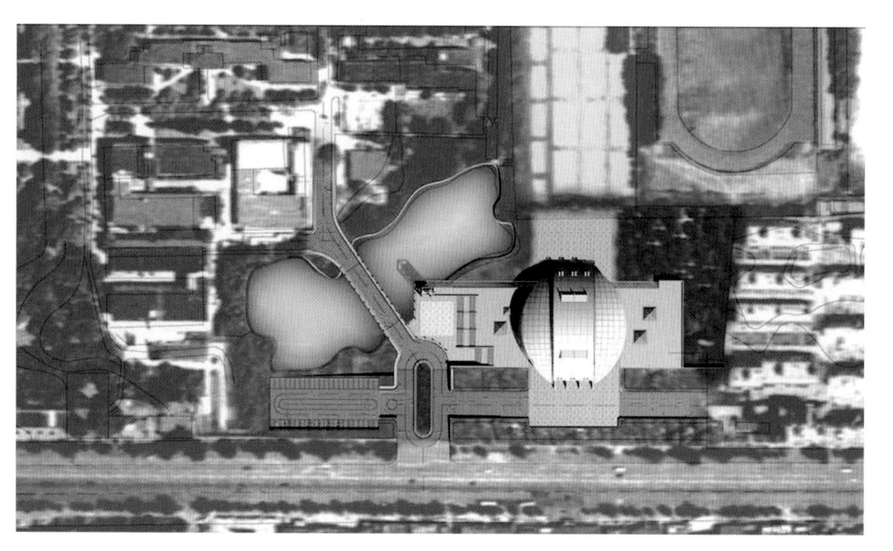

上：基地条件
下：业主对体育馆和道路的期望
右：总体布局方案
Up: Site conditions
Down: Owners' expectations for gymnasium and road
Right: Overall layout plan

图（8）-2　总体布局构思过程
Fig.（8）-2　Ideation process of the overall layout

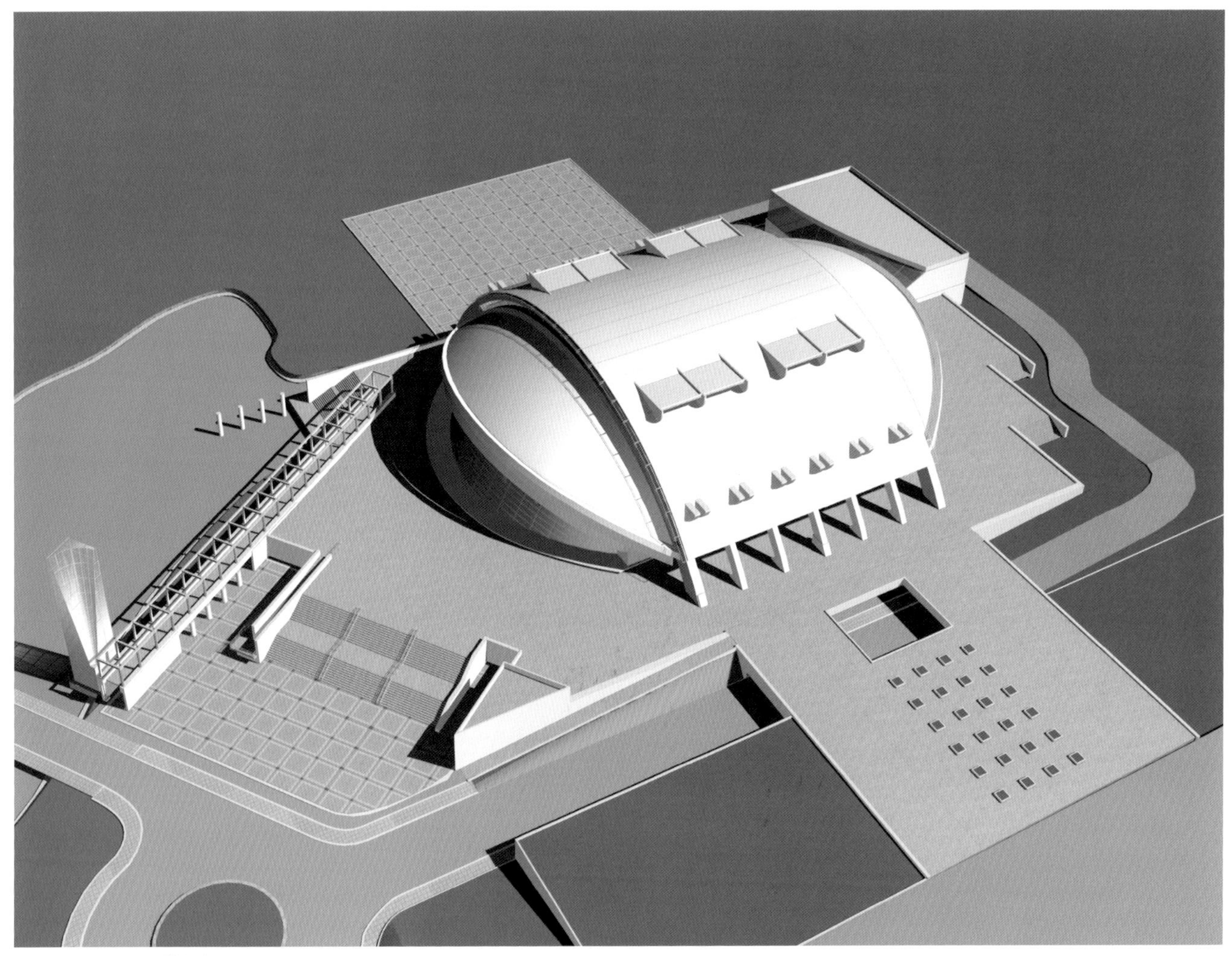

图（8）-3　设计方案总体鸟瞰图

Fig.（8）-3　Overall aerial view of the design plan

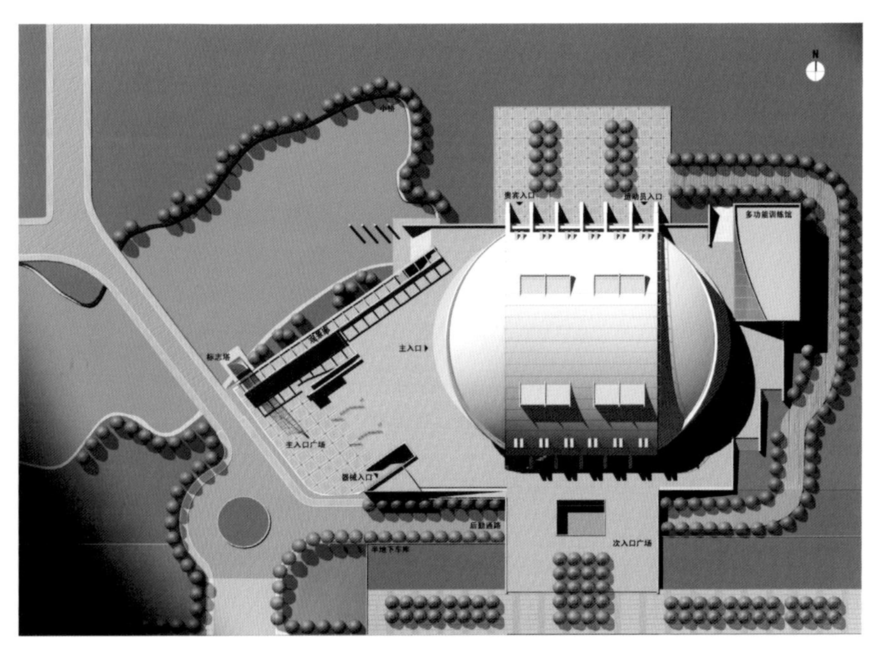

图（8）-4　设计方案总平面图

Fig.（8）-4　General layout of the design plan

图（8）-5　室外照片 1
Fig.（8）-5　Outdoor photo 1

造型设计

本设计试图利用最为成熟的结构体系，创造新颖独特的建筑形象。通过球形网架的高差处理，不但可解决馆内的采光问题，且创造了轻快、飘逸的形体。平台四周的草坡使裙房与场地融为一体，减弱了建筑的体量，利于建筑与环境的融合。

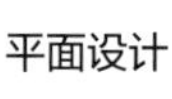

平面设计

首层平面比赛场地选择尺寸 40m × 70m，可布置国际体操比赛场地、手球场、篮球场等。教学和训练时灵活性更大，场地数量可增加。附属用房包括运动员、贵宾、记者、裁判用房、广播、电视转播用房和办公、会议用房。多功能用房包括健身、训练房，比赛期间可供比赛使用，平时可对外开放，供群众锻炼和学生上课使用，同时可达到以馆养馆之目的。设备用房包括空调、变配电用房、水泵房。均按实际需要的建筑面积予以配置。多功能用房、设备用房、办公用房围绕主馆布置，与水面形成了不同层面的开敞空间。为增加场地内绿化，使体育馆形象更为亲切，裙房的周边设有倾斜的草坡，使裙房在视觉上与场地的绿化融为一体。残疾人坡道结合草坡布置。

图（8）-6　室内照片 1
Fig.（8）-6　Indoor photo 1

Form

The design of the structure endeavored to create innovative architectural form with matured structural system and technologies. The elevational drops between spherical trusses not only solved the natural lighting issues inside the building, but also contributed to the design of a lightweight and agile exterior form. The elevated grass-slope blended the attached rooms into the site, mitigating the imposing mass of the main structure.

Plans

For the ground floor plan, the field sized 40m×70m, capable for international gymnastics, handball, basketball competitions, etc. In daily teaching and training, the field had even better flexibility so as to accommodate more fields. Auxiliary rooms included rooms for athletes, VIPs, journalists and judges; rooms for the press, television relay; office and conference rooms. Multi-functional rooms including training rooms open to competition management during games, or to public fitness or campus education during daily operation, in order for the gymnasium to self-sustain. Equipment

图（8）-7　施工照片 1
Fig.（8）-7　Photo of the construction 1

rooms including HVAC, power network, and water pump system rooms. Each was sized to the necessary space its function requires. All rooms aforementioned were arranged around the main building, creating open spaces of various levels with the water level of the lake. To increase vegetation rate, making the gymnasium more amiable, grass-slope was designed by the attached buildings to visual blend together the buildings and the site. Accessibility entrance was provided with the help of the grass-slope.

二层设有部分观众席，并包括残疾人座席，供观众使用的休息厅、卫生间、便利店等设施。对外疏散口布置均匀。三层设有观众厅的楼座部分、计时记分、声控灯控。

The second floor consisted of part of the seats, including seats for the disabled, audience lounges, restrooms, and vending facilities. Emergency exits were distributed evenly. The third floor located the seats, the score board and the acoustic and lighting control.

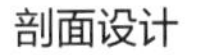

剖面设计

通过屋盖的高差，形成垂直向的天窗，解决馆内的采光问题，同时通过光线的折射解决眩光问题。自然光的引入，不仅形成明亮的室内环境，且可降低体育馆的日常运营的费用。比赛场地最高点净高 25m，不仅能满足体育比赛的需要，还可适应文艺、杂技等多功能使用要求。

Sectional design

The height differences in the roof structure shaped vertical skylights and solved the issue of direct sun. The introduction of natural light lowered the daily operational expense of the venue. The maximum elevation of the floor was 25m, suitable for not only sporting events, but also theatrical performances, acrobatics and other activities.

图（8）-8　施工照片 2
Fig.（8）-8　Photo of the construction 2

结构设计

结构上采用技术最成熟的网架和拱架组成高低起伏，又具有天然采光功能的屋面结构体系，使建筑造型活泼明朗、构思新颖、个性鲜明。其他如空调、灯光、音响等设施以及无障碍设计等，都采用了最新的技术。与此同时，对自然通风和天然采光，也都作了相应的安排，力求最大限度地减少管理运营费用。

Structure design

Structurally, the roof support consisted of steel truss arches of changing heights, so as to assist the introduction of natural lighting and the overall design of the agile and stylish form. Facilities such as HVAC, lighting, stereos and accessibility had adopted cutting-edge technologies. Treatments for natural lighting and ventilation also contributed to the lowest operational cost objectives.

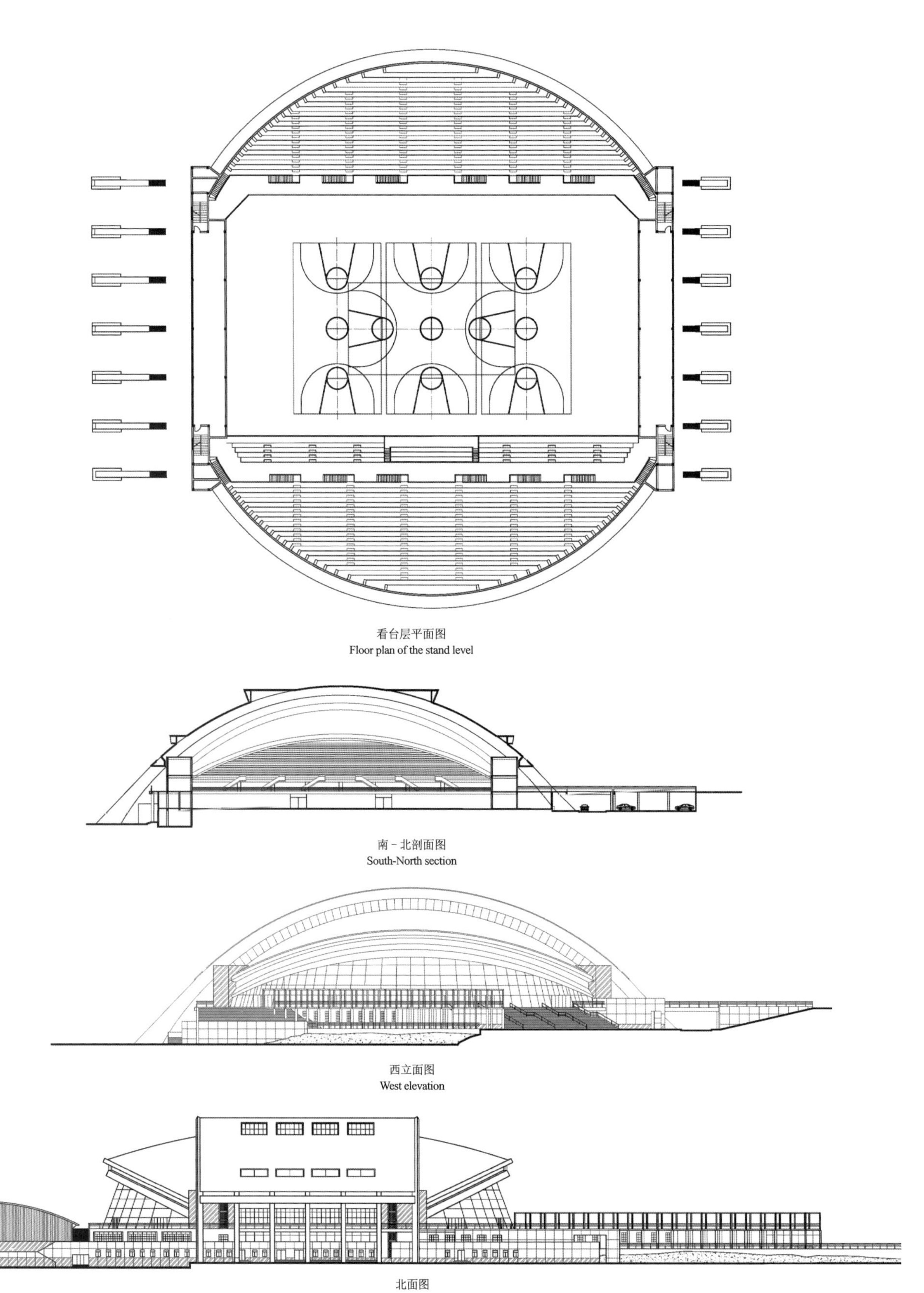

看台层平面图
Floor plan of the stand level

南－北剖面图
South-North section

西立面图
West elevation

北面图
North elevation

图（8）-9　主要技术图
Fig.（8）-9　Main technical digram

图（8）-10　室外照片 2
Fig.（8）-10　Outdoor photo 2

图（8）-11　室外照片 3（上）
Fig.（8）-11　Outdoor photo 3 (up)

图（8）-12　室外照片 4（下）
Fig.（8）-12　Outdoor photo 4 (down)

图（8）-13　室内照片 2
Fig.（8）-13　Indoor photo 2

2011年深圳大运会场馆（宝安体育场）
中国，广东，2008

2011 SHENZHEN UNIVERSIADE VENUES (BAO'AN STADIUM)
GUANGDONG, CHINA, 2008

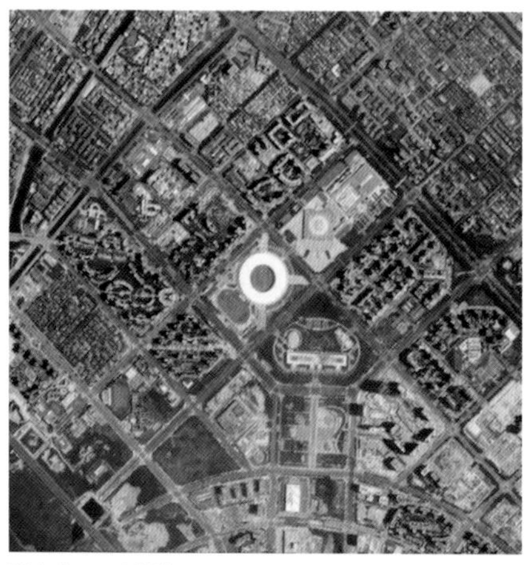

图（9）-1　区位图

Fig.（9）-1　Location map

本项目的总体目标是将大运会的精神与深圳的区域特征结合起来，为宝安区树立一座标志性的建筑。深圳地处亚热带，举步皆为园林；而园林中的葱郁竹林则是南中国的代表性植物，它在这里演变为建筑设计的灵感。竹林素以其曼妙、通透和优雅之态而著称，且随阳光照射的方向可形成不断变化的光影图案。在宝安体育馆，环绕大厅的绿柱正好演绎了这样不规则的光影图案，并支撑了屋顶和上部看台的荷载。观众从绿柱之间渗入体育场，抵达较低处的看台，将整个体育馆的壮观景象尽收眼底。

The ultimate goal for the project was to erect an iconic building that embodied the spirit of the Universiade and regional characteristics of Shenzhen. The inspiration of the design concept was drawn from the bamboo forest of the typical sub-tropical gardens of Shenzhen. Bamboo forest was renowned for its gracefulness and translucency, especially the shadow effect it created under direct sunlight. The green columns surrounding the hall of Bao'an Stadium was a reflection of the bamboo forest imagery. The columns also supported the roof and the upper grandstand. Visitors flow through the columns to the lower grandstand, where they could take into the majestic scenery of the entire Stadium.

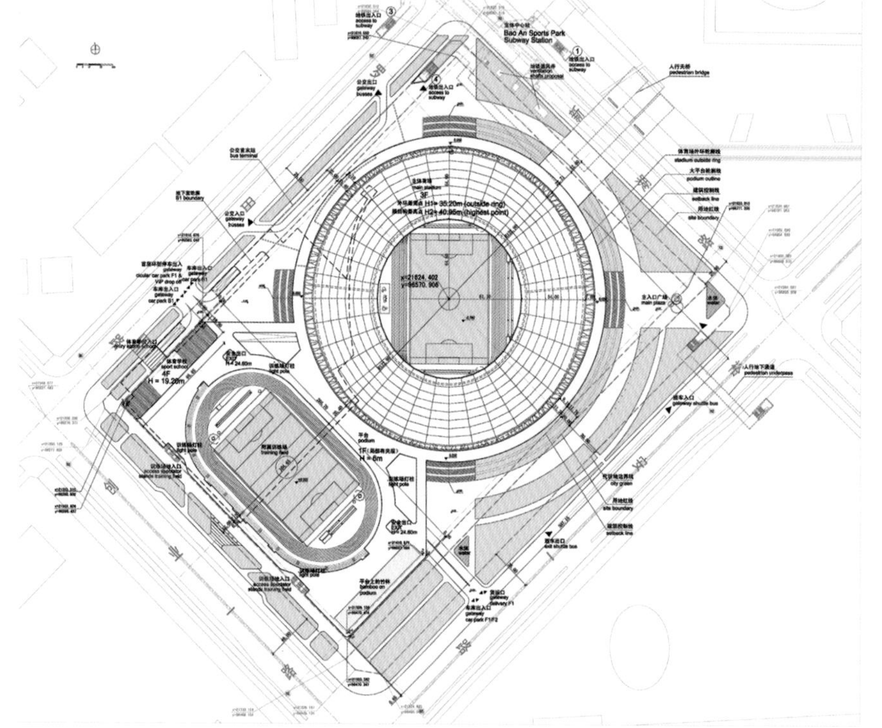

图（9）-2　设计方案总平面图

Fig.（9）-2　General layout of the design plan

图（9）-3　设计方案总体鸟瞰图

Fig.（9）-3　Overall aerial view of the design plan

图（9）-4　总体鸟瞰图
Fig.（9）-4　Overall aerial view

体育场的屋盖构造遵循了上部看台起伏的曲线造型，这一造型在南面和北面较低，而东面和西面则较高。由于体育场屋盖采用宽大的膜结构、以及该构造所具备的高度塑性，因此整个屋盖看上去犹如一朵巨大的浮云升腾于竹林之上；由此塑造了独特的建筑外观，同时又呼应了 2011 年这一欢乐盛事的主题。

The roof structure, responding to the curve of the upper grandstand, was shaper taller at the east and the west, shorter at the north and the south. The roof applied a light membrane structure that made it look like a huge cloud floating on the bamboo forest.

在地面层，所有与足球比赛机构相关的功能均设在西看台之下。

On the ground floor, all programs related to football matches and football associations were set beneath the west grandstand.

立面由两层构成，外层是层次丰富的竹林立面，它们均由空心钢立柱组成，直径不等。高度上有每 4m 一段和每 8m 一段进行划分的拼接。表面为绿色喷涂，凸显竹林的概念。整体构建轻巧美观。竹林后层除了各辅助功能房间、楼梯等自然形成的铝板表面立面外，在 3 层的上方至看台底板设置了环绕的膜结构的第 2 立面，其材料为 PVC 膜，其设置统一解决了 3 层顶部众多空调室外机、设备设施、管道等安置遮挡的问题，白天时其纯粹的形态更加突出了主体育场竹林的概念；夜晚，结合灯光的巧妙设计，人们可以远远地看到立面上灯光投射出的影像、广告、活动信息等，更加强调了该建筑的独特标志性。

The elevation had two layers. Firstly, the hollow steel columns varying in diameter and member lengths (4m or 8m between connections) are painted green to resemble bamboos. Secondly, behind the ‘bamboos’, aside from auxiliary rooms and staircases was a membrane wrapper structure stretched from the top of the third floor to the bottom of the grandstand. The membrane, made of PVC, resolved the problem of the previously-exposed air conditioning units, mechanics and pipes. In daytime, the pure form of the membrane magnified the effect of the bamboo-like bottom structure; at night, with proper arrangement of lighting equipment, projections of advertisement and event information could be seen on the façade from afar, which emphasis the position and the iconicity of the architecture.

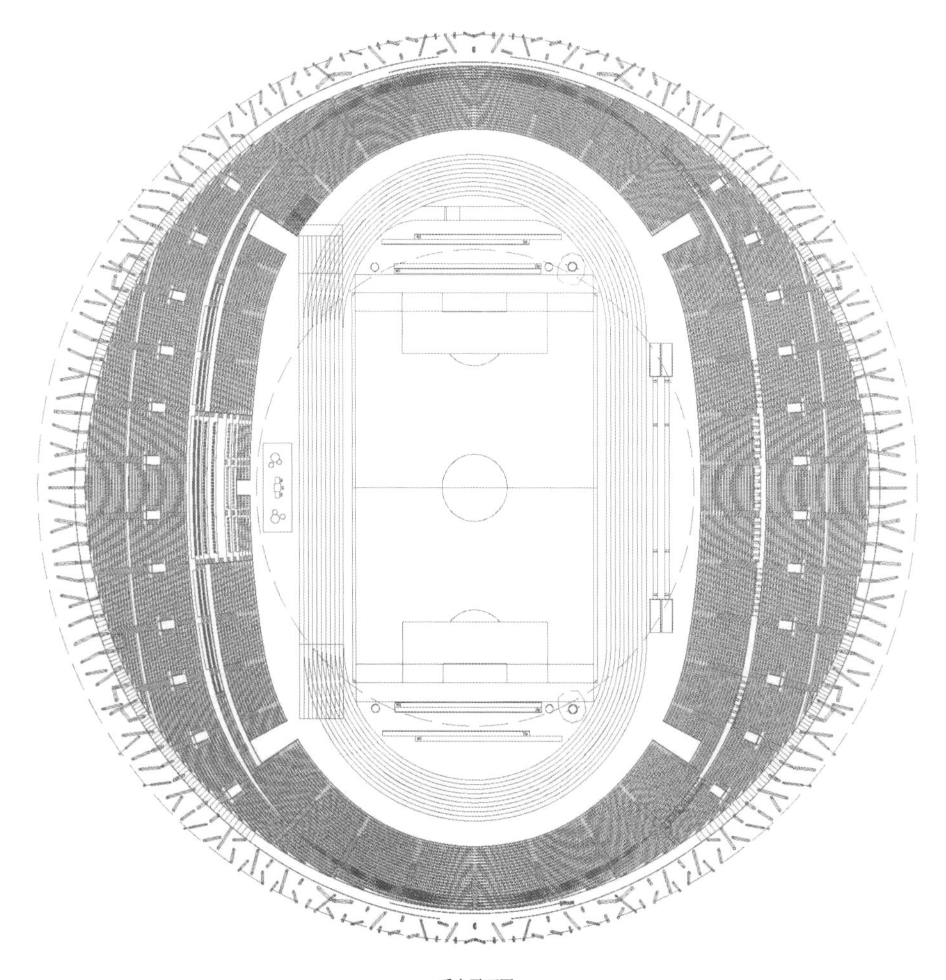

看台平面图
Floor plan of the stand

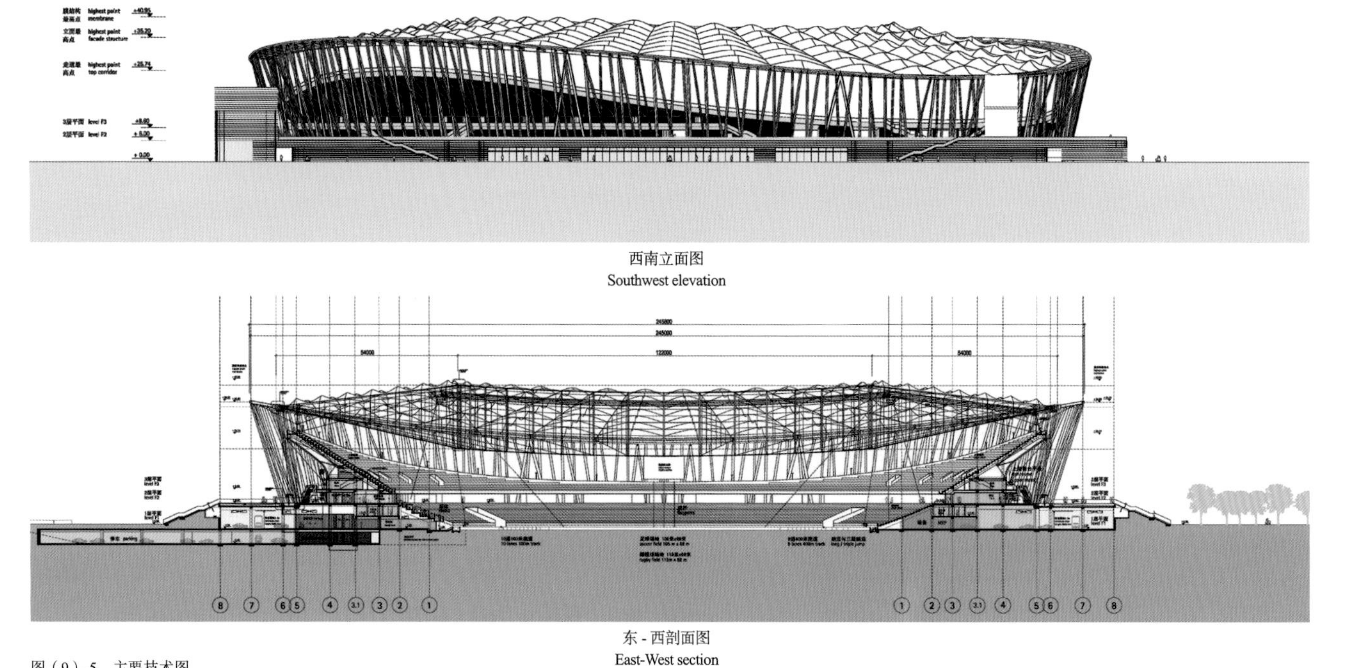

西南立面图
Southwest elevation

东 - 西剖面图
East-West section

图（9）-5　主要技术图
Fig.（9）-5　Main technical diagram

图（9）-6　室内照片 1
Fig.（9）-6　Indoor photo 1

屋面主要结构是由 PTFE 膜、钢、铸钢构成。屋面排水沟设在膜结构屋面外沿的凹处，水经由各段天沟集水，通过立面的空心的“竹子”排到大平台面层下方至通常排水系统。

设计原创团队：德国 GMP 国际建筑设计有限公司，施工图与实施团队：华南理工大学建筑设计研究院。

（本案例的文字稿主要根据合作团队设计文件整理）

The roof was mainly of PTFE membrane, steel and cast steel. Drains were provided along the edge of the roof, where water would be carried through the lower ‘bamboo’ columns to the drainage system beneath the platform.

Architectural design: German GMP International Architectural Design Co., Ltd.Construction drawing and implementation: Architecture Design & Research Institute of SCUT.

图（9）-7　屋面细部照片
Fig.（9）-7　Detailed photo of the roof

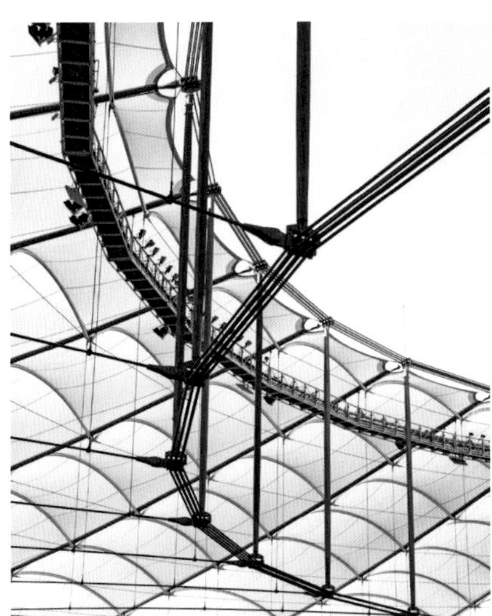

图（9）-8　室外照片 1
Fig.（9）-8　Outdoor photo 1

图（9）-9　室外照片 2（上）
Fig.（9）-9　Outdoor photo 2 (up)

图（9）-10　室内照片 2（下）
Fig.（9）-10　Indoor photo 2 (down)

佛山世纪莲体育中心
广东，中国，2003、2010

CENTURY LOTUS STADIUM GUANGDONG, CHINA, 2003, 2010

图（10）-1 区位图
Fig.（10）-1 Location map

佛山世纪莲体育中心坐落在佛山市东平河畔，为广东省第12届省运会主会场，可容纳3.6万名观众，于2004年6月开始动工，2006年9月建成。体育中心总面积42hm^2，包括体育场、游泳跳水馆、室外热身场地（含400m跑道运动场、足球训练场）、室外水上中心、能源中心及附属配套服务设施。因其体育场外形酷似莲花，故被命名为“世纪莲”。

体育场建筑面积78 193m^2，直径305m，高50m，膜投影面积53 421m^2，由屋盖索膜结构、下部钢筋混凝土支撑结构、看台及附属用房等部分组成，拥有观众座位36 686个，能够满足国际赛事的要求。

屋盖结构坐落在40根折线形钢筋混凝土柱顶，这些折线形柱不仅为屋盖结构提供竖向支承点，还帮助抵抗屋盖结构由于径向收缩产生的强大内力。以受压钢桁架、脊索、谷索、受拉内环索为主，包括膜材和小型索类，共同

Foshan Century Lotus Sports Center locates by Dongping River of Foshan City, serving as the main venue for the 12th Provincial Games of Guangdong. The center, accommodating 36,000 audiences, was constructed from June, 2004 to September, 2006. On a total area of 42hm^2 sited a stadium, a swimming and diving hall, an outdoor warm-up field including a 400m track and a football field, an outdoor swimming center, an energy center and other accessory facilities. The name of the center, Century Lotus, was given due to the resemblance of the stadium's exterior to a lotus flower.

The total construction area of the stadium was 78,193m^2, of which the diameter was 305m , the height 50m, and the roof projection 53,421m^2. The entire building was constituted of a tensile cable-membrane, a reinforced concrete support, a grandstand and attached rooms. The capacity of the stadium reached 36,686, qualified for hosting international competitions.

图（10）-2 总体鸟瞰图
Fig.（10）-2 Overall aerial view

组成了轮辐式的世纪莲屋盖结构。配合每块面积近千平方米的膜材料，巧妙形成了“莲花”造型。

在 2002 年的佛山体育中心国际竞赛投标设计中，我们设计团队同时参加城市设计和体育中心建筑设计的国际竞赛。设计中尝试突破常规的体育中心布局模式，将体育中心的规划与佛山新城市中心区的城市设计相结合。宏观上延续了城市的绿化轴（城市公园）；在总体设计当中，根据具体要求将体育中心的外部空间功能明确化，体育场西面、南面临街广场，配合建筑内部的一些体育、商业功能，北面入口广场硬质铺装广场，为体育赛事提供了疏散空间，同时满足人们日常的体育锻炼对大的开敞空间的需求，如集体舞蹈、舞剑等活动。体育场东面的体育公园提供了一个景观良好的、供人们休憩的空间，东面临街的运动场满足了大众对体育活动的参与。这样的设计并没有牺牲建筑个性的发挥，相反使体育场真正成为人

The roof sit on 40 diagonal fold-line reinforced concrete, which provide both gravitational support to the roof and lateral tension counteracting the radial compression of the roof structure. Pre-stressed steel trusses, ridge and valley cables, inner tension rings, membranes of nearly 1,000m^2 per sheet and minor cables construct the spoke-wheel roof.

In the 2002 international bidding for the Foshan Sports Center, our design team submitted works for both the competition of urban design and architectural design. Our designs attempted to break away from conventional plan for sports center by combining the design of the sports center with the new urban center of Foshan. The urban vegetation axis was thus maintained and extended (City Park). In the overall design, we tailored the functions of the peripheral spaces to the specific requirements, so that the plazas across the streets in the west and south of the stadium were in coordination with the sports and business functions offered by the interior of the architecture; the north entrance plaza made of hard pavement provided an evacuation space for sporting events and the public's daily physical exercise, such as collective dancing, and sword dancing. The Sports Park in the east of the stadium lent a good view, and the playground in the east across the street supported public participation in sports activities. The overall design, instead of compromising architectural individuality, makes the stadium truly an indispensable part of people's life, fits the stadium into the city, and turns it into an active participator of urban space. In addition, the introduction of the concept of sports park gives the external environment of the sports stadium great flexibility and adaptability, with which the external environment is no longer a prohibitive and lifeless square, but a garden-style public space where people relax and play. That internal and external environment of the stadium becomes a pleasant unity for sporting events and public services.

图（10）-3　总体城市设计投标方案
Fig.（10）-3　Bidding scheme of the overall urban design

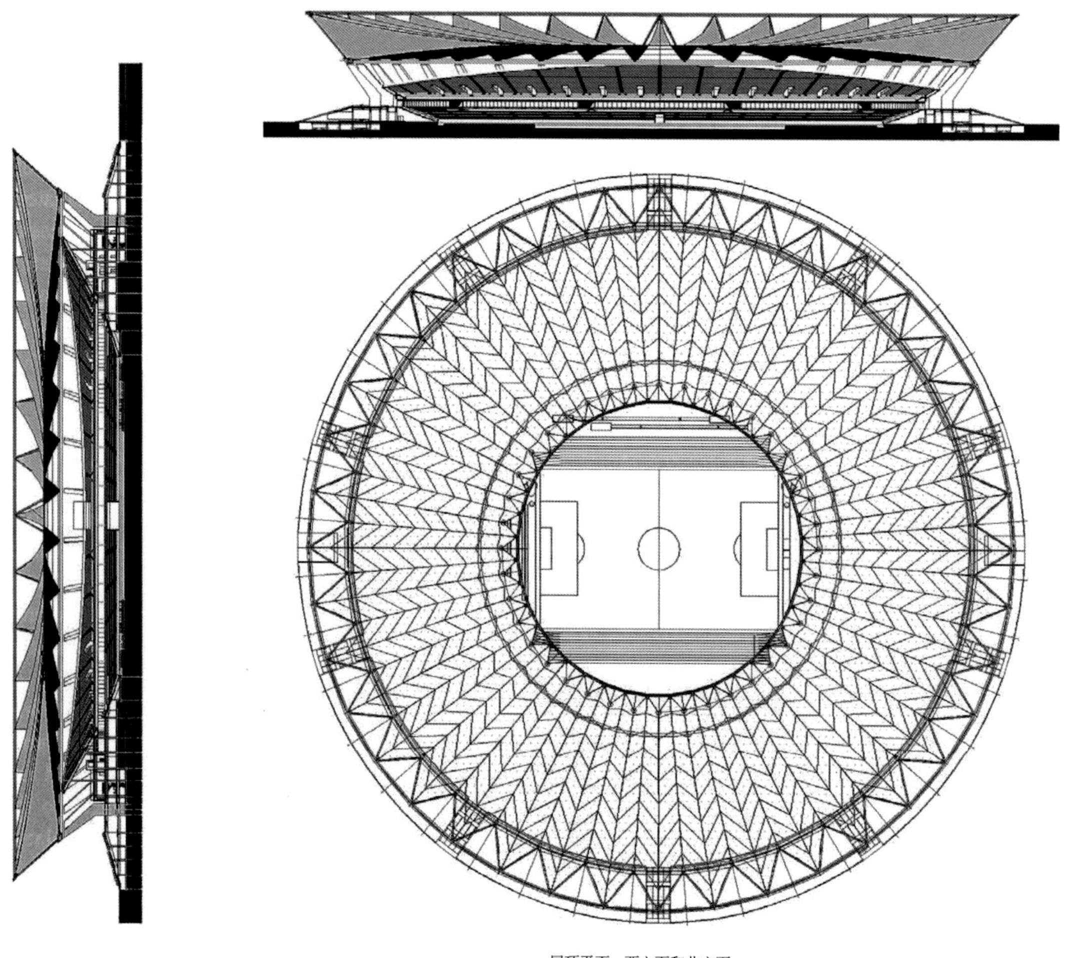

屋顶平面、西立面和北立面
Roof plan, West elevation and North elevation

屋面内部立面
Interior elevation of the roof

图（10）-4　主要技术图
Fig.（10）-4　Main technical diagram

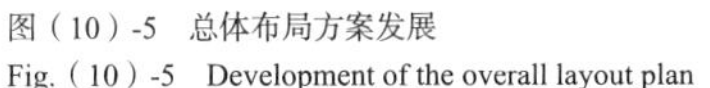
图（10）-5　总体布局方案发展
Fig.（10）-5　Development of the overall layout plan

图（10）-6　建成的总体布局
Fig.（10）-6　Completed overall layout

们生活不可或缺的一部分，使体育场融入城市，成为城市空间的积极塑造者。另外，通过体育公园的概念引入，为体育场馆所处的外部环境创造了很好的灵活适应性。体育场馆的外部环境不再是令人望而却步的、毫无趣味单调的广场，而是利于人们在此休息游玩的园林式公共空间。使得体育场馆的内外部环境成为一个可以轻松愉快地比赛和观赏的场所。

最终，我们的城市设计方案获得采纳，建筑设计则采用了德国 GMP 的设计方案，而我们的团队作为 GMP 的合作单位。接下来的问题在于如何协调建筑师的体育中心布局与城市设计的矛盾之处。德国建筑师的体育中心之所以被最终采纳，得益于其雄伟的标志性设计理念，强烈的中轴对称和富于张力的莲花造型打动了业主。德方迅速修改了总图，但依然是对称布局。最后不得不通过规划局协调。规划局根据城市设计出文，强行规定了体育场距西侧、南侧道路的距离，德方设计师基于良好的专业精神，保留意见的基础上同意在确保建筑设计效果的前提下，将体育场东侧留出尽可能多的公园空间，实现了城市设计的连续步行路径串连城市公共空间的理念。体育中心建成 5 年后，城市增建网球中心，也因为当年体育中心总体布局将东侧土地的空出，新的网球中心得以在东南角建设，从而印证了城市设计理念对可持续发展的重要性。

Ultimately, our urban design and the architectural design of GMP, a German cooperator of ours, were adopted. The next problem was to coordinate the contradiction between the sports center and the urban design. The design of the sports center proposed by the German architect convinced the employers with its iconic magnificence, impressive axial symmetry and the tensile lotus shape. The German party quickly modified its site plan, which nonetheless remained symmetrical. Finally, the coordination issue was referred to the Planning Bureau, which issued indicative documents based on the urban planning and specified the distance of the stadium from the west and south roads. The German designer, in the spirit of professionalism, conservatively agreed to leave as much park space as possible provided the effect of architectural design remains, thus the concept of continuous walking paths connecting with urban public spaces was protected. Five years after the completion of the Sports Center, a tennis center was planned to be built at the south east corner reserved in the original site plan of the Center, which demonstrated the significance of urban design concept for sustainable development.

The roof structures was a full tension structure; its concept was based on a wheel spoke. The structure was composed of powerful upper and lower compression rings and a series of radial valley, ridge and tensioned inner ridge cables, with small suspension cable and PVC film connected

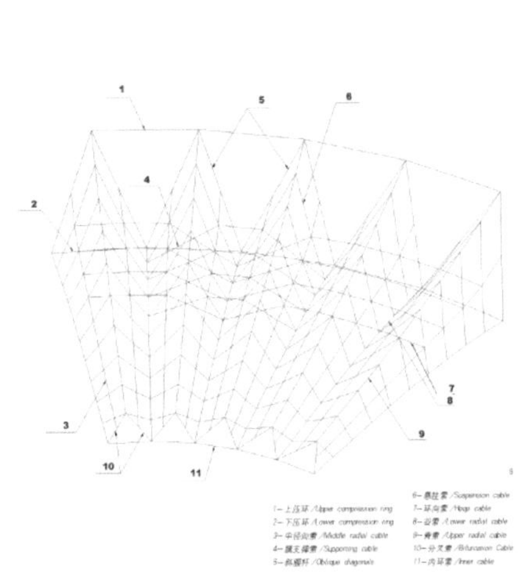

图（10）-7　屋面结构设计图

Fig.（10）-7　Design of the roof structure

图（10）-8　屋面结构照片

Fig.（10）-8　Photo of the roof structure

体育场屋面工程投影为环形，外环直径310m，内环直径为125m，最早的建筑方案中，内环曾设想采用可开启屋面，后因造价的原因取消了活动屋盖，改为现在的半开敞式形式。膜结构屋盖由主体钢结构系统、钢索系统、膜系统3个部分组成，其中主体钢结构系统由上压环、下压环及腹杆组成，分布于场区外圈，下压环标高为29m，上压环标高为49m，上下压环由腹杆连接为整体，为节省用钢量，上压环、下压环及腹杆均采用了钢管混凝土；钢索系统由上径向索（脊索）、下径向索（谷索）、中间径向索、内拉环索、悬挂索、膜支撑索、分叉索组成，其中内拉环索通过上径向索（脊索）、下径向索（谷索）及分叉索与钢结构系统连接成稳定的索网系统，形成空间整体预应力体系，并通过其他钢索的协同作用控制膜结构在不同工况下的位移；膜系统由80个单元膜片组成，膜片四周边界分别与钢结构腹杆、脊索、分叉索及谷索连接。整个膜结构屋盖充分利用了结构基础的整体承载能力，最大限度地发挥了材料力学的效能，具有设计新颖、结构独特、造型美观、雄伟壮观等特点。

设计原创团队：德国GMP国际建筑设计有限公司，施工图与实施团队：华南理工大学建筑设计研究院。

（本案例的文字稿主要根据参考文献[41]整理）

between the valley and the ridge cables. Therefore, the external compression ring, the tensioned inner ring and a variety of cables and films were comprised into a stable structure system capable against dead load and wind load. The roof structure was connected to 40 pre-stressed reinforced concrete inclined columns underneath separately, with 40 spherical hinge points under the compression ring. The structure was then integrated with the lower pre-stressed reinforced concrete structure. The axis diameter of the upper compression ring was 311m, while that of the lower one is 276.15m, with 20m height difference between the upper and the lower compression ring.

Architectural design: German GMP International Architectural Design Co., Ltd. Construction drawing and implementation: Architecture Design & Research Institute of SCUT.

图（10）-9　室外照片（上）
Fig.（10）-9　Outdoor photo (up)

图（10）-10　室内照片（下）
Fig.（10）-10　Indoor photo (down)

1993

中山市体育馆

设计时间：1993 年
建成时间：1997 年
地点：广东中山
规模：建筑面积 19000m²
实施情况：项目竞标第一名并为实施方案
合作单位：-
获奖情况：2000 年获“国家教育部优秀建筑设计三等奖”

Zhongshan Gymnasium

Design: 1993
Completion: 1997
Location: Zhongshan, Guangdong
Scale: construction area of 19,000m²
Implementation: first prize of the competition and for embodiment
Partners: -
Awards: third prize of the outstanding architectrual design, Ministry of Education, 2000

吴江市体育馆

设计时间：1993 年
建成时间：1997 年
地点：江苏吴江
规模：建筑面积 19000m²
实施情况：项目竞标第一名并为实施方案
合作单位：吴江建筑设计院
获奖情况：2004 年获“教育部优秀建筑设计三等奖”

Wujiang Gymnasium

Design: 1993
Completion: 1997
Location: Wujiang, Jiangsu
Scale: construction area of 19,000m²
Implementation: first prize of the competition and for embodiment
Partners: Architectural Design Institute of Wujiang
Awards: third prize of the outstanding architectrual design , Ministry of Education, 2004

1999

广东省国际海上运动基地

设计时间：1999 年
建成时间：2002 年
地点：广东汕尾
规模：建筑面积 48000m²
实施情况：项目竞标第一名并为实施方案
合作单位：-
获奖情况：2001 年获“广东省城市规划优秀设计二等奖”

Guangdong International Marine Sports Base

Design: 1999
Completion: 2002
Location: Shanwei, Guangdong
Scale: construction area of 48,000m²
Implementation: first prize of the competition and for embodiment
Partners: -
Awards: second prize of the outstanding urban planning of Guangdong Province , Ministry of Education, 2001

广东奥林匹克体育场

设计时间：1999 年
建成时间：2001 年
地点：广东广州
规模：建筑面积 14560m²
实施情况：合作实施方案
合作单位：美国 NEB 设计集团
获奖情况：2009 年获“建国 60 周年建筑创作大奖入围奖”；
2002 年获“中国建筑工程鲁班奖”

Guangdong Olympic Stadium

Design: 1999
Completion: 2001
Location: Guangzhou, Guangdong
Scale: construction area of 14,560m²
Implementation: cooperation embodiments
Partners: NEB Design Group of USA
Awards: award finalist of the 60th Anniversary architectural design, 2004;
State Construction Engineering Luban Award, 2002

2002

华中科技大学体育馆

设计时间：2002 年
建成时间：2006 年
地点：湖北武汉
规模：建筑面积 25000m^2
实施情况：中标实施方案
合作单位：华中科技大学建筑设计研究院
获奖情况：2007 年获“湖北省优秀工程勘察设计奖一等奖”

Gymnasium of HUST

Design: 2002
Completion: 2006
Location: Wuhan, Hubei
Scale: construction area of 25,000m^2
Implementation: first prize of the competition and for embodiment
Partners: Institute of Architectural Design of HUST
Awards: first prize of the Outstanding Engineering Investigation & Design of Hubei Province, 2007

2003

佛山世纪莲体育中心

设计时间：2003 年
建成时间：2006 年
地点：广东佛山
规模：建筑面积 13000m^2
实施情况：合作实施方案
合作单位：设计原创团队：德国 GMP 国际建筑设计有限公司
施工图与实施团队：华南理工大学建筑设计研究院
获奖情况：2009 年获“IOC/IAKS 国际体育建筑银奖”

Century Lotus Stadium

Design: 2003
Completion: 2006
Location: Foshan, Guangdong
Scale: construction area of 13,000m^2
Implementation: cooperation embodiments
Partners: Architectural design: German GMP International Architectural Design Co., Ltd.
Construction drawing and implementation: Architecture Design & Research Institute of SCUT.
Awards: Silver prize of the IOC / IAKS international sports architecture, 2009

2004

2008 年奥运会摔跤比赛馆（中国农业大学体育馆）

设计时间：2004 年
建成时间：2007 年
地点：北京
规模：建筑面积 24383m^2
实施情况：项目竞标第一名并为实施方案
合作单位：-
获奖情况：2011 年获“IPC/IAKS 国际体育建筑杰出功勋奖”；
2010 年获“全国优秀工程勘察设计奖银奖”；
2009 年获“全国优秀工程勘察设计行业奖建筑工程二等奖”；
2009 年获“中国建筑学会建筑设计建国 60 周年建筑创作大奖”；
2008 年获“国家优质工程银质奖”；
2008 年获“第五届中国建筑学会建筑创作佳作奖”；
2009 年获“教育部优秀建筑设计一等奖”；
2008 年获“第五届中国威海国际建筑设计大奖优秀奖”

2008 Olympic Wrestling Hall (China Agricultural University Gymnasium)

Design: 2004
Completion: 2007
Location: Beijing
Scale: construction area of 24,383m^2
Implementation: first prize of the competition and for embodiment
Partners: -
Awards: Outstanding Merit Award of IPC / IAKS International Sports Architecture, 2011;
Silver Award of National Excellent Engineering Investigation & Design, 2010;
Second prize of National Excellent Engineering Construction Engineering Survey and Design Industry Awards, 2009;
China Institute of Architectural Design Architectural Creation 60th Anniversary Award, 2009;
Silver Medal of National Quality Engineering, 2008;
Honorable Mentionthe of the creation of the Fifth China Architectural Society, 2008;
First prize of the outstanding architectural design award, Ministry of Education, 2009;
Excellence Award of the Fifth China Weihai International Architectural Design Award, 2008

2008年奥运会羽毛球比赛馆（北京工业大学体育馆）

设计时间：2004年
建成时间：2007年
地点：北京
规模：建筑面积24000m²
实施情况：项目竞标第一名并为实施方案

合作单位：美国NEB设计集团
获奖情况：2009年获“全国优秀工程勘察设计行业奖建筑工程二等奖”；
2009年获“中国建筑学会建筑设计优秀奖”；
2009年获“建筑创作大奖入围奖”；
2008年获“中国土木工程詹天佑奖”；
2009年获“广东省优秀工程设计一等奖”；
2009年获“第五届中国威海国际建筑设计大奖优秀奖”

2008 Olympic Badminton Hall (Beijing University of Technology Gymnasium)

Design: 2004
Completion: 2007
Location: Beijing
Scale: construction area of 24,000m²
Implementation: first prize of the competition and for embodiment
Partners: NEB Design Group of USA
Awards: Second Prize of the National Outstanding Engineering Construction Engineering Survey and Design Industry Awards, 2009;
Excellence Award of China Institute of Architectural Design, 2009;
Award finalistsArchitectural Creation prize, 2009;
China Civil Engineering Zhan Tianyou Award, 2008;
first prize of Guangdong Province excellent design, 2009;
Excellence Award of the Fifth China Weihai International Architecturai Design Award, 2009

2005

武钢体育中心

设计时间：2005年
建成时间：-
地点：湖北武汉
规模：建筑面积26300m²
实施情况：项目竞标第一名并为实施方案，完成初步设计
合作单位：-
获奖情况：-

Wugang Sports Center

Design: 2005
Completion: -
Location: Wuhan, Hubei
Scale: construction area of 26,300m²
Implementation: first prize of the competition and for embodiment, finish the preliminary design
Partners: -
Awards: -

2010年亚运柔道摔跤馆（广州大学城 华工体育馆）

设计时间：2005年
建成时间：2007年
地点：广东广州
规模：建筑面积12783m²
实施情况：项目竞标第一名并为实施方案

合作单位：-
获奖情况：2008年获“国家优质工程银质奖”；
2009年获“教育部优秀建筑设计三等奖”；
2007年获“广东省注册建筑师协会优秀建筑佳作奖”

2010 Asian Games Judo Wrestling Hall (Guangzhou University City SCUT Gymnasium)

Design: 2005
Completion: 2007
Location: Guangzhou, Guangdong
Scale: construction area of 12,783m²
Implementation: first prize of the competition and for embodiment
Partners: -
Awards: National Quality Engineering Silver Medal, 2008;
Ministry of Education, outstanding architectural design prize, 2009;
Guangdong Province Registered Architects Association Award for outstanding architectural masterpiece, 2007

2006

惠州市体育中心

设计时间： 2006 年
建成时间： -
地点： 广东惠州
规模： 建筑面积 93940m²
实施情况： 项目竞标第一名
合作单位： -
获奖情况： -

Huizhou Sports Center

Design: 2006
Completion: -
Location: Huizhou, Guangdong
Scale: construction area of 93,940m²
Implementation: first prize of the competition
Partners: -
Awards: -

东莞市体育馆

设计时间： 2006 年
建成时间： -
地点： 广东东莞
规模： 建筑面积 46927m²
实施情况： 国际竞标优胜入围方案
合作单位： -
获奖情况： -

Dongguan Gymnasium

Design: 2006
Completion: -
Location: Dongguan, Guangdong
Scale: construction area of 46,927m²
Implementation: winning finalists of the international competition
Partners: -
Awards: -

2007

河北省体育中心

设计时间： 2007 年
建成时间： -
地点： 河北石家庄
规模： 建筑面积 24383m²
实施情况： 项目竞标第一名
合作单位： -
获奖情况： -

Hebei Province Sports Center

Design: 2007
Completion: -
Location: Shijiazhuang, Hebei
Scale: construction area of 24,383m²
Implementation: first prize of the competition
Partners: -
Awards: -

2010 年亚运会游泳跳水馆

设计时间： 2007 年
建成时间： 2010 年
地点： 广东广州
规模： 建筑面积 33331m²
实施情况： 项目竞标第一名并为实施方案
合作单位： -
获奖情况： 2011 年获“中国建筑学会创作佳作奖”；
2012 年获“全国优秀工程勘察设计行业奖建筑工程二等奖”；
2011 年获“教育部优秀建筑工程设计一等奖”；
2009 年获“第五届广东省注册建筑师协会优秀建筑佳作”。

2010 Asian Games Swimming and Diving Hall

Design: 2007
Completion: 2010
Location: Guangzhou, Guangdong
Scale: construction area of 33,331m²
Implementation: first prize of the competition and for embodiment
Partners: -
Awards: China Architectural Society of creation Honorable Mention, 2011;
Second Prize, National Excellent Engineering Construction Engineering Survey and Design Industry Awards, 2012;
Outstanding Architectural Engineering Design Award, Ministry of Education, 2011;
fifth Guangdong Province Association of Registered Architects Outstanding Architectural Excellent Work, 2009

2010 年亚运武术馆（南沙体育馆）

设计时间： 2007 年
建成时间： 2010
地点： 广东广州
规模： 建筑面积 30082m²
实施情况： 项目竞标第一名并为实施方案

合作单位： -
获奖情况： 2011 年获“IOC/IAKS 国际体育建筑铜奖”；
2012 年获“第十届中国土木工程詹天佑奖”；
2011 年获“中国钢结构设计金奖”；
2011 年获“第六届中国建筑学会建筑创作奖佳作奖”；
2011 年获“广东省优秀工程勘察设计奖一等奖”；
2011 年获“第三届广东省土木工程詹天佑故乡杯”；
2011 年获“广州市建设工程质量五羊杯奖”；
2011 年获“广东省建设工程质量金匠奖”；
2011 年获“广东省钢结构金奖（粤钢奖）”

2010 Asian Games Wushu Hall (Nansha Gymnasium)

Design: 2007
Completion: 2010
Location: Guangzhou, Guangdong
Scale: construction area of 30,082m²
Implementation: first prize of the competition and for embodiment

Partners: -
Awards: IOC / IAKS international sports architecture Bronze, 2011;
Tenth China Zhan Tianyou Civil Engineering Award, 2012;
China Steel Design Award, 2011;
Sixth China Architectural Society Creation Award Honorable Mention, 2011;
Guangdong Excellent First Prize of Engineering Investigation & Design, 2011;
Third Zhan Tianyou hometown Cup of Civil Engineering in Guangdong, 2011;
Guangzhou Construction Engineering Quality Wuyang Cup Award, 2011;
Guangdong Goldsmith Award for Construction Engineering Quality, 2011;
Guangdong Steel structure Gold Award, 2011

2008

深圳大运会宝安体育场

设计时间： 2008 年
建成时间： 2011 年
地点： 广东深圳
规模： 建筑面积 97712m²
实施情况： 中标并实施

合作单位： 设计原创团队：德国 GMP 国际建筑设计有限公司
施工图与实施团队：华南理工大学建筑设计研究院

获奖情况： 2012 年全国工程建设项目优秀设计成果二等奖；
第四届广东省土木工程詹天佑故乡杯；
2011-2012 年度国家优质工程银质奖；
第十一届中国土木工程詹天佑奖；
2013 年度教育部优秀建筑设计一等奖；
国际体育建筑学会 IOC/IAKS 金奖；
2013 年全国优秀工程勘察设计行业奖 一等奖

Universiade Shenzhen Bao'an Stadium

Design: 2008
Completion: 2011
Location: Shenzhen, Guangdong
Scale: construction area of 97,712m²
Implementation: first prize of the competition and for embodiment

Partners: Architectural design: German GMP International Architectural Design Co., Ltd.
Construction drawing and implementation: Architecture Design & Research Institute of SCUT.

Awards: outstanding design Achievement Award National Construction Projects, 2012;
Fourth Guangdong Province, the hometown of Civil Represented Cup;
2011-2012 National Quality Engineering Silver Medal;
Eleventh China Zhan Tianyou Civil Engineering Award;
The Ministry of Education, annual outstanding architectural design award, 2013;
International Society of Sports Building IOC / IAKS Gold;
National Outstanding Engineering Investigation & Design Industry Awards first prize, 2013

2009

江苏淮安市体育中心

设计时间：2009 年
建成时间：2013 年
地点：江苏淮安
规模：建筑面积 168520m^2
实施情况：项目竞标第一名并为实施方案
合作单位：-
获奖情况：-

Jiangsu Huai'an Sports Center

Design: 2009
Completion: 2013
Location: Huaian, Jiangsu
Scale: construction area of 168,520m^2
Implementation: first prize of the competition and for embodiment
Partners: -
Awards: -

安徽铜陵市体育中心

设计时间：2009 年
建成时间：-
地点：安徽铜陵
规模：建筑面积 104256m^2
实施情况：投标方案
合作单位：-
获奖情况：-

Anhui Tongling Sports Center

Design: 2009
Completion: -
Location: Tongling, Anhui
Scale: construction area of 104,256m^2
Implementation: the bidding program
Partners: -
Awards: -

平远县文化体育中心

设计时间：2009 年
建成时间：2012 年
地点：广东东莞
规模：建筑面积 18464m^2
实施情况：委托项目
合作单位：-
获奖情况：-

Pingyuan Culture and Sports Center

Design: 2009
Completion: 2012
Location: Dongguan, Guangdong
Scale: construction area of 18,464m^2
Implementation: commissioned project
Partners: -
Awards: -

长安体育中心

设计时间：2009 年
建成时间：2013 年
地点：广东东莞
规模：建筑面积 22991m^2
实施情况：委托项目
合作单位：-
获奖情况：-

Changan Sports Center

Design: 2009
Completion: 2013
Location: Dongguan, Guangdong
Scale: construction area of 22,991m^2
Implementation: commissioned project
Partners: -
Awards: -

芜湖市奥体公园二期

设计时间：2011 年
建成时间：-
地点：江苏芜湖
规模：建筑面积 127750m^2
实施情况：项目竞标第一名
合作单位：-
获奖情况：-

Wuhu Olympic Park

Design: 2011
Completion: -
Location: Wuhu, Jiangsu
Scale: construction area of 127,750m^2
Implementation: first prize of the competition
Partners: -
Awards: -

广东梅县体育中心

设计时间：2011 年
建成时间：2012 年
地点：广东梅州
规模：建筑面积 26366m^2
实施情况：中标并实施
合作单位：-
获奖情况：2013 年度广东省优秀工程勘察设计奖二等奖

Guangdong Meixian Sports Center

Design: 2011
Completion: 2012
Location: Meizhou, Guangdong
Scale: construction area of 26,366m^2
Implementation: first prize of the competition and for embodiment
Partners: -
Awards: Annual Guangdong Province Engineering Investigation & Design excellence award Second Prize, 2013

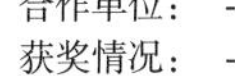

内蒙古党校文体活动中心

设计时间：2011 年
建成时间：2013 年
地点：内蒙古自治区呼和浩特市
规模：建筑面积 12980m^2
实施情况：竞赛中标并实施
合作单位：-
获奖情况：-

Inner Mongolia Party School sports center

Design: 2011
Completion: 2013
Location: Hohhot, Inner Mongolia
Scale: construction area of 12,980m^2
Implementation: first prize of the competition and for embodiment
Partners: -
Awards: -

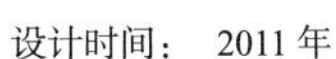

江苏省泰州市奥体中心

设计时间：2011 年
建成时间：-
地点：江苏泰州
规模：建筑面积 175980m^2
实施情况：全国竞赛优胜方案
合作单位：-
获奖情况：-

Olympic Sports Center of Taizhou, Jiangsu Province

Design: 2011
Completion: -
Location: Taizhou, Jiangsu
Scale: construction area of 157,980m^2
Implementation: first prize of the national competition
Partners: -
Awards: -

山西交通大学体育馆

设计时间： 2011 年
建成时间： -
地点： 山西太原
规模： 建筑面积 12580m^2
实施情况： 合作实施方案
合作单位： -
获奖情况： -

Shanxi Jiaotong University Gymnasium

Design: 2011
Completion: -
Location: Taiyuan, Shanxi
Scale: construction area of 12,580m^2
Implementation: cooperation embodiments
Partners: -
Awards: -

2012

江门滨江体育中心

设计时间： 2012 年
建成时间： -
地点： 广东江门
规模： 建筑面积 202500m^2
实施情况： 竞赛中标并实施
合作单位： -
获奖情况： -

Jiangmen Riverside Sports Center

Design: 2012
Completion: -
Location: Jiangmen, Guangdong
Scale: construction area of 202,500m^2
Implementation: first prize of the competition and for embodiment
Partners: -
Awards: -

蓬莱市体育中心

设计时间： 2012 年
建成时间： -
地点： 山东蓬莱
规模： 建筑面积 39212m^2
实施情况： 竞赛中标
合作单位： -
获奖情况： -

Penglai Sports Center

Design: 2012
Completion: -
Location: Penglai, Shandong
Scale: construction area of 39,212m^2
Implementation: first prize of the competition and for embodiment
Partners: -
Awards: -

常州工学院体育馆

设计时间： 2012 年
建成时间： -
地点： 江苏常州
规模： 建筑面积 12850m^2
实施情况： 委托项目
合作单位： -
获奖情况： -

Changzhou Institute of Technology Gymnasium

Design: 2012
Completion: -
Location: Changzhou, Jiangsu
Scale: construction area of 12,850m^2
Implementation: commissioned project
Partners: -
Awards: -

山东大学青岛校区体育中心

设计时间：2012 年
建成时间：-
地点：山东青岛
规模：建筑面积 94000m^2
实施情况：投标方案
合作单位：-
获奖情况：-

Shandong University Qingdao Campus Sports Center

Design: 2012
Completion: -
Location: Qingdao, Shandong
Scale: construction area of 94,000m^2
Implementation: the bidding program
Partners: -
Awards: -

2013

武汉大学体育馆

设计时间：2013 年
建成时间：-
地点：湖北武汉
规模：建筑面积 29850m^2
实施情况：中标并实施
合作单位：-
获奖情况：-

Wuhan University Gymnasium

Design: 2013
Completion: -
Location: Wuhan, Hubei
Scale: construction area of 29,850m^2
Implementation: first prize of the competition and for embodiment
Partners: -
Awards: -

湖北省奥林匹克体育中心

设计时间：2013 年
建成时间：-
地点：湖北武汉
规模：建筑面积 61997m^2
实施情况：投标方案
合作单位：-
获奖情况：-

Hubei Olympic Sports Center

Design: 2013
Completion: -
Location: Wuhan, Hubei
Scale: construction area of 61,997m^2
Implementation: the bidding program
Partners: -
Awards: -

图表来源

图 1-3　1936 年柏林奥运会的德意志帝国体育场
　来源：alamy.com.
图 1-4　1960 年罗马奥运会城市与建筑
　来源：金磊编 . 建筑师看奥林匹克 [M]. 机械工业出版社 , 2004.
图 1-5　神宫外苑奥运体育场构想
　来源：Japan: The Official Guide, 1941.
图 1-6　驹泽体育公园
　来源：Vintage Japanese Postcard Museum（1900-1960）的网站 .
图 1-8　1953 年建成的广州越秀山体育场和 1951 年建成的重庆大田湾体育场
　来源：http：//www.ycwb.com/ePaper/ycwb/html/2007-10/26/content_48970.htm.
1951 年建成的重庆大田湾体育场
　来源：http：//www.cq.xinhuanet.com/2014-08/27/c_1112235978_3.htm.
图 1-9　1990 年北京亚运会场馆
来源：新华社照片（新华社记者郭大岳摄）.
图 1-10　北京奥林匹克运动会的体育设施
　来源：http：//www.china.com.cn/sports/zhuanti/2008ay/txt/2007-01/05/content_7609935.htm.
图 1-11　北京奥运主体育场瘦身
　来源：NcFBBS 网贴 .
图 1-12　“鸟巢”里的文艺演出
　来源：http：//yangxiaowen0822.blog.163.com/blog/static/13589799320123301059123l5/.
图 1-14　北京奥运中的 6 个高校体育场馆
　来源：左下 - http：//blog.sina.cn/dpool/blog/s/blog_13462002d0102v7jt.html?md=gd；　右　上 - http：//www.bit.edu.cn/xxgk/xysz/tyb53/44173.htm；右 中 - http：//news.hexun.com/2008-08-04/107880951.html；右下 - http：//news.hexun.com/2008-08-04/107880951.html.
图 1-15　东莞体育馆
　来源：http：//bbs.zhulong.com/101010_group_201810/detail10125063.
图 1-18　沈阳五里河体育场爆破
　来源：Source：http：//news.sina.com.cn/photo/360/2007/0212/3.html.
图 1-19　北京五棵松体育中心中标方案
　来源：《建筑创作》杂志社 . 建筑师看奥林匹克 [M]. 机械工业出版社，2004.
表 1-3　广州大学城体育场馆建设项目及投资
　来源：李传义 . 广州大学城体育场馆规划与建设回顾 [J]. 城市建筑，2007(11)：21-24.

图 2-5　一个与城市结合的体育建筑案例
　来源：由 sasaki 事务所提供 .

图 3-5　建于 1932 年的广东省人民体育场
图 3-6　建于 1987 年的广州天河体育中心
　来源：google 地图 .
图 3-8　被停车场包围孤立的比赛场馆
　来源：The New York Times，由 nytimes.com 下载 .
图 3-9　美国克利夫兰市体育建筑案例总体布局
　来源：由 sasaki 事务所提供 .
图 3-10　美国克利夫兰市体育建筑案例总体鸟瞰
　来源：sasaki 事务所提供 .
图 3-11　传统都市形式
图 3-12　现代都市形式
图 3-13　克莱尔的理想城市模型
　来源：Roger Trancik 找寻失落的公共空间——城市设计的理论 [M]. 北京：中国建筑工业出版社，2008.
图 3-14　诺里图底关系图
　来源：安德鲁斯· 杜安伊等著 . 新城市艺术与城市规划元素 [M]. 大连：大连理工大学出版社，2008.
图 3-16　第 26 届奥林匹克运动会新建的主体育场
　来源：由 sasaki 事务所提供 .
图 3-17　萨克拉门托国王队新球馆和“铁路广场”项目土地使用计划
　来源：林昆 . 体育娱乐区与城市中心再发展——以萨克拉门托国王队新球馆与“铁路广场”项目为例 [J]. 城市规划，2010(10)：93-96.
图 3-22　北京奥林匹克中心原城市设计方案
　来源：由 sasaki 事务所提供 .
图 3-23　萨萨基公司中标实施的北京奥林匹克中心方案
　来源：由 sasaki 事务所提供 .
图 3-24　佛山体育中心实施布局演进（右图）
　来源：google 地图 .

图 4-9　华中科技大学体育馆的 40m × 70m 场地
　来 源：http：//photo.iyaxin.com/content/2009-06/25/content_1090709.htm.
图 4-12　悉尼奥林匹克公园内皇家农业协会展馆赛时改造为篮球等比赛场地
图 4-13　向公众开放的悉尼水上中心赛时为游泳跳水比赛场地，赛后向公众开放
　来　源：Official Report. Sydney Organizing Committee for the Olympic Games. 2001：386-391.
图 4-14　北京国家游泳中心奥运会赛时剖面
图 4-15　北京国家游泳中心奥运会赛后剖面
　来源：郑方，杨奇勇 . 从体育场馆到公共中心——水立方赛后设计与运营 [J]. 世界建筑，2013(08)：52-59, 128-129.
图 4-16　伦敦奥运水上中心主体空间与临时看台一体化设计
图 4-17　伦敦奥运水上中心赛后将拆除临时看台
　来源：http：//jst-cn.com.

图 5-3　佛罗伦萨体育场
　来源：https：//www.douban.com/note/264531596/.
图 5-4　东京代代木游泳馆
　来源：http：//t.zhulong.com/u9469151/worksdetail4448792.html.
图 5-5　日本岩手县体育馆
　来源：梅季魁 . 现代体育馆建筑设计 [M]. 哈尔滨：黑龙江科学技术出版社，1999.
图 5-6　伦敦奥运游泳馆剖面

来源：http：//www.xiuhome.com/article/show.asp?id=6152.

图 5-7　伦敦奥运游泳馆屋顶

来源：http：//i.ifeng.com/news/news?vt=5&aid=104490307&all=1.

图 5-8　日本东京 2020 奥运主体育场基地现场

来源：日本体育馆委员会公开资料 .

图 5-9　扎哈· 哈迪德事务所方案

来源：http：//i.ifeng.com/news/news?vt=5&aid=104490307&all=1.

表 5-2　自然采光方式和采光策略的适用性分析

来源：根据建筑设计资料集和体育建筑照明设计手册等资料整理 .

其余图表均来源于作者及其团队项目资料。

参考文献

[1] 吉慧 . 公共安全视角下的体育场馆设计研究 [D]. 华南理工大学，2013.

[2] 林昆 . 公共体育建筑策划研究 [D]. 华南理工大学，2011.

[3] 杨威 . 基于使用后评价的体育中心外部公共空间设计准则研究 [D]. 华南理工大学，2016.

[4] 李昕旻 . 低碳视角下江门滨江体育中心体育场游泳馆设计策略 [D]. 华南理工大学，2015.

[5] 任晓璐 . 低碳视角下体育馆及会展建筑设计研究 [D]. 华南理工大学，2015.

[6] 徐子文 . 体育会展综合功能建筑研究 [D]. 华南理工大学，2014.

[7] 易照曌 . 广州亚运部分体育馆节能技术实效验证与分析研究 [D]. 华南理工大学，2014.

[8] 骆乐 . 城市空间视角下的体育中心设计研究 [D]. 华南理工大学，2014.

[9] 吴剑玲 . 我国场地自行车场馆建设探究 [D]. 华南理工大学，2013.

[10] 章艺昕 . 游泳馆建筑设计研究 [D]. 华南理工大学，2013.

[11] 黄燕 . 广东体育馆建筑功能模块的适应性研究 [D]. 华南理工大学，2012.

[12] 周超然 . 中型体育场建筑研究 [D]. 华南理工大学，2013.

[13] 邹林 . 我国中小型体育馆的设计策略与方法 [D]. 华南理工大学，2011.

[14] 谢东彪 . 时代背景下的赛会体育馆创作研究 [D]. 华南理工大学，2011.

[15] 林耀阳 . 基于大学生行为模式的高校体育馆主空间设计 [D]. 华南理工大学，2011.

[16] 周琮 . 北京高校奥运场馆赛后利用比较研究 [D]. 华南理工大学，2011.

[17] 曹黛 . 基于开敞空间的城市体育场所防灾设计研究 [D]. 华南理工大学，2010.

[18] 赵薇薇 . 应对城市防灾避难的体育馆建筑设计初探 [D]. 华南理工大学，2010.

[19] 孙一民 . 责任与艰辛 [J]. 城市建筑，2017(28)：3-4.

[20] 侯叶，孙一民，杜庆 . 启蒙——近代中国体育建筑的内化演变 [J]. 新建筑，2017(05)：83-87.

[21] 孙一民 . 从形式探索到逻辑追寻：走向精明营建 [J]. 建筑师，2015(02)：101-110.

[22] 孙一民，马国馨，崔愷，郭明卓，庄惟敏 .“走向精明营建的体育建筑”系列访谈 [J]. 城市建筑，2015(25)：6.

[23] 孙一民 . 走向精明营建的体育建筑 [J]. 城市建筑，2015(25)：3.

[24] 孙一民，吉慧 . 大空间体育建筑防火疏散设计研究——以广州亚运会游泳跳水馆为例 [J]. 新建筑，2013(02)：104-107.

[25] 孙一民，孟可，刘慧，谭泽阳，吕强，付毅智，肖辉 .“体育建筑设计及建设之思”主题沙龙 [J]. 城市建筑，2013(17)：6-14.

[26] 孙一民 . 可持续性，体育建筑的恒久主题 [J]. 城市建筑，2013(17)：3.

[27] 孙一民，汪奋强 . 基于可持续性的体育建筑设计研究：结合五个奥运、亚运场馆的实践探索 [J]. 建筑创作，2012(07)：24-33.

[28] 孙一民，吉慧 . 平灾结合，应时而变——体育场馆的防灾避难设计对策 [J]. 土木工程学报，2012，45(S2)：113-116.

[29] 孙一民 . 建筑创作的理性途径 [J]. 建筑技艺，2012(05)：150-155.

[30] 孙一民，林隽，邬尚霖 . 走向公民城市——亚运广州城市与建筑 [J]. 建筑学报，2010(10)：40-42.

[31] 孙一民，冷天翔，申永刚，陶亮 . 云水禅心——广东奥林匹克游泳跳水馆设计 [J]. 建筑学报，2010(10)：58-59.

[32] 孙一民，叶伟康，汪奋强，陶亮，谢冠一 . 海天一色——广州亚运武术馆设计 [J]. 建筑学报，2010(10)：73-74.

[33] 何镜堂，孙一民，杨适伟 .2010 年亚运会游泳跳水馆设计 [J]. 中国建筑装饰装修，2010(07)：150-153.

[34] 何镜堂，孙一民，杨适伟 .2010 年广州

亚运会武术馆设计 [J]. 中国建筑装饰装修，2010(07)：144-149.
[35] 孙一民 . 体育建筑 60 年，科学理性新起点 [J]. 城市建筑，2010(11)：3.
[36] 孙一民，王璐 . 重大体育赛事与新城建设发展——广州亚运村建设研究 [J]. 建筑学报，2009(02)：38-41.
[37] 孙一民，汪奋强，叶伟康 . 公共体育场馆的建设标准刍议 [J]. 南方建筑，2009(06)：4-5.
[38] 孙一民 . 产、学、研，密切结合，相互促进的体育建筑实践 [J]. 南方建筑，2009(05)：24-27.
[39] 孙一民 . 北京奥运摔跤馆 [J]. 建筑创作，2009(12)：36-37.
[40] 孙一民 . 体育建筑何处去 ?[J]. 城市建筑，2009(11)：3.
[41] 孙一民 . 佛山世纪莲体育中心，广东，中国 [J]. 世界建筑，2009(10)：48-51.
[42] 孙一民 . 立足城市建筑 弘扬科学理性 [J]. 城市建筑，2009(10)：6.
[43] 孙一民 . 体育场馆适应性研究——北京工业大学体育馆 [J]. 建筑学报，2008(01)：94-97.
[44] 孙一民 . 体育建筑——期待科学理性的回归 [J]. 城市建筑，2008(11)：6.
[45] 孙一民 . 回归基本点：体育建筑设计的理性原则——中国农业大学体育馆设计 [J]. 建筑学报，2007(12)：26-31.
[46] 何镜堂，孙一民，汪奋强，叶伟康，姜文艺，燕雨生 . 简洁内敛的理性探索：2008 年北京奥运会摔跤比赛馆 [J]. 建筑创作，2007(07)：119-127.
[47] 何镜堂，孙一民，江泓，姜文艺，燕雨生，王欣斌 . 城市、校园、体育：2008 年北京奥运会羽毛球及艺术体操比赛馆 [J]. 建筑创作，2007(07)：140-145.
[48] 孙一民，江泓 . 城市空间与体育建筑的契合——北京奥运会羽毛球馆建筑创作 [J]. 城市建筑，2004(02)：31-33.
[49] 孙一民，郭湘闽 . 从城市的角度看体育建筑构思——谈新疆体育中心方案设计 [J]. 建筑学报，2002(09)：27-29.
[49] 孙一民，张春阳 . 走向成熟的城市——九运与广州 [J]. 时代建筑，2002(03)：27-29.
[50] 孙一民 . 基本问题的解决与思考——中山体育馆设计反思 [J]. 华中建筑，1999(03)：59-61.
[51] 孙一民 . 多功能· 综合性 试论我国高校体育“场”. 馆的规划设计 [J]. 华中建筑，1993(04)：47-49.
[52] 邓芳 . 广州大学城华工体育馆可持续设计创作实践 [J]. 建筑创作 , 2012(7):34-45.
[53] 孙一民 , 汪奋强 , 叶伟康 , 等 . 梅县体育中心，梅州，2010-2012[J]. 建筑创作 , 2012(7):46-63.

图书在版编目（CIP）数据

精明营建：可持续的体育建筑 / 孙一民著 .—北京：中国建筑工业出版社，2015.6
ISBN 978-7-112-17860-5

I.①精… II.①孙… III.①体育建筑—建筑设计—英、汉 IV.①TU245

中国版本图书馆 CIP 数据核字（2015）第 037741 号

责任编辑：李 鸽 毋婷娴
责任校对：李美娜 姜小莲

精明营建：可持续的体育建筑
孙一民 著
*
中国建筑工业出版社出版、发行（北京海淀三里河路 9 号）
各地新华书店、建筑书店经销
北京京点图文设计有限公司制版
天津图文方嘉印刷有限公司印刷
*
开本：889×1194 毫米 1/16 印张：17 字数：451 千字
2019 年 2 月第一版 2019 年 2 月第一次印刷
定价：198.00 元
ISBN 978-7-112-17860-5
（29043）